Biographies in the History of Physics

Christian Forstner · Mark Walker
Editors

Biographies in the History of Physics

Actors, Objects, Institutions

 Springer

Editors
Christian Forstner
Ernst-Haeckel-Haus
Friedrich-Schiller-University
Jena, Germany

Mark Walker
Union College
Schenectady, NY, USA

ISBN 978-3-030-48508-5   ISBN 978-3-030-48509-2   (eBook)
https://doi.org/10.1007/978-3-030-48509-2

© The Editor(s) (if applicable) and The Author(s), under exclusive licence to Springer Nature Switzerland AG 2020
This work is subject to copyright. All rights are solely and exclusively licensed by the Publisher, whether the whole or part of the material is concerned, specifically the rights of translation, reprinting, reuse of illustrations, recitation, broadcasting, reproduction on microfilms or in any other physical way, and transmission or information storage and retrieval, electronic adaptation, computer software, or by similar or dissimilar methodology now known or hereafter developed.
The use of general descriptive names, registered names, trademarks, service marks, etc. in this publication does not imply, even in the absence of a specific statement, that such names are exempt from the relevant protective laws and regulations and therefore free for general use.
The publisher, the authors and the editors are safe to assume that the advice and information in this book are believed to be true and accurate at the date of publication. Neither the publisher nor the authors or the editors give a warranty, expressed or implied, with respect to the material contained herein or for any errors or omissions that may have been made. The publisher remains neutral with regard to jurisdictional claims in published maps and institutional affiliations.

This Springer imprint is published by the registered company Springer Nature Switzerland AG
The registered company address is: Gewerbestrasse 11, 6330 Cham, Switzerland

Contents

1 Introduction 1
  Christian Forstner and Mark Walker

Part I Individuals

2 The 'Invisible Hand' of Carl Friedrich Gauß—Retracing the Life
  of Moritz Meyerstein, a 19th century Instrument Maker
  and Universitäts-Mechanicus............................. 13
  Klaus Hentschel

3 The Personal is Professional: Margaret Maltby's Life
  in Physics 37
  Joanna Behrman

4 Erwin Schrödinger in the Second Spanish Republic,
  1934–1935 59
  Enric Pérez

5 Ludwig Prandtl: Pioneer of Fluid Mechanics and Science
  Manager 75
  Michael Eckert

6 Rudolf Tomaschek—An Exponent of the *Deutsche Physik*
  Movement 89
  Vanessa Osganian

7 The 'Better' Nazi: Pascual Jordan and the Third Reich 111
  Dieter Hoffmann and Mark Walker

8 Relativity and Dialectical Materialism: Science, Philosophy
  and Ideology in Hans-Jürgen Treder's Early Academic
  Career 129
  Raphael Schlattmann

9 Biography and Autobiography in the Making of a Genius:
   Richard P. Feynman 145
   Christian Forstner

Part II Collectives

10 A Biography of the German Atomic Bomb 163
    Mark Walker

11 The Multiple Lives of the General Relativity Community,
    1955–1974 179
    Roberto Lalli

12 Whose Biography Is It Anyway? Shared Biographies
    of Institutions, Leaders, Instruments, and Self 203
    Catherine Westfall

Part III Objects

13 Lost in the Production of Time and Space: The Transformation
    of the Airy Transit Circle from a Working Telescope
    to a Museum Object 221
    Daniel Belteki

14 Scientific Instruments Turning into Toys: From Franklin's
    Pulse Glass to Dipping Birds 237
    Panagiotis Lazos

Part IV Limitations

15 I'm Not There. Or: Was the Virtual Particle Ever Born? 261
    Markus Ehberger

16 Biography or Obituary? The Historiographical Value
    of the Death of the Ether 281
    Jaume Navarro

17 The Meaning, Nature, and Scope of Scientific
    (Auto)Biography 301
    Thomas Söderqvist

Index 319

Chapter 1
Introduction

Christian Forstner and Mark Walker

On Biography

Many historians of science who have written on biography are themselves biographers with direct experience of the advantages and difficulties presented by this genre. Thomas Hankins, who arguably began a historiographic discussion of the role of biography in the history of science with an essay entitled "In Defense of Biography," noted that: "The biographer of a scientist tends to be drawn either to the personal life of his subject or to the technical details of his subject's scientific work. It is difficult to bring these two different aspects together in a harmonious way." (Hankins 1979, p. 2). Like many biographers who have followed him, and with some justification, Hankins also bemoaned the fact that mainstream history of science did not give biography much respect. At that time biographical approaches were often associated with the hagiographic tradition of the 19th century and therefore rejected. Scholarship that focused on the social determinants of science had become more influential, breaking with traditional biography and yielding new insight for the history of science. However, in the meantime, the genre of biography has made a comeback.

Mott Greene provided a blueprint for a good biography, arguing that "… to write a biography is almost inevitably to write a historical novel, albeit a historical novel constructed according to a demanding set of rules." Such a biography would follow the rules of the nineteenth-century *Bildungsroman*, portraying a "striving hero" who overcomes obstacles and achieves goals in the service of their self-development

---

C. Forstner (✉)
Friedrich-Schiller-Universität Jena, AG Wissenschaftsgeschichte - Ernst-Haeckel-Haus, Kahlaische Str. 1, 07745 Jena, Germany
e-mail: Christian.Forstner@uni-jena.de

M. Walker
Department of History, Union College, Schenectady, NY 12308, USA
e-mail: walkerm@union.edu

© The Editor(s) (if applicable) and The Author(s), under exclusive licence to Springer Nature Switzerland AG 2020
C. Forstner and M. Walker (eds.), *Biographies in the History of Physics*,
https://doi.org/10.1007/978-3-030-48509-2_1

(Greene 2007, pp. 730, 746). However, Margit Szöllösi-Janze cites Pierre Bourdieu's warning against the "biographical illusion," whereby the chronological narrative of the *Bildungsroman* can suggest more coherence and meaning in a life that was actually there (Szöllösi-Janze 2000, p. 30). According to Helmuth Trischler, the main task of biography, "… describing individual lives, is not… an end in itself, rather it lies in the connection of the general with the particular, in the mediation between individuality and sociability by means of person-centered research" (Trischler 1998, p. 46).

Mary Jo Nye proposes three principal forms of biography for the history of science. First is the 'life of the scientist,' where the biographer studies a scientist as the "chief protagonist in a broad sweep of historical events" that transcend that scientist's own work. Examples include several biographies of J. Robert Oppenheimer and Albert Einstein, Ruth Lewin Simes' book on Lise Meitner, David Cassidy's study of Werner Heisenberg, and Szöllösi-Janze's biography of Fritz Haber (Nye 2015, p. 284). Here one could add Crosbie Smith's and Norton Wise's work on Lord Kelvin (Kragh 2015, p. 275). Next is a 'scientific biography,' interested mainly in the scientist's contribution to scientific knowledge. These include Helge Kragh's study of Paul Dirac, Kostas Gavroglu's book on Fritz London, and Frederic Lawrence Holmes' biographies of Antoine Lavoisier, Hans Krebs, and Claude Bernard. Finally, Nye describes a 'biography of scientific collaboration,' which of course focuses on individuals, but does not "distort the processes of science" by placing excessive emphasis on their roles. Examples include Sam Schweber's history of the creation of quantum electrodynamics, Deborah R. Coen's study of three generations of scientists in Vienna, and Nye's own work on Michael Polanyi (Nye 2015, p. 284).

One of the strongest advocates for biography in the history of science, Thomas Söderqvist, breaks this down into seven subgenres:

(1) Biography as contextual history of science (*ancilla historiae*);
(2) Biography as a means for understanding the construction of scientific knowledge;
(3) Scientific biography and the popular understanding of science;
(4) Scientific biography as belles-letters;
(5) Scientific biography as public commemoration (eulogy);
(6) Scientific biography as private commemoration (labor of love);
(7) Scientific biography, research ethics, and the 'good life.'

For Söderqvist, biography "is not just history by other means." It allows scientific results to be understood in the context of the scientist's "motivations, ambitions, ideas, feelings, personality traits and personal experiences." Biographies can have an intrinsic aesthetic value and can pay respect to the deceased by erecting a "symbolic gravestone." Perhaps most striking is Söderqvist's description of an "existential biography," which focuses both on the life and the writing of the biography as achievements in themselves.

> … one could argue that it is a good thing to write about recent life scientists in order to understand their work and their lives, but it is an equally good thing to write about them as a way of practicing the care of one's own scholarly self. Writing the history of

the life sciences and writing βιοι of contemporary life scientists are thus ways by which historians, biographers, and scientists alike can explore the perennial question of how to craft a worthwhile life-course out of talent and circumstances. (Söderqvist 2011, pp. 636, 637, 642, 647)

Scholars have challenged the traditional definition of biography by seeking to expand it beyond studies of an individual or collections of people to material and immaterial things. In an article on the "Cultural Biography of Objects," the anthropologists Chris Gosden and Yvonne Marshall argue that human and object histories influence and inform each other: "... as people and objects gather time, movement and change, they are constantly transformed, and these transformations of person and object are tied up with each other," giving both of them a 'biography.' As an example, they discuss the role played by tabua, whole whale's teeth, in nineteenth century Fiji. These circulated as part of a ritualized currency exchange between gods, chiefs, and people. Tabua were cradled in the hand, becoming darker in color over time from the oils of the many people who held them and accumulating the power of their successive chiefly owners. "The depth of a tabua's colour, as indicator of a lengthy biography, is a primary determinant of a tooth's value" (Gosden and Marshall 1999, p. 169).

With the turn towards material things, the biography of objects has also found its way into the history of science. Researchers working with scientific instruments have chosen this approach in particular to explore the different levels of meaning for an object in changing environments. David Pantalony, for example, demonstrates these levels in a medical radiation device, the Canadian Theratron Junior (Pantalony 2011). Samuel Alberti has outlined a history of museums written through biographies of objects in their collections (Alberti 2005). Katherine Anderson, Mélanie Frappier, Elizabeth Neswald, and Henry Trim use an artifact to reveal rich and complex networks of narratives (Anderson et al. 2013).

Lorraine Daston and Otto Sibum expand the genre of biography in a different way by examining the scientific persona: "Intermediate between the individual biography and the social institution lies the persona: a cultural identity that simultaneously shapes the individual in body and mind and creates a collective with a shared and recognizable physiognomy." Here the authors in Daston's and Sibum's edited book work are more like botanists than biographers, piecing together a type specimen that represents a class instead of an individual. Persona are rare, and not to be confused with professionalization or institutionalization: "to be a pastry chef or for that matter an inorganic chemist is to follow a profession, but not to embrace a persona. To achieve a persona presupposes a certain degree of cultural recognition, as well as a group physiognomy that can be condensed into a type..." J. Robert Oppenheimer embodied the persona of a modern theoretical physicist by combining elements of the theorist, teacher, administrator, and advisor (Daston and Sibum 2003, pp. 2, 5).

The true bogeyman for current biographers of scientists is undoubtedly Theodore Arabatzis and his book *Representing Electrons: A Biographical Approach to Theoretical Entities*. Helge Kragh, in a chapter in a different book co-edited by Arabatzis, went so far as to say: "Although admitting that this is biography only in a metaphorical sense, he [Arabatzis] maintains that it is a legitimate and useful notion. The question is whether it is biography at all" (Kragh 2015, p. 279). Söderqvist, in his contribution to our book, makes his skepticism clear:

> ... mental constructs are not persons (or assemblages of persons) and do not have any of the properties of personhood; a concept does not literally have consciousness, memory or emotions, and thus does not have a life of its own. Arabatzis' and other historical studies of concepts and theoretical entities can therefore not be called a biographical study in any meaningful way, unless the terms 'life' and 'life course' are defined so broadly that the denotation of 'biography' includes the description and analysis of the change of all kinds of mental constructs over time. (Söderqvist, p. 304)

Arabatzis explains how he came to the genre of biography through the concept of representation. He had assumed that representation was a "plastic resource" that physicists and chemists could manipulate at will. But when he examined what had happened when a new property, spin, was attributed to the electron, Arabatzis realized that the existing representation of the electron did not tolerate the newly suggested property and indeed led to a violation of the special theory of relativity. Physicists were then forced to modify the representation of the election to accommodate spin. This suggested that theoretical entities like the electron, defined by Arabatzis as constructions from experimental data, were "active agents whose internal dynamic transcends the beliefs, abilities, and wishes of human actors and acts as a constraint on the development of scientific knowledge. Even though they are the products of scientific construction, they have a certain independence from the intentions of their makers; that is, they have a life of their own..." Here Arabatzis makes clear that he is using the term 'biography' in a metaphorical sense, and recognizes that every metaphor has its limits. The history of the electron resembles a biography in some respects, but not others. "In particular, I do not want to attribute intentionality to the representation of the electron, or to imply that it had wishes or other anthropomorphic features" (Arabatzis 2005, pp. 35, 46).

These three examples, Gosden and Marshall, Daston and Sibum, and Arabatzis, all explicitly use the term 'biography' in unconventional ways, but arguably do not stray that far from its essence. The whale teeth do not have a 'life' of their own, rather only in the interaction with the people who desire and exchange them. A scientific persona would be meaningless without the many lives of the scientists that this concept embodies. Finally, in a sentence from Arabatzis' book that his critics rarely quote, he makes clear that "... this book is a biography of the representation of the electron and not of the electron itself." (Arabatzis 2005, p. 49) In other words, Arabatzis, like Gosden and Marshall, and Daston and Sibum, can be seen as being engaged in a specific form of collective biography: how have the representations of the electron, each created by one or more individual scientists, changed over time?

This Book

This book brings together both biography in Nye's and Söderqvist's senses and approaches that attempt to stretch this genre. The first section includes 'classical'

1 Introduction

biographies of individuals. Klaus Hentschel reconstructs the life of Moritz Meyerstein, instrument maker and *Mechanius*, a fine toolmaker or specialist in the maintenance of apparatus and instruments. Meyerstein was an 'invisible hand' in nineteenth-century science, overlooked both by contemporaries and historians. Hentschel's sophisticated and subtle account reveals many different facets of Meyerstein: his career as a nineteenth-century instrument maker; his innovations in instrument technology; the scientific experiments performed using his instruments; a member of the petite bourgeoisie; and "a baptized native Jew who [succeeded] against considerable odds." (Hentschel, p. 22)

Joanna Behrman's essay on Margaret Maltby, one of the pioneering women in American physics, also sheds new light on her subject's life. This is done, not by focusing on what Behrman calls 'traditional' markers of a scientific biography, rather by blurring the line between the professional and the personal. Behrmann unearths a personal scandal that might well have derailed a woman scientist's career, but Maltby instead navigated in a personally fulfilling and professionally successful way. Historians can, Behrmann argues, "wield an incredible power over what is remembered, and how. Is this one of those stories in need of telling?" (Behrman, pp. 54–55)

The visits Erwin Schrödinger paid to Spain during the 1930s is the subject of Enric Pérez's chapter. These took place during the so-called "Silver Age," when Spanish culture and especially science flourished. In contrast to Albert Einstein, who also visited Spain between the two World Wars, Schrödinger was not a celebrity, so that only physicists and physics students were aware of his visits. At the time, Schrödinger was developing his critique of the Copenhagen interpretation of quantum mechanics. Because there was a dearth of theoretical physicists in Spain at the time, "quantum mechanics was arriving in Spain via Schrödinger's views." (Pérez, p. 71) This was all cut short by the Spanish Civil War.

Michael Eckert's chapter investigates the career of Ludwig Prandtl, a pivotal figure in German science, aeronautics, and science policy during the first half of the Twentieth Century. Prandtl was a father-figure to a group of scientists and engineers who were attempting to provide a solid scientific basis for applied mechanics. Whether Prandtl was studying the challenging subject of turbulence, founding world-class, innovative scientific institutions, or serving as a goodwill ambassador for the Nazi regime, he always saw himself as unpolitical. Indeed as Eckert explains, the "myth of the unpolitical scientist relies on the separation of scientific merits from political attitude." (Eckert, p. 85)

Rudolf Tomaschek, a German experimental physicist during the Third Reich, is the subject of Vanessa Osganian's chapter. Contemporaries saw his appointment in Munich as a victory for the 'Aryan Physics' (*Deutsche Physik*) movement, but Osganian shows that this story is more complicated by reconstructing and analyzing Tomaschek's diverse connections to a complex network of conservative and *völkisch* (racist-nationalist) scientists. Indeed this complexity may have allowed Tomaschek to be successful, both within and outside the 'Aryan Physics' movement, "without taking a distinctive stand or clearly marking his position." (Osganian, p. 106)

Dieter Hoffmann and Mark Walker examine the career of Pascual Jordan during the Third Reich. Although perhaps most German scientists had to accommodate themselves to the Nazi regime, Jordan stands out because he combined in one person traits that usually would be considered contradictory. He was both a co-founder of quantum mechanics and a vigorous Nazi propagandist. Despite being a fanatic Nazi, Jordan clashed with and triumphed over his 'Aryan Physics' opponents because in the end he was much more useful to the tyrannical National Socialist state, especially through his involvement in technical military projects like rocket development. "His biography reveals a symbiotic relationship between some of the most modern developments in science and the reactionary ideology of National Socialism." (Hoffmann and Walker, p. 126)

Hans-Jürgen Treder, the subject of Raphael Schlattmann's chapter, initially must have appeared to the East German Communist authorities as the ideal young scientist: both very talented in physics and "consciously biased, anti-fascist, committed and capable; merging the politically intended world view with modern science." (Schlattmann, p. 141) However, Treder's attempts to become a teacher of Marxist-Leninist philosophy were ironically redirected by his own academic mentors in the Socialist-Unity-Party and science bureaucracies, who told him to abandon philosophy and dialectics as a profession and instead devote himself to physics. Indeed Treder abstained from philosophical topics until 1963, after the construction of the Berlin Wall, by which time he was fully integrated into the East German science system.

Christian Forstner scrutinizes the popular perception of Richard Feynman as a 'genius.' As a young, aspiring physicist, "Feynman was enthusiastic about finally having arrived at the frontline of contemporary research, meaning he was thrilled by the questions specified by the thought collective and thought style." (Forstner, p. 150) His famous argument, that the path to the goal did not matter, so long as it was supported by experimental results, reflected the pragmatism of the physics community in the United States at the time. Forstner breaks up the physicist's perceived 'genius' into three images: Feynman as the 'magician'; Feynman removed from the world; and Feynman as a 'revolutionary man.' All of these were reinforced by the physicist's own popular writings.

Although less common than individual biographies, there is also a long tradition of collective biography. Mark Walker examines how the German atomic bomb was perceived by different groups, and how this perception changed over time. It "is still influential today, not really because of what happened, but instead because of what might have been." (Walker, p. 176) The discovery of nuclear fission in 1939 appeared to be an amazing source of tremendous energy. After the war began, outside of Germany the potential of nuclear fission became the specter of powerful weapons in ruthless Nazi hands. Once the war turned against Germany, scientists were able to keep themselves away from the front lines by suggesting that these weapons might be able to stave off the looming defeat. In the postwar period, the German atomic bomb, which had never actually existed, was used both as evidence that, on one hand, the German scientists had been incompetent, and on the other, that German scientists had resisted the Nazis by denying nuclear weapons to Hitler.

1 Introduction

Roberto Lalli examines the formation of a community of scientists. Albert Einstein's theory of General Relativity had been marginalized in comparison to the rapidly-developing nuclear and solid state physics since the 1920s, but by the 1950s had returned to the mainstream of physics. Physicists and mathematicians came together to form an informal elite organization called the International Committee on General Relativity and Gravitation (ICGRG). Lalli approaches the history of this committee from a biographical perspective, as the "living embodiment of a scientific domain." Biography is used "both as a metaphor and an analytical tool" (Lalli, pp. 180, 198) to understand how the early conferences developed into a mature professional organization, despite being affected by the turbulence of the Cold War.

Two founders of premier scientific institutions, Robert Wilson of Fermilab and Hermann Grunder of Jefferson Lab, are compared and contrasted by Catherine Westfall in what she calls a 'shared biography.' This is a study of what the lives of the founding directors, the laboratories, and the instruments, which were the reason for building the laboratories in the first place, had in common. Such an analysis requires knowledge of all that surrounds and interacts with the laboratory, including "the scientific communities that provide standards, elite decision makers, and outside accelerator users; local communities that shape and are shaped by the presence of the laboratory; and political establishments that provide financial and cultural support." (Westfall, p. 216)

As noted above, far more controversial than a collective biography are the biographies of objects. Daniel Belteki examines the Airy Transit Circle, a combination of transit instruments (to measure when a celestial body crossed the meridian, or the point of transit) and mural circles (to measure the angle at which the transit occurred). The history of this instrument "emerges not as one that passively gave into the wishes of its users, but rather, as a sensitive instrument that through its ever-changing character resisted the physical control of its users." (Belteki, p. 229) The data gathered and published from the Airy Transit Circle demonstrates that the instrument was not a passive executor of human commands, but instead had its own 'personality.'

Panagiotis Lazos' contributes a different type of essay on Franklin's pulse glass, which developed from a toy made by artisans to a scientific instrument. The toy came into the hands of Benjamin Franklin, Count Rumford, and other natural philosophers who adapted it in order to extend the frontiers of research on the propagation of thermal radiation. Eventually these devices were replaced in research, but continued to be used in scientific instruction in schools up until the twentieth century. Today the pulse glass has completed "an almost circular path" (Lazos, p. 237) and can be found in the form of toy dipping birds.

Finally, three authors examine the limitations inherent in the extension of biography beyond individuals or groups. Markus Ehberger questions the application of biographical metaphors to objects by asking whether it makes sense to speak of the 'birth' of the virtual particle? In the course of his investigation, Ehberger considers mathematical techniques, physical ideas and processes, diagrammatic methods, empirical evidence, and terminology. Assuming that a 'birth' means something completely new is formed, Ehberger argues that this would "force us to make a distinct judgement when a specific conception or theoretical representation could be called a virtual particle, at the expense of treating the framework in which the actors worked and thought on their own terms." (Ehberger, p. 277)

In contrast to Ehberger, Jaume Navarro's study of the ether is more open to the idea of a biography of an object. By the early 1950s, most physicists had abandoned the ether, but only a generation before it had been very much 'alive.' The ether "worked as a kind of interstitial concept that bridged the classical and the modern, the cultural and the scientific, the material and the spiritual, the British and the Continental." (Navarro, p. 282) For a time the development of technologies like radio appeared to save the ether. It was obviously real, the argument went, because without it wireless would be impossible. When scientists declared the ether dead in an 'obituary,' Navarro shows, they also reformulated the object in order to justify its demise.

The final chapter comes from Thomas Söderqvist, the biographer of Niels Jerne. (Söderqvist 2003) Söderqvist has two aims: "to remind historians of science that the genre of biography, including scientific biography, is about people, not institutions, concepts, or objects; and, secondly, to bring autobiography and memoir into the discussion." After analyzing the use of 'biography' for historical studies of scientific institutions, theoretical entities, and material objects, Söderqvist concludes that this concept should be restricted to studies of the lives of individual persons. However, he also argues that the value of autobiography has been underappreciated and could "make future discussions about the relation between scientific auto/biography and the history of science more varied and interesting." (Söderqvist, pp. 301, 315)

This book grew out of a workshop we held at the very accommodating Physics Center in Bad Honnef with the generous support of the Wilhem and Else Heraeus Foundation and the German Physical Society. We are very grateful to both the Society and Foundation. The inspiration for this book came from the work of Dieter Hoffmann, who is also a contributor to it. Dieter's love of biography and generous support of colleagues have made important contributions to the historiography of nineteenth- and twentieth century physics. We dedicate this book to Dieter.

References

Alberti, S. L. (2005). Objects and the museum. *Isis, 96,* 559–571.
Anderson, K., Frappier, M., Neswald, E., & Trim, H. (2013). Reading instruments: Objects, texts and museums. *Science & Education, 22,* 1167–1189.
Arabatzis, T. (2005). *Representing electrons: A biographical approach to theoretical entities.* Chicago: University of Chicago Press.
Daston, L., & Otto Sibum, H. (2003). Introduction: Scientific personae and their histories. *Science in Context, 16*(1/2), 1–8.
Gosden, C., & Marshall, Y. (1999). The cultural biography of objects. *World Archaeology, 3*(1/2), 169–178.
Greene, M. T. (2007). Writing scientific biography. *Journal of the History of Biology, 40*(4), 727–759.
Hankins, T. (1979). In defense of biography: The use of biography in the history of science. *History of Science, 17,* 1–16.

Kragh, H. (2015). On scientific biography and biographies of scientists. In T. Arabatzis, J. Renn, & A. Simões (Eds.), *Relocating the history of science. Boston studies in the philosophy and history of science 312*, (pp. 269–280). Cham: Springer.

Nye, M. J. (2015). Biography and the history of science. In T. Arabatzis, J. Renn, & A. Simões (Eds.), *Relocating the history of science, Boston studies in the philosophy and history of science 312* (pp. 281–296). Cham: Springer.

Pantalony, D. (2011). Biography of an artifact: The Theratron junior and Canada's atomic age. *Scientia Canadensis, 34,* 51–63.

Söderqvist, T. (2003). *Science as autobiography: The troubled life of Niels Jerne.* New Haven: Yale University Press.

Söderqvist, T. (2011). The seven sisters: Subgenres of 'βίοι' of contemporary life. *Journal of the History of Biology, 44*(4), 633–650.

Szöllösi-Janze, M. (2000). Lebens-Geschichte–Wissenschafts-Geschichte. Vom Nutzen der Biographie für Geschichtswissenschaft und Wissenschaftsgeschichte. *Berichte zur Wissenschaftsgeschichte, 23,* 17–35.

Trischler, H. (1998). Im Spannungsfeld von Individuum und Gesellschaft. In W. Füßl & S. Lintner (Eds.), *Biographie und Technikgeschichte* (pp. 42–57). Opladen: Leske & Budrich.

Part I
Individuals

Chapter 2
The 'Invisible Hand' of Carl Friedrich Gauß—Retracing the Life of Moritz Meyerstein, a 19th century Instrument Maker and Universitäts-Mechanicus

Klaus Hentschel

The Instrument Maker and *Mechanicus* as a Profession

The term 'mechanic' was originally minted during antiquity, coming from the Greek μηχανικός, and began to be used sporadically in German as '*Mechanicus*' or in English also as 'mechanist' since the mid-sixteenth century. (OED, 2nd ed., vol. 9, pp. 534f.) Its more frequent usage occurs toward the end of the eighteenth century, often in parallel with the designation '*opticus*'. Such opticians rather specialized in such instruments as telescopes and microscopes as well as in the grinding of lenses or polishing of glass surfaces generally. The professional field of a *Mechanicus* signified more than the later mechanic, a fine toolmaker or specialist in the maintenance and repair of apparatus and instruments (*Kluges Etymologisches Wörterbuch*, Hirsch 1985, p. 47, Weil 2000, §1). *Mechanicus* referred to a wide palette of instrument makers and model builders, from the 'weather-glass maker' to a lecturer's assistant in performing demonstration experiments. (The importance of clockmakers and eyeglass grinders is discussed in Loewenherz 1882, pp. 216–219, Behrendsen 1907, p. 95, de Solla Price 1984, p. 11.) Depending on the local circumstances of the clientèle—be they princes, academies, universities or private scholars and professors—the mechanic (then nearly always male) also made mathematical, astronomical, geodetic, meteorological or surgical instruments. The professional differentiations arose out of the guilds for toolmakers and locksmiths, compass-, tool-, and clock-makers as well as eyeglass grinders. By the time Meyerstein was born, this process had already come to a close. However, the tasks performed by this professional field were rapidly changing with the technical developments. The universal instrument maker with his

---

Translated into English by Ann M. Hentschel, a freelance translator specialized in the history of science.

K. Hentschel (✉)
University of Stuttgart, Stuttgart, Germany
e-mail: klaus.hentschel@hi.uni-stuttgart.de

© The Editor(s) (if applicable) and The Author(s), under exclusive licence to Springer Nature Switzerland AG 2020
C. Forstner and M. Walker (eds.), *Biographies in the History of Physics*,
https://doi.org/10.1007/978-3-030-48509-2_2

broad array of products touching almost all areas of the experimental sciences became increasingly specialized in an individual sector, such as, just in mathematical tools or just in optical astronomical instruments. A transition took place from the skilled crafting of unique pieces to serial production with the division of labor. For instance, toward the end of the nineteenth century the university and court mechanic Carl Zeiss (1816–1888) made the transition into industrial manufacturing in bulk, marrying science with technical expertise by engaging suitable scientists—in his particular case, the physicist Ernst Abbe (1840–1905). A similar coupling occurred in the close cooperation between Moritz Meyerstein and Carl Friedrich Gauß (1777–1855).

German industrialization occurred smack in the middle of the life of our principal figure, almost a full century later than in Great Britain. It effectuated a boom in the German instrument-making sector, particularly in precision engineering and optics, which could regain its lead against the hitherto stronger British competition in the art dating back to its early-modern heights in Nuremberg and Augsburg. This also applied to electrical engineering during the second half of that century. Faraday's discovery of electromagnetic induction in 1831 or the technical development of electrical telegraphy by Gauß and Weber were converted by others into practically operable telegraph systems, electric generators and motors as well as communications networks spanning the continents.

The founding of the German customs union, *Deutsches Zollverein* in 1833, which Hanover also joined in 1854 along with other members of the Northern German tax union (*Steuerverein*), as well as the founding of the German Empire in 1871, led to intensified trade throughout the region. The demand for compatible and accurate weights and measures rose accordingly for the new state struggling toward standardization. The imperial authority on standard calibration, the *Kaiserliche Normal-Aichungs-Commission* was established in that same year and the German bureau of standards soon followed in 1877. The task of the *Physikalisch-Technische Reichsanstalt* (PTR) in Charlottenburg on the outskirts of Berlin was to test and certify scientific instruments, make available metrological standards, and develop new instruments. (Brachner 1985, pp. 121f.; Cahan 1989; Weil 2000, §15; Terry Shinn in Joerges and Shinn, eds. 2001; Hoffmann and Witthöft 1996; cf. the many papers by Dieter Hoffmann on the PTR and its coworkers.) Scientific and technical associations such as the Bavarian *Polytechnische Verein für das Königreich Bayern* founded 1815, the Prussian association for the promotion of commerce, *Verein zur Beförderung des Gewerbefleißes in Preußen* founded in 1821, or the Hanoverian trade association *Gewerbeverein*, founded 1828 and reorganized 1834, offered practitioners in the field the possibility to present newly developed instruments or technical processes in their publishing organs and to discuss them personally during annual conventions and expositions. The section about trade fairs in my Meyerstein biography shows that such visits, to the World Exposition in Paris in 1855 for instance, inspired our protagonist to implement new ideas on a small scale in activities by the Göttingen trade association, which he presided over for a while. Meyerstein's death date was nearing when the local professional trade associations united nationwide in 1879 to form the *Deutsche Gesellschaft für Mechanik* with 97 founding members—*und Optik* being appended to its official name in 1881, with 133 instrument makers counting among its growing membership besides 62 others. The periodical *Central-Zeitung für Optik und Mechanik*, appearing from 1880 on, along with this society's own organ *Zeitschrift für Instrumentenkunde* from 1881, was also the place to publish company notices and obituaries of major and minor master instrument makers. Ambitious artisans could dare to submit articles about newly developed instruments to other reputable journals, such as Poggendorff's *Annalen der Physik*, Schumacher's *Astronomische Nachrichten*, or Henle's *Zeitschrift für rationelle Medicin*. Longer contributions were better placed in the *Repertorium für physikalische Technik* edited from 1866 by the Munich experimental physicist and instrument maker Philipp Carl. A glance at lists of instruments by Meyerstein from 1845 and 1860 shows that he was quite successful in placing such reports about his own instruments. Twice he even managed to have a full listing appear of his deliverable instruments—this was actually a form of free advertisement. These Meyerstein price lists are reprinted in my Meyerstein book (Hentschel 2005, pp. 268–288; on fair trades see Sect. 2.11).

The Difficult Situation of the Sources

The 'hero' of our biography was not famous. This means that not many letters, documents, nor scraps of writing of his had been carefully saved up on the consideration that it might well prove to be valuable. The University of Göttingen had been Meyerstein's workplace for decades, but its archive had kept no staff file bearing his name. Unlike for professors, the files of a mere mechanic were clearly not deemed worthy archival material. That doesn't necessarily mean that there was nothing to be found in this and other archives. What little had survived simply had to be ferreted out from other nooks and crannies in their holdings. They include a general file on *Mechanici*, for instance, as well as some academic department files, such as a dossier on the *Mathematisch-physikalische Seminar*, founded 1850; the records of the *Magnetische Verein*, along with log journals by its participating scientists; inventory lists for various instrument collections; documentation on honorary degrees; municipal files; parish records; and the '*Copulations-Register*' (i.e., marriage lists). The research techniques needed for this kind of investigative work are those of a social historian, who must consult city directories, church books, tax rolls and other serial sources in order to be able to attribute to a given individual sometimes known only by name, a world of work, living conditions and a residence, a sphere of influence and a context, and perhaps ultimately something like an identity. What has carried down to posterity is much more selective than is usually the case with 'great minds', because usually only aspects of legal or economic import in the lives of the 'little folk' find their way into an archive: certificates of birth, marriage and death. In our present case we also have his official establishment as a mechanic, which was certainly not a simple step to take due to major resistance by local artisans (Hentschel 2005, pp. 61–77); his engagement as '*Universitäts-Mechanicus und Maschinen-Inspector*' at the University of Göttingen even necessitated, among other things, his conversion to

Christianity. There is also Meyerstein's interaction with renowned clients from all over the world, which was by no means limited to business concerns.

Paradigmatic studies by Natalie Z. Davis on three female figures of early modernity served as my model in delving into potential source material. They show how one can succeed in reconstructing the life and works of a relatively unknown individual notwithstanding a difficult source situation and how exciting such a historical 'manhunt' can be. As yet, the history of science has little to offer of comparable quality, unfortunately. (Davis 1995; on instrument makers: de Clerq 1985; Hirsch 1985; Williams 1994; Clifton 1996, the thoroughly researched series of miniatures by Paolo Brenni on a number of French instrument makers in *Bulletin of the Scientific Instrument Society*, Hentschel 1993 on Rowland's laboratory assistant Lewis E. Jewell, as well as K. and A. Hentschel 2001 on Dulos, a French copper engraver and illustrator.) The biography—including the 'scientific biography' in the history of science—is one of the most popular genres on the book market. Even so, biographical studies on scientific instrument makers are extremely rare. Museums have compiled some elaborate catalogues on a few instrument makers from the eighteenth and nineteenth centuries, with particular care given to illustration quality for their instrument exhibits; but from this perspective the lives and research contexts of the focal artisans tend to be given marginal treatment. (Brachner et al. 1983 on G. F. Brander; Kirchvogel 1938; Breithaupt 1962; von Mackensen 1987 on the Breithaupt dynasty; Turner 1989 on the revolution in the manufacture of precision instruments based on the case of E. Lenoir.) Comparable monographs exist on some British and American instrument makers (Millburn 1976, 2000 on Benjamin Martin and Adams in the 18th c., Warner 1968 on Alvan Clark & Sons in the early 20th c.), on the Berlin instrument maker Carl Philipp Heinrich Pistor (1778–1847), and on the Hamburg master fireman and mechanic Johann Georg Repsold (1770–1830), as well as on Fraunhofer's successors, Georg and Sigmund Merz in Munich (Weil 2000; Koch 2000, 2001; Kost 2015; further Loewenherz 1887, pp. 208–215; Repsold 1914, pp. 18ff., 59ff.)—These monographs are comparable as regards topic and completeness, albeit not as regards the focal issue, narrational form or sociohistorical background. Otherwise, if at all, only brief obituaries or memoir articles have appeared about this group of persons. Another parallel among those mentioned above exists as well: a common involvement in the establishment of standard lengths, weights and measures (*Etalons*)—Pistor for Prussia, Repsold for the Danish, and Meyerstein for the Kingdom of Hanover (Hentschel 2005, pp. 104ff., 2007).

The Documentary Biography

In writing my book about Meyerstein from 2005 it was a priority for me to preserve as much of the flair of his time as possible. That is why I chose the form of a documentary biography which cites heavily from the surviving primary literature. Such material was available in a dozen archives and other collections. This form of scientific biography, more familiar under the label 'the life and letters of X', has a long tradition going back to David Brewster's biography of Newton in the early nineteenth century. (On the history of biography and problems of a methodical nature see, e.g., Engelberg and Schleier 1990; Shortland and Yeo, eds. 1996; Füssl and Ittner 1999; for more recent biographical approaches in social history see Gestrich et al. eds. 1988). The special attraction of the present study is that our X is in the lower case—in other words, the figure is not as renowned, which adds a very different quality to the cited primary sources. Moreover, Meyerstein was not only an exceptionally gifted instrument maker, setting him apart from other members of his guild. He was also one of the few Jews to succeed prior to 1850 in transgressing the cramping legal and social limits set for this group of the population. His grandfather and his father had been confined within the trade sector; he was able to advance beyond, into the artisanal field of scientific instrument making, which due to a variety of reasons was the domain of his Christian contemporaries. As a consequence, the first chapters of my book focus on describing his family background and marriage situation as well as the fates of his many siblings, which makes it easier to understand his own career choices and personal decisions.

Two of his elder brothers received financial support from their father to study medicine or chemistry at the local university, the *Georgia Augusta* in Göttingen. Moritz himself being the youngest son was less fortunate. He probably had to leave school prematurely because the family could not afford to support another son academically. But even if this had been possible, there was no suitable vocational school in the area for someone with his interest in technical subjects. Compared to other major German-speaking regions, the Kingdom of Hanover lagged behind in opening its own trade school. Vienna founded its first in 1815, Berlin in 1820, Karlsruhe in 1825, Munich in 1827, Dresden in 1828, and Stuttgart in 1829. The *Hannoversche höhere Gewerbeschule*, established in 1831, was later renamed *Polytechnische Schule* and was subsequently raised to a *Technische Hochschule*, in 1879.[1] As regards the training options for aspiring technicians and mechanics during Meyerstein's youth, the German provinces could only look up to such models as *Gresham College* in London or the *École Polytechnique* in Paris. (On the gradual improvement during the 19th c., see Lundgren 1973, 1988; more specifically on the "phase of a beginning practical enlightenment of municipal craftsmanship in the Kingdom of Hanover" see Huge 1990 which treats esp. the pedagogical activities of the Hanoverian *Gewerbeverein*. For a survey of the local technical educational institutions, including statistics on teachers and students, see *Mittheilungen des Gewerbe-Vereins für das Königreich Hannover*, 1855, esp. cols. 169f.) Moritz was left the only option of apprenticeship under a master of his own choosing. As a journeyman afterwards, he then went to Munich. The fact that this aspiring *Mechanicus* also attended lectures at the university there speaks for his thirst for higher education.

---

[1] The *Höhere Gewerbeschule* run by Karl Karmarsch in Hanover opened in 1831 with a pupil count of only 64; by 1875 it had risen to 868.

The Broad Array of Meyerstein Instruments

Meyerstein's product line resembled Pistor's, Repsold's or Nathan Mendelsohn's, for instance—another rare Jew in this business in Berlin: It included geodetic instruments for the land surveying projects being initiated everywhere at the time; precision balances for chemists and apothecaries; and astronomical instruments for the modern observatories. Meridian circles became the precision instruments they were since the eighteenth century thanks to important innovations by Duc de Chaulnes, Ramsden, and Reichenbach. First-class circular and longitudinal dividing machines were essential for reliable ruling of their scales, which all of the above-mentioned persons were able to master with differing degrees of success. For the quality control of their new instruments, refined calibration apparatus was indispensable, such as the lever gauge, adjustment tools such as the bubble level, and reading aids such as lenses and verniers, etc. It was possible to find out how Meyerstein acquired some aspects of his special technical expertise, his skills. (Hentschel 2004, 2005, pp. 50–60, part 3 on his scientific instruments, and Appendix 2 there for a list of surviving pieces in museums and collections worldwide.) Apart from his regular training as an apprentice under masters in Göttingen, Kassel, and Munich, he also spent some time in Scandinavia where he received instructions in the weighing arts, in geodesy and circular graduation as well as perhaps in the handling of steam engines which were already being employed extensively in mine-rich Sweden.

Meyerstein's products were not decorative 'pretty' instruments like the ones destined for a royal collection such as George III's or those Medici princes' enthusiastic about science, for display purposes. With a few exceptions, they were not demonstration apparatus for scientific instruction either. They were carefully designed and continually improved prototypes, research instruments of interest only to serious experimental scientists because high precision was their strength, not ease of handling. As historians of science have already shown, over the course of the first half of the nineteenth century, precision in the exact sciences had acquired cultural value of the first order (Wise, ed. 1995) in many national contexts, particularly in France, Great Britain and the German-speaking realm (Hoffmann and Witthöft, eds. 1996). The causes were multilayered and ranged from the restructuring of basic scientific disciplines such as chemistry and physics (Cahan 1985, or Olesko 1991),[2] to the establishment of modern institutional structures such as physics curricula focusing not only on instruction but also on research, to upheavals studied by the history of mentality (Lepenies 1976 and Stichweh 1984 discuss the end of natural history and the disciplinary breaks of the *Sattelzeit*). The aims of eighteenth-century natural history included encyclopedic synthesis of knowledge and classificatory system for large amounts of data. In the German-speaking realm, particularly between 1780 and 1820, there was a high appreciation for romantic natural philosophy with a strong mix of characteristic mental motifs like polarity, analogy and metamorphosis. Imaginative invention often took the upper hand in this philosophizing, though. This

---

[2] For example, Franz Neumann's seminer in Königsberg served as Wilhelm Weber's model for the establishment of his seminar for mathematical physics at Göttingen.

approach also had some heuristic potential, however, in the discoveries of what we now call ultraviolet radiation by Johannes Ritter in 1800 or electromagnetism by Hans Christian Ørsted in 1820. There were few advocates of it at the enlightened University of Göttingen, though; Carl Friedrich Gauß counted as an outright opponent of this fashionable romantic natural philosophy, defended so enthusiastically in Jena, Berlin and Munich.

Space constraints prevented my documentary biography of a peripheral figure from going into the details of Gauß's own role, that 'Titan of science' as he has been justly called. Reference must be made instead to the ample literature[3] on this '*princeps mathematicorum*' (Dunnington 1955 and Michling 1887 provide readable introductions to Gauß's life and works; Schering 1881, 1887; Rühlmann 1885, pp. 289–290; Schaefer 1929 and Schimank 1971 go into his physical research; Oesterley 1838, pp. 444–447 lists contemporary awards and his most important publications.) His extensive work in mathematics, astronomy, geodesy and physics was only brushed upon insofar as it pertained immediately to Moritz Meyerstein's life, activities and instrumentation. The same applies to the experimental physicist Wilhelm Eduard Weber (1804–1891); see Anon. 1842, p. 158f.; Riecke 1892; Weber 1893; Wiederkehr 1967, and the introduction to Weber and Kohlrausch 1968, pp. 5–13), who was appointed to Göttingen in 1831 at Gauß's recommendation and assisted his senior colleague by almost thirty years in practical experimentation (Schimank 1936). Despite the great differences between them, Gauß and Weber got on very well personally. They designed many instruments that Meyerstein converted into working prototypes and after thorough testing and often numerous improvements were produced in low numbers for interested clients around the world. The magnetometer of the association on magnetism, *Magnetischer Verein*, stands out among those instrument designs. This association's magnetic observatory, initially erected in the observatory's garden in summer 1833 (later transferred to the garden of the University of Göttingen's geophysics institute on the city margins where it can still be visited), had no iron parts—no nails or door hinges, for instance—and Gauß recommended that it be closely studied when planning to build a similar observatory. As a result, in the 1830s scientists began to come to Göttingen from all over the world to view it. They also became customers of the university mechanic Meyerstein, placing orders for the magnetic instruments that Gauß and Weber had designed. (Schering 1887, p. 54; Schaefer 1929, pp. 38, 47; Wiederkehr 1964, pp. 173f.; Hentschel 2005) The association on magnetism founded by Gauß and Weber thus did its part in increasing the demand in this specialized market (Gauss and Weber 1837, 1840; Wiederkehr 1967, pp. 53ff.; Schröder and Wiederkehr 2000).

---

[3] Many other aspects of his life and environs have already been documented in the *Mitteilungen* periodically published by the Gauß-Gesellschaft, and are cited here where relevant. A comprehensive scientific biography of Gauß satisfying the standards of modern history of science still remains to be written.

In 1837 Weber and six other scholars were dismissed from their chairs for having signed a protest against a breach of the state constitution by the new Hanoverian king Ernst August (1771–1851, Weber 1893; Kück 1934; Wiederkehr 1964, pp. 191ff., 1967, pp. 70–84; Gresky 1985).[4] Unlike three of these 'Göttingen seven', Weber was fortunately not banished from the kingdom and was able to stay in Göttingen as a private scholar and continue to publish other volumes of the 'Results of the Observations by the Magnetic Association' until 1842. But his dismissal put an end to the association's activities. The aging Gauß was unable to maintain it after Weber left for Leipzig in 1843 to fill another chair. Even when it became possible for Weber to return to his tenureship at Göttingen in 1849, Gauß had already moved away from the topic to purely mathematical problems and Weber had to look for other collaborators. Meyerstein was still there as *Universitäts-Mechanicus* and continued to supply the Göttingen physicists with instruments up to the end of his life.

I see an important overall pattern behind Meyerstein's collaboration with these two Göttingen professors. It corroborates a statement about Weber that Gauß himself made in a letter to a trusted friend (to Olbers on 19 Apr. 1838, cited after Wiederkehr 1967, p. 82): "In fact, our researches were so intertangled and enmeshed as is only possible in the friendliest of relationships." It is sometimes extremely hard to keep apart the theoretical contributions by Gauß and Weber in thematically related work. The same applies to the materialization of their metrological ideas as concrete prototypes: Weber and Meyerstein often cooperated just as intensely in their construction. Unfortunately for us historians, this interaction between experimenter and instrument maker occurred almost exclusively in direct conversation while they were looking over each other's shoulders at the work bench. Precious little direct evidence exists about Meyerstein's cooperation with Gauß and Weber, but there are some indirect hints in the form of mentions in their publications. (Hentschel 2005, Sect. 2.5) We can be glad to have these mentions at all, though. It had recently been the general convention to suppress such acknowledgements completely. That's why Stephen Shapin (1989) referred to the operators of Robert Boyle's air pumps as "invisible technicians", with the great seventeenth-century experimenter himself merely giving directions as suited his rank as a nobleman. (For other examples of nameless mechanics, laboratory assistants and other persons behind the scenes, see Hentschel 1998, p. 40, 2009) It is a sign of character that Gauß and Weber praised the achievements of their mechanic publicly so often and acted on behalf of this coworker of theirs otherwise as well (see Gauß's opinion supporting a petition for a raise in Meyerstein's compensation, Hentschel 2005, pp. 144–146). Other contemporary researchers had more qualms about acknowledging the contributions of subordinates outright.

---

[4]The '*Göttinger Sieben*' protested against the reintroduction of the civil law of 1819 in place of the new state constitution from 1833 by the new Hanoverian king, who ascended the throne on 20 June 1837.

There is more behind this than personal vanity or the ceaseless public fascination about 'great' men. A very particular image of science is involved: that of busy and adept assistants implementing the ingenious idea of a charismatic genius. This genius is mostly also a theoretician and, in this distorted picture of theoretical science, experimenters and technicians act as mere agents. If any attention is given to an experimentally occupied scientist, then it is to imagine him as a virtuoso soloist laboring in solitude in his laboratory. Scientific *practice* looks entirely different, however, with exceptions being few and far between. Figures like Gauß, Weber, and Bessel were talented theorists as well as being versed practitioners not hesitant about performing extremely tedious and lengthy series of measurements and if need be struggling with the most obtuse errors generated by their instruments. Yet even they were not alone in completing their experiments and building the necessary apparatus. Practitioners were first in broaching many of the fundamental problems in metrology, such as how to subdivide angles and lengths with utmost precision, or how to make a perfect plane, cylinder or sphere and check them. They would continue to perfect their method in conformance with a centuries-long tradition in the mechanical arts. For example, a clockmaker constructed the first angular dividing engine. The inventor of the telescope was not Galileo (who simply improved it effectively) but a Dutch optician. No place may have been granted to the 'mechanical arts' at universities in that period. Nevertheless, they played a substantial part in the achievements of empirical research since early modern times. This point is made for angular graduation and the 'mechanical arts' in Loewenherz (1882–1887), Ambronn (1919, 1921), de Solla Price (1980) and Herbst (1996); for case studies in astrophysics, spectroscopy and metrology of the 19th and 20th c.: Hentschel (1998, 2002).

Even if the guiding ideas behind the novel electrical measurement instruments unquestionably did *not* originate with Meyerstein but with the two professors Gauß and Weber, it must be granted to Leopold Loewenherz that "numerous designs [were] based on a very simple idea, whose implementation is, however, extraordinarily difficult". As early as 1882 it was thus possible to regard it as "unrisky to make mention of the support of the resourceful mechanic Meyerstein in research even by the great master Gauss" (Loewenherz 1882, p. 215) This happens extremely rarely nonetheless. Even someone like the experimental physicist Clemens Schaefer, who had delved most thoroughly into the material for his overview article about Gauß's physical research in the final volume of the famed scholar's collected works, only mentions Meyerstein in passing and does not even list his name in the detailed index to his contribution. The same blind spot appears in the otherwise so thoroughly researched history of nineteenth-century theoretical physics by Jungnickel and McCormmach (1986, e.g., vol. 1, p. 76: "They [Gauß and Weber] perfected their instruments with one purpose in mind: to aid in the investigation of the mathematical laws"—entirely overlooking Meyerstein's crucial practical involvement in this fine-tuning.) Gauß's biographers Michling and Dunnington do not go into his *Mechanicus* at all either, perhaps because this satellite figure would have skewed their portrait of a 'Titan of science'. I was able to show that Gauß himself had to struggle against this widespread distorted view of science as the contrivance of but a few great minds. In 1836 the

Hanoverian ministry contacted Privy Councillor Gauß, wanting to reduce the uncertainty about the definition of the Hanoverian pound and its conversion into other standard weights. When he tried to delegate some of the intricate weighing work to his mechanic Meyerstein, the ministry protested. It insisted on Gauß's personal involvement and direct oversight, as the Hanoverian Land Survey had also done on a previous occasion. In the ministry's view, the reliability of the measurements was warranted not by the precision of the method or the quality of the scales used but solely on Gauß's personal authority.[5] (See the physics file no. 26 among the Gauß papers, Olesko in Wise, ed. 1995, pp. 119f.) I was able to show from texts about Meyerstein's metrological research how substantial his role in fact was in the production and testing of the Hanoverian standard weights and measures (Hentschel 2005, Sect. 2.7, 2007).

Jewish Emancipation

Another facet of the story is religion. The case of Moritz Meyerstein involves a baptized native Jew who against considerable odds succeeded in a way that would have been difficult even for the son of a middle-class tradesman of Christian roots to achieve. (On the hopes and fears, career strategies and existential threats faced by tradesmen and craftsmen in the German-speaking realm during the 19th c., Blackbourn 1984; on the social structure specifically in Göttingen at the turn into that century, Spörer 1980; Sachse 1986, 1987; Koch 1958; Wellenreuther 2001.) Exceptional talent was an important element of this success, of course, along with a considerable willingness to assimilate, purposeful career planning that took advantage of the training options available in more liberal countries abroad along with the contacts made there with high-ranking members of modern science, and not least the unusual liberal mood and unconditional acknowledgement of Meyerstein's accomplishments at Göttingen, where the level of anti-Semitism among academic circles was relatively low at that time. The decades after 1780 are known as the period of Jewish emancipation. It began in German-speaking lands—depending on location—shortly before the French revolution and ended about three generations later with the constitutional guarantee of legal and political equality for Jews within the German Empire in 1871. My documentary biography reflects many facets of this development, such as the gradual gentrification and assimilation of Jews to the norms and behaviors of the Christian majority, as well as some quite considerable resistance to this social integration and equality in society. (Hentschel 2005, Sect. 2.1)

The starting point of this development was the stratified society of the seventeenth and early eighteenth centuries, when the official authorities merely tolerated the presence of most Jews inside the country. They were prohibited from being landowners

---

[5] I am indebted to Katherine Olesko for referring me to this point. References to the Staats- und Universitätsbibliothek Göttingen, where the Gauß papers are held, are henceforth abbreviated as SUB.

or farmers, from joining the guilded trades, from taking positions in the civil service or aspiring toward higher ranks in the military—in short, they were left outside of the social order in a way only comparable to the vagrant gipsy nations. About 10 per cent of the most impoverished among them, the 'beggar' Jews, relied on the charity of fellow members of their faith and were compelled to move from place to place because they were prohibited from staying longer in any one town. About three quarters of the Jewish population, having managed to make a modest career as a pedlar or shop assistant, for instance, although not allowed to marry were able to gain a temporary permit. Only a narrow middle class of so-called *Schutzjuden*, 'protected Jews', were explicitly tolerated by the regional authorities in most of the territorial states and free cities during the eighteenth century. This residential permit (*Schutzbrief*) had to be bought. It was conditional on the payment of a considerable annual duty within a complicated system of fees that wealthy Jews were expected to pay out to the local authorities. Jews without means had little hope of obtaining such a permit.

Only the local sovereign had the power to issue a regularly renewable permit, but the advice of the magistrate of the pertinent town was normally sought in advance. (Michaelis 1910; Schütz 1994; Sabelleck 1994) A marriage certificate was generally issued only to one or two sons provided that the family was quite wealthy, in order to limit the natural growth of the local Jewish community. Influx from elsewhere was checked by restricting the right to move around, which most states granted to Jews only around 1848 or later. (Richarz 1976 offers the general social history; on the local situation in the Kingdom of Hanover, which joined Prussia in 1866: Wilhelm 1973; Schütz 1994, and Gerdes in Bertram et al 1999. Compare Kieckbusch 1998 pp. 189ff., 213ff. on the employment and income statistics of Jews in neighboring Holzminden.) During the first half of the nineteenth century there was a strong professional shift even so. In Prussia the percentage of Jews employed as merchants or innkeepers in 1813 dropped from 91.8 to 43% in 1843, only 19% were craftsmen, 14% were day laborers or servants and 2.7% were employed in academic or artistic professions (statistics from Richarz 1974, p. 87: in cities such as Berlin the last group came to about 6% of the Jewish population, in rural regions such as Posen (Poznán) only 1.5%, and about half were academics, for instance, 49 academics in Frankfurt in 1844, or 252 artists and scholars in Berlin in 1852).

The Meyer family from which our protagonist originated also had to pay such residence duties in Einbeck, a town thirty kilometers north of Göttingen, during the eighteenth and nineteenth centuries. It had to write petitions for permission to marry and struggle with countless other special duties and restrictions affecting the daily lives of followers of the Jewish faith. Our case can serve as exemplary of the dynamics and difficulties attached to the gradual social ascent and professional reorientation of one member of a discriminated minority, with the fate of the family figuring as a kind of backdrop to Moritz Meyerstein's own impressive trajectory. The living conditions of this Jewish family of tradesmen in a small-town setting could be reconstructed on the basis of archival documents (Freise 1975; Gerdes 1999, p. 18 and 21–25 on the town of Einbeck since 1673, with a facsimile of the first page of the *Schutzbrief* issued to our mechanic's grandfather Elias Meyer in 1786; on Göttingen: Wilhelm 1973, 1978) along with a complete family tree (Hentschel 2005, p. 34).[6]

Meyerstein's Workshops

Like his master, the mechanic Rumpf before him, Meyerstein did not keep a shop of his own. We read in the trade journal *Monatsblatt des Gewerbe-Vereins für Hannover* (no. 1, January 1866, p. 3, category for mechanics and opticians, no. 15): "Meyerstein, as University Mechanic, supplies almost exclusively astronomical and physical instruments and works only to order." According to contemporary testimonies, he employed in his workshop on Weender Straße "on average 5 or 6 assistants and 6–8 apprentices. He was very strict with his subordinates, not lavish with praise or compensation. By nature ambitious, he always threw his entire energy into realizing his plans, taking little heed of others in doing so."[7] Up to four of his apprentices received temporary 'room and board' in the Meyerstein household.[8] This large number of apprentices and assistants places Meyerstein's workshop above the statistical average for craftsmen of the period. (Blackbourn (1984, p. 56) mentions for 1800 an average of 0.54 apprentices per master craftsman. Karl-Heinrich Kaufhold (in Abel, ed. 1978, pp. 58f.) mentions manpower figures—including the master—for medium-sized businesses for the same period as between 1.26 in Southern Germany and 1.84 in Silesia, with the larger businesses tending to be in cities and metalworking operations—including scientific instrument makers—at an average size between 1.6 and 2.3.) Only the instrument maker Friedrich Apel, who specialized in the manufacture of simpler instruments in larger numbers, employed more staff than Meyerstein, with about 18 assistants and 10 apprentices in 1844, for example. (Behrendsen 1901, pp. 18–21; Koch 1958, p. 186) A glance at the commercial statistics gathered by the city of Göttingen from a census from the beginning of December 1861 counting the numbers of masters, assistants and apprentices in various craftsman workshops in the city confirms that the sector of "mechanics for mathematical, optical, [and] physical objects" lay far above the average with a total of 34 assistants and apprentices working for just 6 masters. It was a booming business. Apel and Meyerstein, and later also Sartorius, dominated this sector of scientific instrument making so much

---

[6] I thank in particular Ms. Susanne Gerdes in Einbeck for her preliminary work and for many helpful pointers without which this genealogical reconstruction would not have been possible.

[7] Behrendsen (1907, p. 137) does not indicate where he got this information about Meyerstein's character from or the number of his employees, but I presume that in 1907 it was still possible to speak with people who had experienced that time, perhaps even former apprentices.

[8] Holdings of the Göttingen municipal archive, *Stadtarchiv Göttingen*, are abbreviated as StAG. See the taxation file there Kä 48, Klassensteuerrolle 1869, running nos. 2876–2879 on the apprentices Carl Engelbert, Gustav Kornmüller, Georg Bartels and August Müller, and no. 2880 on another tenant in the household: Adolphine Neuß, the widow of a city councillor who was classified under tax bracket 1 (at an annual income of 250 thalers); all apprentices fell under bracket 2 and paid only 2 groschens tax yearly.

that it was very hard for their own trainees to open and maintain their own workshops there.

Biographical descriptions of other instrument makers reveal that their workshops and living quarters were generally modest and cramped. Pistor's workshop was in the rear wing of his home and later crept further into other parts of the house (cf. Weil 2000, p. 5, see Sect. 17 about the typical equipment such as turning lathes). Meyerstein's business, by contrast, was spacious (Fig. 2.2). It bordered on the city bulwark, with his residence facing the busy thoroughfare Weender Straße (Fig. 2.1). The parlor and kitchen were probably on the ground floor, with the bedrooms of his small family of three and a few tenants in the storey above, and the apprentices and household staff members roomed in the upper floors.[9] The Göttingen directory (*Adreß-Buch*) for 1865 indicates that Meyerstein was the registered owner of the corner residence on Weender Straße (no. 4, later renumbered 8, then 82, currently no. 102), right next to the city gate *Weender Tor,* which existed until 1855. His almost cubic corner property with three main stories, was 10 meters broad, with a basement and an attached 19-m long workshop. It was officially classified as a '*Kothaus*'—which normally describes poor housing, but according to the Grimm's *Deutsches Wörterbuch* from 1873, in Göttingen rather means all buildings "not being breweries or having a brewing license" (vol. 5, reprint vol. 11, col. 1896). No. 1 across the street was the gatekeeper's house, no. 2 was the residence of a hauler of goods and other neighbors included a superior court director Baron von Bobers, the lithographer Georg Honig, some shoemakers, carpenters and professors, the university's fencing master, porter and curator. (Adreß-Buch 1857, 1865, p. 44; on their incomes: StAG, Kä 48, Klassensteuer-Rollen 1856–1869) The university's riding hall used to be diagonally across the street. It was typical of small university towns like Göttingen for various classes of residents to live right next to each other. In larger cities, socially segregated living quarters usually emerge. In Göttingen that started to happen when the settlement spread beyond the city mound 'before the gates'; and the eastern quarter became the residential area for professors, the suburb across the stream Leine, called Grone, turned into the industrial district, and the university expanded westwards into other areas.

In 1874 Meyerstein sold his corner house and workshop on Weender Straße to the instrument maker August Becker (born 1838) for an allegedly "very high price" (Behrendsen 1901, pp. 34f., 59–65; Dunnington 1955, p. 272). This native of Göttingen had trained under Wilhelm Apel (not under Meyerstein, as is erroneously contended in Jenemann 1988, p. 200) and then after various jobs in Göttingen and Gotha, he opened his own mechanical and optical shop. After the founding of the German empire he sold this shop and moved back to Göttingen where he initially

---

[9]The file on Weender Straße 102 at the Göttingen building authorities (Bauordnungsamt) only contains blueprints from later times (after 1875). The corner building with an attached workshop in the rear was first occupied by a carpenter and then, after major renovation, by an apothecary for many decades.

Fig. 2.1 The workshop and residence that Meyerstein took over from Rumpf in Göttingen. This undated historical image taken by Arno Stanke (Bildarchiv Häuser, No. IX, W 65) shows the workshop with the arched gateways on the left and the (since modified) residence on the right with the main facade toward Weender Straße, photographed after 1855

worked as a mechanic in the center of town.[10] When purchasing Meyerstein's workshop Becker had hoped to be able to assume his clientèle and complete product line as well, but he was badly mistaken (Behrendsen 1901 p. 35):

> Meyerstein's contracts were certainly already diminishing, the number of anticipated existing orders shrank to a small figure and because Meyerstein even refused to convey incoming orders to the purchaser of his shop, Becker found himself in a difficult position, particularly when Meyerstein, having continued to remain in business for a while, even set up a new workshop and joined Becker's competition.

This new workshop was in the basement of a house that the architect Rathkamp had drawn up for Meyerstein and built 1876 at Bürgerstraße no. 6A (currently no.

---

[10] According to the communal tax rolls: StAG, Communalsteuerrolle 1873 (one of the last years before the introduction of the mark), running no. 1075: Aug. Becker, mechanic, Lange Geismarstr. no. 9, paid a class and general tax of 1 thaler (Th) 18 groschen (gr); for comparison: Prof. Wilhelm Weber's total tax obligation in 1873 was (per no. 2682) 33 Th 18 gr, with 19 gr in addition for his large garden, 6 Th 6 gr for military quartering, and 7 gr property tax along with other smaller items coming to a total tax of 44 Th 4 gr. Comparative figures for Meyerstein unfortunately are unavailable because only the first volume of the communal tax roll, ordered by street, for this one year has survived. Looking at another example, in 1873 the mechanic Wilhelm Apel (no. 363) paid only 12 Th 24 gr, 4 Th 18 gr for military quartering and approx. 8 Th for his house and garden, totaling 27 Th 6 gr; and the mechanic Heistelhagen, who had recently moved into Untere Masch no. 21 from elsewhere paid as little as 8 Th (supplementary vol. 1873, postscript no. 50).

Fig. 2.2 Meyerstein's first workshop in Göttingen by the city mound, photographed by the author around 2000

Fig. 2.3 Meyerstein's later workshop in Göttingen on Bürger Straße (near the later Gauß-Weber-memorial), photographed by the author around 2000

48).[11] It is just beyond the city mound surrounding the old town of Göttingen—incidentally now situated diagonally across from the Gauß-Weber memorial unveiled 1899, and very close to the plot which the University of Göttingen used in 1906 to accommodate their expanding physics and mathematics institutes.[12] In this new house (Fig. 2.3) with a workshop in the basement Meyerstein continued to fill orders for his existing clients as long as he lived, albeit in considerably lower numbers than before. The problems with a diminishing demand that Meyerstein and other instrument makers experienced toward the end of the nineteenth century was part of a more general industrialization trend in the German national economy which drove businesses in the crafts into a variety of trades falling under the labels: production, repair services and retail and only granted secondary importance to the original manufacturing sector (Rößle 1956, p. 44; Voigt 1956, p. 28; Zorn 1956; Aßmann and Stavenhagen 1969). The workshop of Carl Diederichs, a former apprentice of Meyerstein, who had opened business in 1875, was incorporated 1898 into the large precision engineering firm of Spindler and Hoyer, founded in Göttingen 1892. It specializes in high-sensitivity seismographic, physiological and meteorological instruments (Behrendsen 1901, p. 99–119; Ambronn 1919 p. 33; Brachner 1985, pp. 138, 145–148). Göttingen's culture of instrument making, for which Apel, Rumpf and Meyerstein figure as such prominent examples since the first half of the nineteenth century, was very fruitful (on this longue-durée perspective: von Saldern 1976 and Witthöft 1979, 1990). It has remained alive to this day in the case of the Sartorius firm, and among experts the university town of Göttingen has the reputation of being Northern Germany's 'measurement valley'.

Meyerstein's many achievements were acknowledged during his lifetime and were famed for their originality. One author viewed this "very competent and resourceful man" as "among Germany's most important mechanic artisans" (Behrendsen 1901, pp. 23–25). Another author found it noteworthy that Meyerstein was among the first expert instrument makers not to think it essential to spend his journeyman years in England (Olesko 2002, p. 40). A third historian (Koch 1958, p. 186) considered him "the first Jew to attain a position of repute among the Göttingen bourgeoisie." Meyerstein was apparently also awarded many distinctions, besides the University of

---

[11] See, e.g., the file on land and building tax receipts, StAG, Mappe Einnahmen aus Grund- und Gebäudesteuern ab 1877, no. 1762 (roll 1109, previously 1355) M. Meyerstein at Bürgerstr. 6A: no land tax and 24 marks (M) 75 pfennigs (Pf) building tax (for 1879/81) and 39 M (from 1880/81). A. Becker and spouse Berta, née Gerlach, at Weenderstr. no. 4 (running no. 1544, roll 949, prev. 925) paid 21 M 60 Pf (1877/78), 16 M 20 Pf (1879/80) and 57 M (from 1880/81). Sartorius von Waltershausen (running no. 40, roll 1111, prev. 609) paid 5 M 80 Pf land tax and 54 M, later 78 M 90 Pf building tax. Wilhelm Weber (no. 1260, roll 503, prev. 1191) with his small gardening cottage remodelled into a home at Jüdenstr. 40, was charged just 2 M 20 Pf for land tax and 12 M for the building.

[12] Cf. *Die physikalischen Institute der Universität Göttingen 1906*. The following is based on the file for Bürgerstr. 48 at the Göttingen building authorities, which Ms. Ströbig kindly made available to me upon the approval of Ms. Lemmerich of the land registry office (*Liegenschaftsamt*) as the current owner. Meyerstein's last workshop stands exactly across from the former auditorium of the university's Third Institute of Physics and right next to the building formerly occupied by the Institute of Physical Chemistry (now by anthropologists).

Göttingen's honorary doctorate in 1863 mentioned above. According to one obituary (E. H. 1882, pp. 241f.), he also received the gold medal for science, art and industry issued by König Ludwig II in August 1872, the silver and gold medal of the Hanoverian *Gewerbeverein*, and the Gauß medal conferred by the Göttingen Academy of Sciences, albeit I was unable to locate any primary sources confirming these last four awards.[13]

If one looks back on Meyerstein's professional trajectory from the commercial perspective, the various stages of his life indicate the typical developmental phases in the market for scientific instruments that economists have worked out for product cycles generally (Schmidt 1978, p. 278; Schmoller 1870, pp. 653ff. on the crisis experienced by metal craftsmen and businesses; Huge 1990) During the first years since establishing his business in Göttingen, we have a pronounced **experimental phase**. Meyerstein worked closely with Gauß and Weber on developing numerous prototypes of innovative scientific instruments as well as collaborating on their practical testing in experimental series of analyses (esp. within the context of the Magnetic Association). The broadening and internationalization of his reputation toward the close of the 1830s and 1840s signaled the **expansion phase**. Meyerstein was able to deliver tried-and-true models in larger numbers around the world. The detailed descriptions of many of his instruments, published in articles that he submitted to *Carl's Repertorium für physikalische Technik*, the *Annalen der Physik*, and some manuals of the late 1850s and 1860s, marked the third phase, **maturity**. The spread of the product also began to slow down then. The sale of his workshop in the center of town in 1874 coincided with the **stagnation phase** (as the last longer quotation clearly indicates). His products began to sense the limits of the market's ability to absorb them as well as the increasing pressure by competing industrialized producers. Meyerstein's success as a small businessman in Göttingen also depended on his clearly recognizing each of these phases and making the utmost out of each of them. Of greater weight than this remarkable social and economic rise was his work for the science of his day. It proved to be extremely useful. Meyerstein always placed economic considerations of profit-making from his instruments second against the relentless exigency for ever higher precision. Other instrument makers (such as Wilhelm Apel in Göttingen) produced large quantities of low quality. Meyerstein preferred to position himself in the small market of high-precision instrumentation. Mass production was not his thing. His case exemplifies what some social historians have described as a particular phase that started in the first half of the nineteenth century: the 'practical enlightenment of city artisanry' which is clearly distinguishable from conventional academic scholarship and academic enlightenment, but also surmounted the old corporate order by social rank in the craft sector and embodied a new type of knowledge and activity. Meyerstein's close cooperation with scientists gave him a decisive lead against his

---

[13] Dr. Reinicke's attempt to verify the award of the Gauß medal among the documents in the archive of the *Akademie der Wissenschaften zu Göttingen* yielded no reference to Meyerstein (email from Ms. S. Henschen dated 12. Aug. 2002), which casts doubt on the truth of this last distinction. The archive inspector Ms. Piller at the *Bayerische Hauptstaatsarchiv* was unable to find any documentation about the award of the Bavarian *Ludwigsmedaille* to Meyerstein among their holdings (Außenministerium-Ordensakten 1872–1882).

Fig. 2.4 The sole surviving portrait of Moritz Meyerstein, from his obituary (E. H. 1882)

competitors who were still producing according to the traditional rules of the trade (Fig. 2.4).

Moritz Meyerstein died on April 30th, 1882, at eight o'clock in the evening. His death was reported to the official at the magistrates office of the city of Göttingen on the following day by the university's senior janitor, Heinrich Gaßmann, who "had been instructed about the above case of death by his own information"[14] (*Göttinger Zeitung*, XIX, no. 5612, Saturday 6 May 1882, indicates his age as 73 years, 10 months, 14 days; see also the modest death notice by Betty and Emilie Meyerstein, ibid., Monday, 1 May 1882, no. 5607, p. 3). The funeral took place on May 4th, 1882 at three in the afternoon in the municipal cemetery on Kasseler Landstraße which Göttingen had just opened the year before. I found the following entry in Prof. Benedikt Listing's diary[15]:

---

[14]The death certificate, no. 199, is on file at the Göttingen registry office and was signed by the civil magistrate G. Fröhlich. I thank Ms. Anders for sending me a copy of it. Court proceedings regarding Meyerstein's estate are not locatable in the name catalogue of the Göttingen district court, according to a message by Herr Justizobersekretär Capelle on 13 Feb. 2002.

[15]See Listing's papers at the Niedersächsische Staats- und Universitätsbibliothek, Göttingen, no. 12 Diarium 1877–1882, entry inadvertently under May 3rd with the note: "all applies to May 4th". Evidently, Listing had learned of Meyerstein's death on May 2nd. It appears that Meyerstein was already not feeling well in mid-January, because one of Listing's diary entries under Jan. 13th 1882 reads: "[medic.] association met here at home (7 heads) mech[anic] and Meißner absent." Unfortunately, his grave was leveled in 1973 but we can gather from the burials roster and other

sweltering hot, humid summer day. 6 h darkening, evening stormy, quite esp. hefty during the night. 3 h Dr. Meyerstein's funeral. Sermon by Pastor Rettig (jun), I follow on foot all the way out [to] the new cemetery. Quite exhausted by the mud and puddles.

Thus the life of one of the most important German instrument makers of the nineteenth century came to an end. His name is now forgotten to all except for a few specialists. Reconstructing such biographies is nevertheless indispensable in order to acquire an adequate understanding of the scientific practice of that period. Sociohistorical research on such persons without presorted archival collections of their papers is difficult and requires training—the result is highly instructive, though, which does make the trouble worthwhile.

References

Allgemeines Adreß-Buch für Göttingen. (1857, 1865). Göttingen: Vandenhoeck & Ruprecht.
Abel, W. (ed.) (1978). *Handwerksgeschichte in neuer Sicht*, Vol. 1 of Göttinger Beiträge zur Wirtschafts- und Sozialgeschichte. Göttingen: Schwartz.
Ambronn, L. (1919). Beitrag zur Geschichte der Feinmechanik. *Beiträge zur Geschichte von Technik und Industrie, 9*, 1–40.
Ambronn, L. (1921). Die Beziehungen der Astronomie zu Kunst und Technik. In *Die Kultur der Gegenwart: Astronomie*, (part III, Sect. 3, Vol. 3, pp. 568–597). Leipzig: Teubner.
Anon. (Oppermann, H.A., & Bock). (1842). *Die Universität Göttingen*, (a) in *Deutsche Jahrbücher für Wissenschaft und Kunst*; (b) offprint, Leipzig: Wigand.
Aßmann, K., & Stavenhagen, G. (1969). *Handwerkereinkommen am Vorabend der industriellen Revolution*. Göttingen: Schwartz.
Behrendsen, O. (1901). *Die mechanischen Werkstätten der Stadt Göttingen, ihre Geschichte und ihre gegenwärtige Einrichtung. Denkschrift, herausgegeben von den vereinigten Mechanikern Göttingens*. Melle in Hannover: F. E. Haag.
Behrendsen, O. (1907). Zur Geschichte der Entwicklung der mechanischen Kunst. *Deutsche Mechaniker-Zeitung* no. 10: 93–96, no. 11: 101–107, no. 12: 115–121, no. 13: 129–137, no. 15: 160–165.
Bennett, J. A. (1987). *The divided circle. A history of instruments for astronomy, navigation and surveying*. Oxford.
Bertram, F., Gerdes, S., Mosler-Christoph, S., Kirleis, W., & Prieß, W. (1999). *Verloren aber nicht vergessen. Jüdisches Leben in Einbeck*. Oldenburg: Isensee Verlag.
Beuermann, G., & Görke, R. (1983/1984). Der elektrische Telegraph von Gauß und Weber aus dem Jahre 1833. *Gauß-Gesellschaft—Mitteilungen* (Vol. 20/21, pp. 44–53).
Beuermann, G., & Werner, Th. (1986). *Die historische Sammlung des I. Physikalischen Institutes der Georg-August-Universität Göttingen. Ausstellungskatalog anläßlich der 250-Jahrfeier der Georg-August Universität im Jahre 1987*. Göttingen: Goltze.
Blackbourn, D. (1984). Between resignation and volatility: The German petite bourgeoisie in the nineteenth century. In G. Crossick & H. G. Haupt (Eds.), *Shopkeepers and master artisans in nineteenth-century Europe* (pp. 35–59). London and New York: Methuen.

---

administrative documents at the cemetery that his grave had been in the former sector E near the cemetery wall to the right of the main entrance and that at his burial, two other burial places had been purchased for his wife and unmarried daughter. They continued to live in Göttingen for a few years and moved to Hanover in 1884. Meyerstein's widow and daughter were laid to rest two or three days after their deaths, on either side of Moritz in their family grave on the Göttingen *Stadtfriedhof*.

Brachner, A. (1985). German 19th-century scientific instrument makers. In: de Clerq, ed. 1985 (pp. 117–157).

Brachner, A., et al. (1983). *G. F. Brander 1713–1783, wissenschaftliche Instrumente aus seiner Werkstatt.* Munich: Deutsches Museum.

Breithaupt, G. (1962). *Friedrich Wilhelm Breithaupt. Kurhessischer Hofmechanikus und Münzmeister.* Kassel: Firma F.W. Breithaupt & Sohn.

Cahan, D. (1985). The institutional revolution in German physics, 1865–1914. *Historical Studies in the Physical Sciences, 15*(2), 1–65.

Cahan, D. (1989). *An Institute for an Empire. The Physikalisch-Technische Reichsanstalt 1871–1918.* Cambridge University Press.

Clifton, G. (1996). *Directory of british scientific instrument makers 1550–1851.* London: Zwemmer.

de Clerq, P. R. (Ed.). (1985). *Nineteenth-century scientific instruments and their makers.* Amsterdam: Rodopi.

Davis, N. Z. (1995). *Women on the margins: Three seventeenth century lives.* Harvard University Press.

Dunnington, G. W. (1955). *C.F. Gauß, Titan of Science* (1st ed.), New York: Exposition Press (2nd ed.), Haftner, 1960.

E.H. (1882). Dr. phil. Moritz Meyerstein [Obituary]. In *Central-Zeitung für Optik und Mechanik 3,* November 1 (no. 21, pp. 241–242).

Engelberg, E., & Schleier, H. (1990). Zur Geschichte und Theorie der historischen Biographie. *Zeitschrift für Geschichtswissenschaft, 38,* 195–217.

Freise, I. (1975). Nachrichten zur Geschichte der Juden in Niedersachsen. *Norddeutsche Familienkunde, 10,* 245–254.

Füssl, W., & Ittner, S. (1999). *Biographie und Technikgeschichte.* Opladen: Leske & Budrich.

Gauß, C. F. (1870–1929). *Werke, herausgegeben von der Göttinger Akademie der Wissenschaften* (11 Vols.). Göttingen: Vandenhoeck & Ruprecht.

Gauß, C. F., & Weber, W. (Eds.). (1837). *Resultate aus den Beobachtungen des magnetischen Vereins im Jahre 1836,* (Vol. 1). Göttingen: Dieterich.

Gauß, C. F., & Weber, W. (Eds.). (1840). *Resultate aus den Beobachtungen des magnetischen Vereins im Jahre 1839,* (Vol. 4). Leipzig: Weidmann.

Gauß, C. F., Weber, W., & Goldschmidt, C. W. B. (1840). *Atlas des Erdmagnetismus nach den Elementen der Theorie entworfen. Supplement zu den Resultaten aus den Beobachtungen des magnetischen Vereins unter Mitwirkung von C.W.B. Goldschmidt,* Leipzig: Weidmann (with 18 charts).

Gerdes, S. (1999). Juden in Einbeck im 19. Jahrhundert. In Bertram et al. 1999 (pp. 17–72).

Gestrich, A., Knoch, P., & Merkel, H. (Eds.). (1988). *Biographie—sozialhistorisch. Sieben Beiträge,* Göttingen: Vandenhoeck & Ruprecht.

Gresky, W. (1985). Zwei bisher unveröffentlichte Briefe Benjamin Goldschmidts über die Entlassung der Göttinger Sieben. *Gauß-Gesellschaft—Mitteilungen, 22,* 45–52.

Hentschel, K. (1993). The discovery of the redshift of solar Fraunhofer lines by Rowland and Jewell in Baltimore around 1890. *Historical Studies in the Physical Sciences, 23*(2), 219–277.

Hentschel, K. (1998). *Zum Zusammenspiel von Instrument, Experiment und Theorie. Rotverschiebung im Sonnenspektrum und verwandte spektrale Verschiebungseffekte von 1880 bis 1960.* Hamburg: Kovač.

Hentschel, K. (2002). *Mapping the spectrum. Techniques of visual representation in research and teaching.* Oxford: Oxford University Press.

Hentschel, K. (2004). Leben und Werk des Instrumentenmachers Moritz Meyerstein (1808–82). *Einbecker Jahrbuch, 49,* 157–184.

Hentschel, K. (2005). *Gaußens unsichtbare Hand: Der Universitäts-Mechanicus und Maschinen-Inspector Moritz Meyerstein.* Vol. 52 of Abhandlungen der Akademie der Wissenschaften zu Göttingen. Göttingen: Vandenhoeck & Ruprecht.

Hentschel, K. (2007). Gauß, Meyerstein and Hanoverian metrology. *Annals of Science, 64,* 41–72.

Hentschel, K. (Ed.). (2008). *Unsichtbare Hände. Zur Rolle von Laborassistenten, Mechanikern, Zeichnern und anderen Amanuenses in der physikalischen Forschungs-und Entwicklungsarbeit.* Stuttgart: GNT-Verlag.

Hentschel, K. (2009). Unsichtbare Hände in der Wissenschaft. Auf der Suche nach den ungewürdigten Helfern der Forschung. *Physik-Journal, 8*(1), 37–40.

Hentschel, K., & Hentschel, A. (2001). An engraver in nineteenth-century Paris: The career of Pierre Dulos. *French History, 15,* 64–102.

Herbst, K.-D. (1996). *Die Entwicklung des Meridiankreises 1700–1850. Genesis eines astronomischen Hauptinstrumentes unter Berücksichtigung des Wechselverhältnisses zwischen Astronomie, Astro-Technik und Technik,* Stuttgart: GNT-Verlag.

Hirsch, O. (1985). Die Kunst, geodätische Instrumente zu bauen, und die Mechaniker, die dies konnten. In H. Junius (Ed.), *Vermessungswesen und Kulturgeschichte* (pp. 47–65). Stuttgart: Wittwer.

Hoffmann, D., & Witthöft, H. (Eds.). (1996). *Genauigkeit und Präzision in der Geschichte der Wissenschaften und des Alltags* (Vol. 4). Brauschweig: PTB-Texte.

Huge, W. (1990). Gewerbeförderung und Handwerkerfortbildung im Königreich Hannover. Zum pädagogischen Wirken des 'Gewerbevereins für das Königreich Hannover' (1828–1866). *Technikgeschichte, 57,* 211–234.

Jenemann, H. R. (1988). Die Göttinger Präzisionsmechanik und die Fertigung feiner Waagen. *Göttinger Jahrbuch, 36,* 181–201.

Joerges, B., & Shinn, T. (Eds.). (2001). *Instrumentation between science, state and industry.* Dordrecht: Kluwer.

Jungnickel, C., & McCormmach, R. (1986). *Intellectual mastery of nature. Theoretical physics from Ohm to Einstein.* Chicago: Univ. of Chicago Press.

Kieckbusch, K. (1998). *Von Juden und Christen in Holzminden, 1557–1945: Ein Geschichts- und Gedenkbuch.* Holzminden: Verlag Jörg Mitzkat.

Kirchvogel, P. A. (1938). *Der Hofmechanikus Johann Christian Breithaupt.* Kassel: F.W. Breithaupt.

Koch, D. (1958). *Das Göttinger Honoratiorentum vom 17. bis zur Mitte des 19. Jahrhunderts. Eine sozialgeschichtliche Untersuchung mit besonderer Berücksichtigung der ersten Göttinger Unternehmer.* Göttingen: Vandenhoeck & Ruprecht.

Koch, J. (Ed.). (2000). *Der Briefwechsel von Johann Georg Repsold mit Carl Friedrich Gauß und Heinrich Christian Schumacher. Kommentierte Übertragung der Brieftexte.* Hamburg: Libri books on demand.

Koch, J. (2001). *Der Hamburger Spritzenmeister und Mechaniker Johann Georg Repsold (1770–1830), ein Beispiel für die Feinmechanik im norddeutschen Raum zu Beginn des 19. Jahrhunderts.* Dissertation, Univ. Hamburg. Norderstedt: Libri.

Kost, J. (2015). *Wissenschaftlicher Instrumentenbau der Firma Merz in München (1838–1932).* Vol. 40 of Nuncius Hamburgensis – Beiträge zur Geschichte der Naturwissenschaften. Hamburg: tredition.

Kück, H. (1934). *Die Göttinger Sieben. Ihre Protestation und ihre Entlassung im Jahre 1837.* Aachen: Edition Herodot im Rader Verlag.

Lepenies, W. (1976). *Das Ende der Naturgeschichte: Wandel kultureller Selbstver-ständlichkeiten in den Wissenschaften des 18. und 19. Jahrhunderts.* Munich: Hanser.

Loewenherz, L. (1882–1887). Zur Entwicklung der mechanischen Kunst. *Zeitschrift für Instrumentenkunde* 2 (1882), pp. 212–219, 254–260, 275–285; 3 (1883), pp. 52–55, 99–103; 6 (1886), pp. 405–419; 7 (1887), pp. 208–215.

Lundgren, P. (1973). *Bildung und Wirtschaftswachstum im Industrialisierungsprozeß des 19. Jahrhunderts.* Berlin: Colloquium Verlag.

Lundgren, P. (1988). *Bildungschancen und soziale Mobilität in der städtischen Gesellschaft des 19. Jahrhunderts.* Göttingen: Vandenhoeck & Ruprecht.

von Mackensen, L. (1987). *Feinmechanik aus Kassel. 225 Jahre F.W. Breithaupt & Sohn. Festschrift und Ausstellungsbegleiter.* Kassel: Wenderoth.

Meyerstein, M. (1845). *Verzeichnis astronomischer und physikalischer Instrumente*. Göttingen: H.C. Seemann.

Meyerstein, M. (1860). Preisverzeichnis der astronomischen und physikalischen Werkstätte von M. Meyerstein, Universitäts, Instrumenten- und Maschinen-Inspector in Göttingen. *Astronomische Nachrichten, 53*, cols. 155–160, 263–272, 301–304.

Meyerstein, M. (1866/67). Preisverzeichnis der astronomischen und physikalischen Werkstätte von M. Meyerstein, Universitäts-Instrumenten-und Maschinen-Inspector in Göttingen. *Astronomische Nachrichten, 68*, cols. 125–128, 139–144, 159–160, 175–176.

Michaelis, A. (1910). *Die Rechtsverhältnisse der Juden in Preußen seit dem Beginne des 19. Jahrhunderts: Gesetze, Erlasse, Verordnungen, Entscheidungen*, Berlin: Lamm.

Michling, H. (1887). *Carl Friedrich Gauß—Episoden aus dem Leben des princeps mathematicorum*. Göttingen: Verlag Göttinger Tageblatt, (a) 1st ed., 1976; (b) 2nd exp. ed., 1982; (c) 3rd reprinted ed., 1997.

Millburn, J. R. (1976). *Benjamin Martin: Author, instrument-maker and 'country showman'*. Leiden: Nordhoff.

Millburn, J. R. (2000). Adams of fleet street: Instrument maker to King George III. Aldershot: Ashgate.

Oesterley, G. H. (1838). Geschichte der Universität Göttingen in dem Zeitraume vom Jahre 1820 bis zu ihrer ersten Säcularfeier im Jahre 1837. Göttingen: Vandenhoeck & Ruprecht.

Olesko, K. M. (2002). Training generations: Social change and Kohlrausch's ethos of practice, 1850–1870, manuscript.

Olesko, K. M. (1991). *Physics as a calling: discipline and practice in the Königsberg seminar for Physics*. Ithaca and London: Cornell University Press.

Repsold, J. A. (1914). *Zur Geschichte der Astronomischen Meßwerkzeuge* (Vol. 2): *Von 1830 bis um 1900*. Leipzig: Reinicke.

Richarz, M. (1974). *Der Eintritt der Juden in die akademischen Berufe. Jüdische Studenten und Akademiker in Deutschland 1678–1848*. Tübingen: Mohr.

Riecke, E. (1892). *Wilhelm Weber - Rede gehalten in der öffentlichen Sitzung der Akademie der Wissenschaften*. Göttingen: Dieterich.

Richarz, M. (Ed.). (1976). *Jüdisches Leben in Deutschland. Selbstzeugnisse zur Sozialgeschichte 1780–1871*. Stuttgart: Deutsche Verlags-Anstalt.

Rößle, K. (1956). Handwerksbetrieb. In *Handwörterbuch der Sozialwissenschaften* (Vol. 5, pp. 42–50). Stuttgart: Gustav Fischer et al.

Rühlmann, M. (1844). Über Gewerbe-Ausstellungen im allgemeinen und über die diesjährige Hannoversche insbesondere. *Gewerbe-Blatt für das Königreich Hannover, 3*(4/5), 125–131.

Rühlmann, M. (1885). Vorträge über die Geschichte der technischen Mechanik und theoretischen Maschinenlehre sowie der damit in Zusammenhang stehenden mathematischen Wissenschaften. Leipzig: Baumgärtners Buchhandlung.

Rühlmann, M. (1855/56). Ueber einige technische Neuigkeiten der Pariser Industrie-Ausstellung. *Mittheilungen des Gewerbe-Vereins für das Königreich Hannover*, 1855: 324–332, 1856, no. 1: 146–155.

Sabelleck, R. (1994). Aufenthalt auf Abruf—Zur Praxis der Schutzbriefgewährung im Kurfürstentum und im Königreich Hannover. In R. Sabelleck (Ed.), *Juden in Südniedersachsen: Geschichte, Lebensverhältnisse, Denkmale* (pp. 83–99). Hannover: Hannsche Buchhandlung.

Sachse, W. (1986). Zur Sozialstruktur Göttingens im 18. und 19. Jahrhundert. *Niedersächsisches Jahrbuch für Landesgeschichte, 58*, 27–54.

Sachse, W. (1987). *Göttingen im 18. und 19. Jahrhundert: Zur Bevölkerungs- und Sozialstruktur der Universitäts-Stadt*. Studien zur Geschichte der Stadt Göttingen, no. 15. Göttingen: Vandenhoeck & Ruprecht.

von Saldern, A. (1976). Die Einwirkung der Göttinger Feinmechanik und Optik auf Universität und Wissenschaft im 19. Jahrhundert. In W. Treue & K. Mauel (Eds.), *Naturwissenschaft, Technik und Wirtschaft im 19. Jahrhundert* (Vol. 1, pp. 363–370). Göttingen: Vandenhoeck & Ruprecht.

Schaefer, C. (1929). Über Gauß' physikalische Arbeiten (Magnetismus, Elektrodynamik, Optik). In Gauß *Werke* 1929 (Vol. 11, No. 2).

Schering, E. (1887). Carl Friedrich Gauß und die Erforschung des Erdmagnetismus. *Abhandlungen der königlichen Gesellschaft der Wissenschaften in Göttingen, 34*(3), 1–79.

Schering, K. (1881/1882). Beobachtungen im magnetischen Observatorium. I. Bestimmung der Horizontalintensität; II. Magnetische Inclination und allgemeine Theorie des Erdinductors. *Nachrichten der königlichen Gesellschaft der Wissenschaften zu Göttingen* 1881: 133–176 with plate; 1882: 345–392 and pp. I–II.

Schimank, H. (1936). Gauß und Weber. *Die Großen Deutschen, 3,* 266–279.

Schimank, H. (1971). Carl Friedrich Gauß. *Gauß-Gesellschaft—Mitteilungen, 8,* 6–35.

Schmidt, K. H. (1978). Bestimmungsgründe und Formen des Unternehmenswachstums im Handwerk seit der Mitte des 19. Jahrhunderts. In Abel, ed. 1978 (pp. 241–281).

Schmoller, G. (1870). *Zur Geschichte der deutschen Kleingewerbe im 19. Jahrhundert. Statistische und nationalökonomische Untersuchungen.* Halle: Verlag der Buchhandlung des Waisenhauses.

Schreiber, H. (2000). *Historische Gegenstände und Instrumente im Institut für Geophysik der Universität Göttingen.* Göttingen: Akademie der Wissenschaften, Kommission für historische Instrumente.

Schröder, W., & Wiederkehr, K. H. (2000). Erdmagnetische Forschungen im 19. Jahrhundert. *Sudhoffs Archiv, 84,* 166–183.

Schütz, S. (1994). Das Judenrecht im Kurfürstentum und Königreich Hannover. In R. Sabelleck (Ed.), *Juden in Südniedersachsen: Geschichte, Lebensverhältnisse, Denkmale* (pp. 57–82). Hannover: Hannsche Buchhandlung.

Shapin, S. (1989). The invisible technician. *American Scientist, 77,* 554–563.

Shortland, M., & Yeo, R., (Ed.) (1996). *Telling lives in Science. Essays on scientific biography,* Cambridge University Press.

de Solla Price, D. (1980). Philosophical mechanism and mechanical philosophy. *Annali dell'Istituto e Museo di Storia della Scienza di Firenze, 1,* 75–85.

de Solla Price, D. (1984). The science/technology relationship, the craft of experimental science, and policy for the improvement of high technology innovation. *Research Policy, 13,* 3–20.

Sonne, H. D. A. (1834). *Beschreibung des Königreichs Hannover,* esp. (Vol. 5): *Topographie des Königreichs Hannover alphabetisch geordnet.* Munich: Cotta, 5 vols., 1829–1834.

Spörer, C. (1980). *Einbeck im Jahre 1853. Bürgerliches Leben am Vorabend der Industrialisierung.* Einbeck: Rüttgeroth.

Spörer, C. (1997). *Einbeck im späten 18. Jahrhundert.* Vol. 11 of Studien zur Einbecker Geschichte in 2 vols. Isensee: Einbecker Geschichtsverein.

Stichweh, R. (1984). *Die Entstehung des modernen systems wissenschaftlicher Disziplinen: Physik in Deutschland 1740–1890.* Frankfurt: Suhrkamp.

Turner, A. J. (1989). *From pleasure and profit to science and security: Etienne Lenoir and the transformation of precision instrument-making in France, 1760–1830.* Cambridge: Whipple Museum.

Voigt, F. (1956). Handwerk. In *Handwörterbuch der Sozialwissenschaften* (Vol. 5, pp. 24–35). Stuttgart: Gustav Fischer et al.

Warner, D. (1968). *Alvan Clark & Sons: Artists in optics.* Washington: Smithsonian Institution Press.

Weber, H. (1893). *Wilhelm Weber—Eine Lebensskizze.* Breslau: Trewendt.

Weber, W. (1892–1894). *Gesammelte Werke,* 6 vols. Göttingen: Akademie der Wissenschaften.

Weber, W., & Kohlrausch, R. (1968). *Über die Einführung absoluter elektrischer Maße.* Ostwalds Klassiker, new ser., no. 5, with commentary by K. H. Wiederkehr. Braunschweig: Vieweg.

Weil, H. (2000). *Carl Philipp Heinrich Pistor: Begründer der optisch-mechanischen Kunst in Berlin. Versuch einer Biographie* (3rd ed.) on CD. Berlin: Author's pub.

Wellenreuther, H. (2001). Vom Handwerkerstädtchen zur Universitätsstadt. *Göttinger Jahrbuch, 49,* 21–37.

Wiederkehr, K. H. (1964). Aus der Geschichte des Göttinger Magnetischen Vereins und seiner Resultate. *Nachrichten der Akademie der Wissenschaften in Göttingen, 2. mathem.-physik. Klasse* (pp. 165–205).

Wiederkehr, K. H. (1967). *Wilhelm Eduard Weber—Erforscher der Wellenbewegung und der Elektrizität 1804–1891*. Stuttgart: Wissenschaftliche Verlagsgesellschaft.

Wilhelm, P. (1973). *Die Jüdische Gemeinde in der Stadt Göttingen von den Anfängen bis zur Emanzipation*. Vol. 10 of Studien zur Geschichte der Stadt Göttingen. Göttingen: Vandenhoeck & Ruprecht.

Wilhelm, P. (1978). *Die Synagogengemeinde Göttingen, Rosdorf und Geismar 1850–1942*. Vol. 11 of Studien zur Geschichte der Stadt Göttingen. Göttingen: Vandenhoeck & Ruprecht.

Williams, M. E. W. (1994). *The precision makers. A history of the instruments industry in Britain and France, 1870–1939*. London and New York: Routledge.

Wise, N. (Ed.) (1995). *The values of precision*, Princeton University Press.

Witthöft, H. (1979). *Umrisse einer historischen Metrologie zum Nutzen der wirtschafts- und sozialgeschichtlichen Forschung. Maß und Gewicht in Stadt und Land Lüneburg, im Hanseraum und im Kurfürstentum/Königreich Hannover vom 13. bis zum 19. Jahrhundert*. 2 Vols. Göttingen: Vandenhoeck & Ruprecht.

Witthöft, H. (1990). Längenmaß und Genauigkeit 1660 bis 1870 als Problem der deutschen historischen Metrologie. *Technikgeschichte, 57*, 189–210.

Zorn, W. (1956). 'Zünfte'. In *Handwörterbuch der Sozialwissenschaften* (Vol. 12, pp. 484–489). Stuttgart: Gustav Fischer et al.

Chapter 3
The Personal is Professional: Margaret Maltby's Life in Physics

Joanna Behrman

Introduction

"What I aspired to be/And am not, comforts me," up-and-coming physicist Margaret Maltby wrote in 1897 in a letter to Swedish physical chemist Svante Arrhenius, describing her resolution to remain unmarried and continue research and teaching.[1] In these lines pulled from Robert Browning's poem "Rabbi ben Ezra," Maltby expressed her feelings of acceptance, tinged with loss, over her unmarried status. And yet, Maltby also found certain advantages and even joys in remaining unmarried. For Maltby, as for all female physicists of the early twentieth century, there was no clear demarcation between personal and professional considerations. The two were inextricably linked and the one often shaded into the other. In fact, to rework a phrase which would only emerge decades after Maltby's death, "The personal is professional" (Hanisch 1970).

Using previously overlooked archival records and documents held by her descendants, this paper will use Maltby as a case study to investigate the intermingling of private-personal and public-professional lives of female physicists. Although some aspects of women's personal lives were automatically considered public fodder, Maltby's life gives insight into how women were able to exercise agency over their own privacy and at the same time use their personal identities for professional gain.

Maltby has usually been studied as one individual among many noteworthy female 'firsts'. She is best known as one of the first women to earn a Bachelor of Science

---
[1] Margaret Maltby to Svante Arrhenius, 22 December 1897, Svante Arrhenius Papers, Royal Swedish Academy of Science.

J. Behrman (✉)
Johns Hopkins University, American Institute of Physics, Baltimore, MD, USA
e-mail: jbehrma2@jhu.edu; jbehrman@aip.org

from MIT, the first woman to earn a Ph.D. in physics from any German university, and the first woman to be named as a research assistant at the Physikalisch-Technische Reichsanstalt. To these 'firsts' could be added her long tenure as an influential physics professor at Barnard College, her activism in the American Association of University Women, or her collaboration with many noted physicists such as Walther Nernst, Friedrich Kohlrausch, and Arthur Webster. As with many historical women in science, she appears in numerous compendia of short biographies, yet deserves further attention as a multifaceted individual, rather than as a statistic (Harrison 1993; Kidwell 2006; Wiebusch 1971).

The emphasis on Maltby's professional life, on her as a 'great first', is an artifact of extreme historical simplification. Historians often mark the achievements of great scientists in the form of professional successes: papers published, awards received, ground broken (Shortland and Yeo 1996). Quite frequently, this is all (and sometimes not even this) that can be easily recovered for a woman or for a person of color in science, as their lives and achievements tend to be silenced during their physical and archival lifetimes (Rossiter 1993). This chapter will skip over many parts of her life that might otherwise be included in a traditional biography, in order to push back against the use of scientific papers or other kinds of professional achievements as the main markers of a person's biography.

Researching the life of Margaret Maltby is therefore also a methodological exercise in recovering a lost voice. Because of inadvertent or deliberate actions by herself and others, much of Maltby's life remains in fragments. As Virginia Woolf remarked when she began to write a biography of Roger Fry, "How can one make a life out of six cardboard boxes full of tailors' bills, lovers' letters and old picture postcards?" (Shortland and Yeo 1996, p. 35). To piece together Maltby's life requires not only piecing together those fragments but also interrogating the silences of the archive. The record of Maltby's life reveals her active and ongoing work to manage her personal and professional, or perhaps more properly, her private and public lives. Even years after her death, the formation and maintenance of the archival silences reveal that the same balance achieved in her lifetime is uneasily continued in her historical legacy (Fig. 3.1).

The Girl Who Asked Why

Margaret Eliza Maltby was born in 1860 in Bristolville, Ohio, the youngest of three daughters. Originally named Minnie by her two older sisters (ages 13 and 15 at the time of her birth), she intensely disliked her name and changed it to Margaret in 1889 (Wiebusch 1971).[2] Maltby recalled that she was a frequent questioner of nature: "Why did water boil? Why did an eyedropper suck up liquid? Why did it snow?"

---

[2]Margaret Maltby to A. S. Roof, 17 January 1890, Box 673, Folder: Margaret Eliza Maltby, Series: Graduates and Former Students, RG 28/2 Alumni and Development Records, Oberlin College Archives.

3 The Personal is Professional: Margaret Maltby's Life in Physics

Fig. 3.1 From left to right, Betsy, Margaret, and Martha Jane Maltby. Photograph courtesy Raymond and Jane Gill

(Ferris and Moore 1927, p. 213). Her parents, she recounted in an interview for *Girls Who Did*, a career advice book for girls, encouraged her. Although her high school did not offer science, her parents taught her to be at ease with simple machinery, and her father, Edmund Maltby, particularly encouraged her in mathematics.

Not much else is known about her childhood apart from what is in that book. Her living relatives recall that she was an intensely private person, particularly when it came to her personal history.[3] Maltby's biographical details in *Girls Who Did* are therefore helpful, but simultaneously problematic to use as a source. The context of a career advice book shades Maltby's story into a triumphalist narrative; her story incorporates many tropes of scientific biography. The 'Questioning Child,' like the 'Tinkering Child' who builds a science lab in the basement or takes apart household technologies, is an increasingly common trope of narratives about scientists in the twentieth century (Onion 2016). Childhood is often then taken to be, as it is in Maltby's interview, "a prophecy of her life-long interest" (Ferris and Moore 1927, p. 213). Biographies like these need to be carefully read, and the most trustworthy

---

[3] Author interview with Mary Traynor and Elizabeth Traynor. Guntersville, Alabama. 25 February 2018.

details are often those which deviate from the norm. For example, the lack of laboratory facilities or even science classes at Maltby's high school is mentioned in contrast with the increasing prevalence of science and technology in the daily life of many children and young adults of the twenties.

Of course, the greatest deviation from the norm of scientific biographies is Maltby's gender. The fact that her life story is being read into scientist tropes is an assertion of equality on the part of Maltby and her interviewer. Despite the lack of support and recognition, Maltby and the interviewer acknowledged, for girls who wanted to go into science there were no barriers other than those put into place by society which could limit the achievement of a precocious girl. And even those barriers, they argued, could be surmounted. Thus while this rare glimpse into Maltby's childhood is blatantly an untrustworthy source, in its untrustworthiness it claims power by asserting Maltby's right to a triumphalist scientist's biography.

After Maltby's father died, her mother moved the Maltby family to Oberlin, Ohio, to be closer to educational opportunities. From 1877 to 1882, Maltby attended Oberlin College, taking first a year in the preparatory department before entering the Freshman class.[4] As was mentioned in *Girls Who Did*, science classes were not available to Maltby at her high school, and nor were they much available to her at Oberlin. At the time of her attendance at Oberlin, she remarked, science education remained "in the descriptive stage of development." Maltby would likely have seen physics experiments demonstrated, but not participated in them herself. This was not unusual for the time, as the laboratory method of physics education was only just then emerging in the United States (Kremer 2011). Nevertheless, Maltby credited her classical education at Oberlin with giving her a way to see the broader connections among different branches of science.[5]

Oberlin cultivated her interests in art as well as music. At her college commencement she gave an oration on "Modern Aestheticism," and it was the "collegiate education together with opportunities to hear the best in classical music" along with the "fine character and high ideals of the men and women on the faculty" that she would later recall as being among Oberlin's most important influences (Oberlin Weekly News 1882).[6] Maltby's love of music continued throughout her life. One descendant recalled that she and her brother needed to be quiet when they visited "Aunt Margaret" as soon as the Met Opera came on the radio.[7] Maltby would later go on to develop at Columbia University one of the first courses in the physics of music. She based her course heavily on laboratory work and manual practice with instruments

---

[4] Margaret Maltby short biography, undated, Box 673, Folder: Margaret Eliza Maltby, Series: Graduates and Former Students, RG 28/2 Alumni and Development Records, Oberlin College Archives.

[5] Margaret Maltby, undated, "Autobiography of Margaret E. Maltby, Ph.D., A.A.U.W. European Fellow, 1895–1896," Elizabeth Traynor Private Collection, page 1.

[6] Margaret Maltby response to an alumni survey, undated circa 1916, Box 673, Folder: Margaret Eliza Maltby, Series: Graduates and Former Students, RG 28/2 Alumni and Development Records, Oberlin College Archives.

[7] Author interview with Mary Traynor and Elizabeth Traynor, Guntersville Alabama, 25 February 2018.

Fig. 3.2 MIT women's laboratory, 1888. Ellen Swallow Richards is in the back row on the far left. Margaret Maltby is in the back row, fourth from the left. Image courtesy MIT Museum

of music and precision measurement.[8] In this class it is possible to see two of her major interests come together, but it would be a mistake to assume they were ever very far apart.

After graduation from Oberlin, Maltby studied at the Art Students League in New York City before teaching at the high school level for four years.[9] While teaching, she became drawn to physics and decided to pursue the subject. In 1887 Maltby enrolled at the Massachusetts Institute of Technology, eventually earning a Bachelor of Science in 1891, for which Oberlin awarded her a congratulatory Master of Arts. During and after earning her degree, Maltby worked as a laboratory assistant at Wellesley College and conducted research and postgraduate studies with Charles Cross and Silas Holman at MIT (Fig. 3.2).[10]

While attending MIT, Maltby became very close to Ellen Swallow Richards, and this friendship would continue throughout their lives (Hunt 1912, p. x).[11] Richards,

---

[8]Physics Department Inventory Checks 1903–1914, Box 1, BC 13.42 Physics Department 1900–1940s, Barnard Archives and Special Collections, Barnard College.

[9]1935 Quinquennial Report for Margaret Maltby, Box 673, Folder: Margaret Eliza Maltby, Series: Graduates and Former Students, RG 28/2 Alumni and Development Records, Oberlin College Archives.

[10]Author email with MIT Institute Archivist Dana Hamlin, March 1, 2018.

[11]Margaret Maltby to Ellen Swallow Richards, 6 December 1908, Folder "Maltby, Margaret," Ellen Swallow Richards Papers 1882–1910 MS 130, Series II Correspondence General, Smith Collection, Smith College Libraries Special Collections.

Fig. 3.3 Mary Winston, Grace Chisholm, and Margaret Maltby, circa December 1893. Photograph courtesy Elizabeth Traynor

a pioneer in the fields of home economics, euthenics, and sanitary science, was the only female instructor at MIT at the time, and she also took on a motherly role to the female students who gradually began attending MIT in greater numbers (Stage 1997). With her husband, Richards took these young women into her home at 32 Elliot Street (in the Jamaica Plain neighborhood of Boston), both as visitors and as part of 'the family'. The women could work for board and pay with housekeeping duties in addition to the work that was offered at Richards's laboratory (Hunt 1912, pp. 118–124). Although Maltby did not board with Richards at the time—according to her alumni and student records she lived closer to campus at 331 Columbus Ave and later at 7 Irvington Street[12]—Maltby formed a deep bond with Richards, admiring her wide-ranging passion for the sciences, her keen intellect, and her practical nature (Science 1912, pp. 176–177). This bond would later prove invaluable (Fig. 3.3).

---

[12]Maltby response to Oberlin questionnaire, circa 1889, Box 673, Folder: Margaret Eliza Maltby, Series: Graduates and Former Students, RG 28/2 Alumni and Development Records, Oberlin College Archives; Author email with MIT Institute Archivist Dana Hamlin, March 1, 2018.

Graduate Studies in Germany

From MIT, Maltby went to the University of Göttingen to study physics, chemistry, and mathematics. She earned her Ph.D. there in 1895. While not among the first women to attend lectures, Maltby was one of the first three women admitted to study officially at the university, although they were admitted as exceptions to the general rule.[13] Grace Chisholm and Mary Winston, the other two women admitted at the same time as Maltby, both joined Felix Klein's mathematical research group (Jones 2009, Whitman 1987). There are few materials regarding Maltby's time at Göttingen. It is known that Maltby did write home about her experiences, but only fragments of these letters survive.

To understand Maltby's graduate experience is then to see it through the eyes of her friends, Winston in particular, whose letters home survive. Winston described Maltby as "very tall also and rather stout with round, red cheeks and a pleasant, jolly way. She is considerably older than we [Winston and Chisholm] are".[14] Winston and Chisholm were both in their mid-twenties, about eight to nine years younger than Maltby who was in her early-to-mid thirties. Winston and Chisholm's being in the same research group, along with the age difference, perhaps contributed to the tighter friendship that developed between Winston and Chisholm than between Winston and Maltby. Maltby is often absent from Winston's letters, but her absence is explained as being a result of much time spent in the laboratory or traveling around Europe during the occasional vacation. Winston, Chisholm, and Maltby gathered to celebrate holidays and milestones in their graduate careers. Maltby and Winston also both participated in the activities of the "American Colony" in Göttingen. Within this group the male graduate student who had been the longest in attendance at Göttingen was given the title "The Patriarch". Maltby was able to inaugurate the title of "Matriarch." (Fig. 3.4)[15]

While at Göttingen Margaret Maltby studied with Walther Nernst. (It was also in Germany that she met Svante Arrhenius. Arrhenius, along with Wilhelm Ostwald and Nernst, were among the pioneers of physical chemistry (Crawford 1996).) Originally Maltby had travelled to Leipzig to work with Ostwald, but she was turned away due to overcrowding in his lab. On the strength of her recommendation letters, including one from Arthur Noyes, Ostwald recommended Maltby work with Nernst, and wrote to Nernst suggesting he take her (Zott 1996, p. 62). Nernst, known to his friends as

---

[13] Mary Frances Winston to her mother, 29 October 1893, Mary Frances Winston Papers, MS 213, Folder 10, Family Typed Transcripts 1893–1895 Page 21, Sophia Smith Collection, Smith College Libraries Special Collections.

[14] Mary Frances Winston to Charley, 17 October 1893, Mary Frances Winston Papers, MS 213, Folder 10, Family Typed Transcripts 1893–1895, Page 17, Sophia Smith Collection, Smith College Libraries Special Collections.

[15] Mary Winston, "My Student Days in Germany," undated manuscript circa 1952, Page 13, Mary Frances Winston Papers, MS 213, Folder 10, Family Typed Transcripts 1893–1895 Sophia Smith Collection, Smith College Libraries Special Collections; Mary Frances Winston to Charles, 21 November 1893. Mary Frances Winston Papers, MS 213, Folder 10, Family Typed Transcripts 1893–1895, Pages 34–35, Sophia Smith Collection, Smith College Libraries Special Collections.

Fig. 3.4 Walther Nernst research group, University of Göttingen, 1896. Nernst is in the front row, number 1. Maltby is unnumbered in the back row. Photograph courtesy Elizabeth Traynor

a bit of a philanderer, was not opposed to coeducation, and neither was Eduard Rieke, director of the Physical Institute at Göttingen (Bartel 1989, p. 52; Barkan 1999, p. 51). Maltby worked with Nernst on making better measuring instruments for electrochemistry, as many of the available instruments were made purely for physical applications. For Maltby's dissertation, she made an instrument to measure large electrolytic resistances (Bartel 2007, p. 89; Zott 1996, pp. 95–96; Maltby 1895). After completing her doctorate, Maltby remained in Germany for post-graduate research work until August 1896 at which time she worked on the measurement of high-frequency oscillations (Zott 1996, pp. 92–93; Maltby 1897). The atmosphere of research Maltby experienced in Göttingen was something she would try to foster for the rest of her career (Maltby 1896) (Fig. 3.5).

Returning to the U.S.

Unfortunately, despite a strong interest in research beginning from her time at MIT and continuing through her career, Maltby encountered many obstacles to research work as soon as she got back from Germany. For the first decade that Maltby was back in the United States, research seemed just around the corner, but Maltby became increasingly discouraged. For example, in her second position after graduate school, Maltby taught at Lake Erie College in Painesville, Ohio, and attempted to do research

Fig. 3.5 Margaret Maltby and Philip Randolph Meyer, age 4. Circa 1901. Photograph courtesy of Raymond and Jane Gill

with Edward Morley at Western Reserve University (a little more than 40 miles away from Painesville in Hudson, Ohio). But despite her best efforts, her busy teaching load prevented her from making the journey more often than for part of one day every week.[16] (This collaboration did not lead to any historical record via publication.) For the rest of her career, Maltby's desire to research never quite worked out to her advantage.

And other events soon overshadowed Maltby's success at Göttingen. After only one semester into her time at Wellesley in the fall of 1896, Maltby was forced to withdraw from the college over the winter recess. As she wrote to a friend in December of the next year,

---

[16]Margaret Maltby to Svante Arrhenius, 22 December 1897, Svante Arrhenius Papers, Royal Swedish Academy of Science.

You perhaps know that I met with rather a serious accident last winter shortly after I wrote you. It prevented my return to my school duties last spring and winter.[17]

Despite the "serious accident," Wellesley College felt slightly aggrieved at Maltby's sudden departure and did not offer her a position again the following year. Whether or not Wellesley knew the circumstances behind the "accident" is unclear (Wellesley 1897, pp. 6, 17). To the public eye, Maltby's "accident" remained an unfortunate, but ultimately unremarkable, blip in her otherwise stellar career. However, the revelation of the actual truth would have been far more devastating to her career.

Details of Maltby's life in this time period are few and far between, deliberately obscured by Maltby's friends or Maltby herself. One fragment of this time lingers, for example, in the archived papers of Maltby's close friend Ellen Swallow Richards— the last page of a seven-page letter Maltby had written to Richards. The page survived because it contained notes on discussions in the Association of Collegiate Alumnae— notes which Richards needed to write up as part of a longer article to be published. Hardly would someone think to read much into this single scrap among the larger collection of Richards papers, which were nearly entirely of a professional nature.

By extension from Maltby's usual style, the full letter would have been six pages front and back filled with neat, tightly-spaced handwriting. An immense and intensely private account must have been conveyed in those pages. Even without the previous six pages, Maltby's concluding remarks on this very last page are illuminating. "Grüsse an Herrn Prof.," she wrote. "Greetings to Mr. Prof," a nod both to Richards's husband and to Maltby's relatively recent return from Germany. But the letter ends on a note of sadness: "I shall smile again soon. With my love, Margaret E. Maltby."[18] No one could doubt, reading this fragment, the depth of emotion which must have been conveyed in the previous pages. Knowing the contents of the missing six pages would almost certainly answer the questions raised by Maltby's parting words. However, any hints at an answer would wait decades.

Richards's power, in keeping some papers and not others, ensured that the sensitive details of her friend's suffering were kept out of the hands of not only Richards's heirs, but also future archivists and the occasional enterprising historian. So often the silence of women in the archives is not of their own making, but here the silence is deliberate, an extension of their power over their own legacy.

---

[17] Margaret Maltby to Svante Arrhenius, 22 December 1897, Svante Arrhenius Papers, Royal Swedish Academy of Science.

[18] Margaret Maltby to Ellen Swallow Richards, Circa 1898, Box 1, Folder 47, Ellen Swallow Richards Papers MS 130, Box 1, Folder 47, Sophia Smith Collection, Smith College Libraries Special Collections.

Adoption and Motherhood

In 1902, Maltby wrote again to Svante Arrhenius:

> In addition to teaching, I have also taken certain home cares. I have been housekeeping the past year with my cousin & a small boy – an orphan & the son of a very dear friend – whose education I was to look after.[19]

Maltby looked after more than just the education of this 'orphan and son of a very dear friend,' to him she was a mother in nearly every sense of the word. Maltby adopted Philip Randolph Meyer, a boy of four years, in 1901, almost as soon as she had started her position at Barnard College, where she stayed until retirement in 1933. Philip was to have a profound impact on her life, although his origins would remain shrouded in mystery.

Maltby and Philip had a very close relationship. She gushed over him in letters to her friends, and said that Philip brought an aspect of normalcy and the "human feminine" touch to a life which she had previously found somewhat lonely.[20] However, whatever the degree of her loneliness before she adopted Philip, Maltby had firmly resolved not to marry, and she never did. As she wrote to a former Oberlin classmate, "I had all the pleasures of family life without the disadvantage of supervision by a 'better half.'"[21]

While she found caring for Philip in a way freeing, Maltby found marriage, especially marriage as a female scientist, to be potentially very confining. She argued that both male and female scientists ought to get married only to those who would share and understand their interests, that is to other scientists. But while they should match their partner in knowledge, they also needed to match their partner in personality. A strong-willed woman herself, Maltby thought it would be disastrous if one or the other partner were to be either subsumed or brought down by the influence of a too-strong or too-weak spouse.[22] At one point during her life, Maltby did refuse an offer of marriage from another noted scientist because she did not want her career to be totally subsumed into his.[23]

This contrast between the demands of children and marriage played out at Barnard College in 1906, only six years after Maltby arrived there. She was at that point an adjunct professor of physics with a fair amount of standing with the Columbia

---

[19] Margaret Maltby to Svante Arrhenius, 3 September 1902, Svante Arrhenius Papers, Royal Swedish Academy of Science.
[20] Margaret Maltby to Newton W. Bates, 23 May 1922, Box 673, Folder: Margaret Eliza Maltby, Series: Graduates and Former Students, RG 28/2 Alumni and Development Records, Oberlin College Archives.
[21] Margaret Maltby to Newton W. Bates, 23 May 1922, Box 673, Folder: Margaret Eliza Maltby, Series: Graduates and Former Students, RG 28/2 Alumni and Development Records, Oberlin College Archives.
[22] Margaret Maltby to Svante Arrhenius, 22 December 1897, Svante Arrhenius Papers, Royal Swedish Academy of Science.
[23] Raymond Gill email to author, 25 November 2018.

physics department and the Barnard administration. In 1904 Barnard had hired Harriet Brooks, a promising physicist who had worked with Ernest Rutherford and J. J. Thompson. However, Brooks announced her intention to get married over the summer of 1906 to another physicist who worked at Columbia. Brooks wrote Laura Gill, the Dean of Barnard College, that she hoped not to cause any trouble and would happily continue all her duties. Unfortunately Gill was not so sanguine, and she politely requested, with all the weight of a demand, that Brooks leave the college upon her marriage. The engagement between Brooks and the Columbia physicist was later broken, and Brooks left the college citing her grief about the whole process and the uncomfortable atmosphere (Rayner-Canham and Rayner Canham 1992, pp. 45–51).

Maltby tried to persuade Gill to allow Brooks's continued employment; Brooks was a promising teacher and researcher and Maltby doubted that Brooks was the sort of person to let any duties at college or home slide.[24] But the Dean was unmoved because her objection stemmed from a question of the loyalty of married women to the college. Gill was informed by the Barnard trustees that,

> The College cannot afford to have women on the staff to whom the college work is secondary; the College is not willing to stamp with approval a woman to whom self-elected home duties can be secondary.[25]

There were specific aspects of college work on which Gill believed the crux lay. She wrote to Maltby,

> There are many phases of scientific, educational, and philanthropic work which a married woman can do without possible embarrassment or business disadvantage. Regular classroom work which cannot be interrupted without serious detriment to the class is not a form which is, in my judgment, adapted to young married women.[26]

Concerned that the first priority for a married woman's time would be her husband, rather than her students, Gill's reasoning was rooted in her valuation of loyalty at a woman's college—that a married woman would be more loyal to her husband than the college.

Although a later dean allowed women at Barnard to keep their employment following marriage, it was not unusual at the time for a woman teaching in a college to voluntarily resign or be pushed out of her position because of marriage (Rossiter 1982, pp. 14–16). What *is* unusual is that the dean certainly knew that Maltby had adopted Philip. Adopting a child ought logically to be classified by Gill and the trustees as a 'self-elected home duty', and Maltby did occasionally have to place college work secondary to care for Philip.[27] But a child was not considered to be

---

[24] Margaret Maltby to Laura Gill, 24 July 1906, Box 6, Folder 41, BC 5.1 Dean's Office Correspondence 1906–1908, Barnard Archives and Special Collections, Barnard College.

[25] Laura Drake Gill to Margaret Maltby, 30 July 1906, Box 6, Folder 41, BC 5.1 Dean's Office Correspondence 1906–1908, Barnard Archives and Special Collections, Barnard College.

[26] Laura Drake Gill to Margaret Maltby, 30 July 1906, Box 6, Folder 41, BC 5.1 Dean's Office Correspondence 1906–1908, Barnard Archives and Special Collections, Barnard College.

[27] Margaret Maltby to Laura Gill, 9 September 1906, Box 6, Folder 41, BC 5.1 Dean's Office Correspondence 1906–1908, Barnard Archives and Special Collections, Barnard College.

the same drain on time nor was a working woman's raising a child the same shame to the college that a working, married, woman would be. Maltby was in fact quite fortunate to have household help during Philip's early childhood. In 1905, at the time of Philip's youngest years, Maltby was keeping house with a friend, Mariane Woodhull, and had a servant, Charlotte Penborn.[28] Household help for Maltby made her work at the college and extra-collegiate activism possible, much in the way that many women enabled the work of supposedly 'solitary genius' male scientists.

While the role of heterosexual wife might be frowned on in women's colleges, the role of mother was not only allowed but encouraged. The loyalty of a female professor to the college was not that of a spouse but of a mother to her students, and female college professors frequently took on 'motherly' roles. When Maltby publicly positioned herself as a maternal figure to Philip, it was as his guardian in which position she was in charge of his education. She thus made a parallel between her work as a maternal educator at college and a maternal educator at home.

The Professor as Maternal Figure

Maltby took on many 'motherly' labors at the college. This work included aspects of mentoring—for her students and her colleagues. Described as the "very popular" head of physics, Maltby could be relied on to help put some steel in the spine of any aspiring undergraduate in the sciences (Barnard Bulletin Apr. 1917a; Barnard Bulletin Nov. 1917b). She was very active in college activities, participating in student groups, teas with faculty, and serving as a judge in competitions ranging from Greek Games to Choruses. Her 'motherly' demeanor was pronounced enough to extend into parody, as she was once satirized in a student play as the sort of woman who would serve tea during a mystery investigation (Barnard College Yearbook 1923, p. 40).

In another example, at the 1912 Barnard graduation, Maltby served as a judge for the alumnae parade during Senior Week. Different graduating classes of alumnae dressed in different costumes, and the most recently graduated students dressed as little girls (Barnard Bulletin 1912). With Maltby sitting watch as one of the three matriarchs, the class of 1912 strode forth and sang, "1912 the very day were staid old Seniors /.../ But we've metamorphosized into Babies. /.../ For we've changed from Undergrad to Alumnae".[29] Traditions such as these marking the growth of classes from one stage of life to another were fairly common at women's colleges in the United States at this time. Within these traditions, different classes of female

---

[28] New York State Census Records for Manhattan Borough and Bronx (1905), Election District 47, Assembly District 21, Block M, Page 88, Lines 35–36. FHL microfilm 1,433,097. Accessed via FamilySearch https://familysearch.org/pal:/MM9.1.1/SPF8-VYL. Last updated 21 December 2017.

[29] Eleanor Myers Jewett Scrapbook, vol. 4, 1911–1912, Page 132, Inclusion 1, Barnard College Archives. http://digitalcollections.barnard.edu/islandora/object/BC15-14:1666.

students took on roles as big and little sisters, with the faculty maternally watching over (Bronner 2012, pp. 211–215, 238–241).

Over the course of Maltby's career in the late nineteenth and early twentieth century, female college professors at women's colleges were commonly expected to play 'motherly' roles for the students. At Smith College, for instance, female faculty were expected to live as surrogate mothers with the students in home-like cottages (Horowitz 1984). Although at Barnard the female faculty were never expected to *live* with the students, the faculty did take on emotional labor in caring and mentoring the students. Margaret Maltby doubled-down on extra-curricular mentoring, possibly because it had not been possible to have the research career she had wanted for herself. At Barnard, for instance, Maltby advocated for more part-time positions so women could more easily combine marriage and a career (Barnard Bulletin 1929). She also encouraged women to travel abroad for research, as she herself had done, though her work with the American Association of University Women (Maltby 1929; Talbot 1931). And her appearance in *Girls Who Did* was not her only career-advice publication—she also wrote a chapter on "The Physicist" for a book geared towards secondary-school and college-aged women (Maltby 1920).

A Family Mystery

As children do, Philip Randolph Meyer grew up, and eventually had children of his own. Meyer's son-in-law, Raymond Gill (no relation to Barnard College Dean Laura Gill) is an enthusiastic family geneologist, and what Philip Meyer related to Gill of his origins struck Gill as odd. So Gill went digging, and the puzzling details mounted.

Maltby told Philip that he had been born in Virginia in 1897, that his mother (Elizabeth? Randolph) had died soon after his birth, and that his father (Wilhelm Meyer) was a German who had been killed in China around 1900 during the Boxer Rebellion.[30] But, Philip's birth certificate had never been seen by Philip's descendants or even Philip himself. Acknowledging a German father but no birth certificate caused some trouble when Philip went to enlist in the U.S. army in 1917. Only after Maltby had a private meeting with the recruiting officer was Philip allowed to enlist (Gill 2016) (Fig. 3.6).

Finding documentary evidence to corroborate or provide an alternative explanation for Philip's origins has been difficult both for Raymond Gill and Elizabeth Traynor, Philip's granddaughter. Even with the addition years later of a historian (the author) who has brought to bear all the privileges that having a university affiliation brings, only a few additional gaps have been filled in.

There are very few documents relating to Philip's early life. "Philip Randolph Meyer" does not seem to exist in the 1900 Federal Census, when Maltby was living as a boarder in Worcester Massachusetts while she worked with Arthur Webster at

---

[30]Raymond Gill email to author, 26 November 2018.

Fig. 3.6 Margaret Maltby and Philip Randolph Meyer, 1917. Photograph courtesy Elizabeth Traynor and Raymond and Jane Gill

Clark University.[31] Philip was seven years old at the time of the 1905 New York State Census, and is listed as Maltby's ward; nothing which was not already known.[32] The next official document is a little more illuminating—a 1909 passport application made out by Maltby to take Philip and herself for a sabbatical in England. In this application Philip Meyer is again listed as her ward, with a birthplace of Waynesboro, Virginia.[33] Although this birthplace accords with what Maltby told Philip, no birth certificate for him can be found on record. Maltby also told Philip a fire at Waynesboro had destroyed the records there, making it impossible to obtain a copy of the birth certificate. However, no such fire occurred (Gill 2016, p. 50) (Fig. 3.7).

---

[31] United States Census (1900), Worcester City in Worcester County, Massachusetts, Precinct 2, Ward 6, Enumeration District 1760, Sheet 5B, Family 128, Line 78. NARA microfilm publication T623 (Washington, D.C.: National Archives and Records Administration, 1972.); FHL microfilm 1,240,697. Accessed via FamilySearch https://familysearch.org/pal:/MM9.1.1/M9BD-384.

[32] New York State Census Records for Manhattan Borough and Bronx (1905), Election District 47, Assembly District 21, Block M, Page 88, Lines 35–36. FHL microfilm 1,433,097. Accessed via FamilySearch https://familysearch.org/pal:/MM9.1.1/SPF8-VYL. Last updated 21 December 2017.

[33] Margaret E Maltby Passport Application, May 24–25, 1909. Certificate #6536 from Passport Applications, January 2, 1906—March 31, 1925, NARA Microfilm series M1490, Roll 86 (Washington D.C.: National Archives and Records Administration, n.d.). Accessed via FamilySearch https://familysearch.org/pal:/MM9.1.1/QKDX-BQLN. Updated 16 March 2018.

Fig. 3.7 Front and back of a postcard fragment, written by Margaret Maltby in Germany to her family in Ohio, 27 May 1894. Elizabeth Traynor Private Collection

There was, however, another fire. Margaret Maltby had asked Philip to burn all her personal papers upon her death. He, by all accounts the dutiful son, did so.[34] It is puzzling why Maltby would want to burn her papers, aware as she must have been of her place in history as one of the United States's pioneering female scientists—unless of course she had something to hide. This is one reason why so few records of Maltby's survive, especially from her time in Germany. Interestingly, some of the few personal letters which were not burned and instead passed down in her family were those which had foreign stamps that she then gave to Philip for his stamp collection. Thus a fragment of a postcard Maltby sent to her family from Göttingen survives.[35]

The paper trail in this story would only lead so far it seemed, and yet what records there were, were tantalizing. In the 1909 passport application, Maltby appended a letter in which she wrote:

> My ward, accompanying me, is to all intents "my child," as he is in my care, resides with me, & has done so since infancy, and is dependent upon me solely, since he is an orphan.[36]

---

[34] Author interview with Raymond Gill and Jane Gill, Brooklyn, New York, 18–19 January, 2017; Author interview with Mary Traynor and Elizabeth Traynor, Guntersville, Alabama, 25 February 2018.

[35] Margaret Maltby to her family in Ohio, 27 May 1894, Elizabeth Traynor Personal Collection.

[36] Letter Margaret E. Maltby to the Secretary of State, 24 May 1909. Margaret E Maltby Passport Application, May 24–25, 1909. Certificate #6536 from Passport Applications, January 2, 1906–March 31, 1925, NARA Microfilm series M1490, Roll 86 (Washington D.C.: National Archives and Records Administration, n.d.). Accessed via FamilySearch https://familysearch.org/pal:/MM9.1.1/QKDX-BQLN. Updated 16 March 2018.

3 The Personal is Professional: Margaret Maltby's Life in Physics

However, Philip's descendants strongly suspected that he was Maltby's child in *all* respects. In 2014, Philip's two daughters, Jane Gill and Mary Traynor, took a DNA test via Ancestry.com and compared their results to known relatives of Margaret Maltby. There were many significant matches—Philip Meyer was almost certainly Margaret Maltby's biological son (Gill 2016, p. 51).

With this crucial piece of evidence, new connections could be made. Margaret Maltby withdrew from Wellesley College in late December 1896 on account of her "accident", never clearly described. Philip Meyer was born June 4, 1897. It would have been impossible for Maltby to retain her position at the college with a child born out of wedlock—and it would have been tenuous even had she been married (Rossiter 1982, pp. 14–16). There is no solid evidence of an address for Maltby between December 1896 and when she reappears Fall 1897 at Lake Erie College, but it is likely that she either spent the later months of her pregnancy with Ellen Swallow Richards, or used Richards as a trusted and discrete connection to the outside world by leaving Richards's address as her forwarding address with Wellesley (Wellesley College 1900, p. 12). Maltby then fostered Philip with relatively well-to-do friends who already had young children of their own; Philip's earliest memories were of living with a family he was not related to (Gill 2016, p. 49). Once Maltby obtained a relatively secure position with Barnard College, Maltby then was able to 'adopt' Philip. Of course, discovering that Margaret Maltby was Philip's biological mother immediately raised other questions, particularly with respect to Philip's biological father.

What Maltby had told Philip regarding his father's being German appears to have been true. Philip told Raymond Gill that money came from Germany every month for his care until Philip joined the American forces in 1917 to fight in the First World War. And if June 4, 1897 is an accurate birth date for Philip (and there appears to be no reason why not), then Maltby would have been finishing her work in Germany around the time that Philip would have been conceived. But who then was the father in Germany? Again, Philip's descendants had a suspicion.[37]

Many years ago, Elizabeth Traynor, Philip's granddaughter, was looking at the photograph of the Nernst lab in Göttingen and was instantly drawn to a figure in the front who looked uncannily like her brother John (or his identical twin Philip). Showing the photograph to her siblings and her mother (Philip Meyer's daughter), all of them pointed to Nobel Laureate Walther Nernst, and said, "That looks like John!" Decades later, Traynor was able to use the internet to research and gather more photos, from more angles, of Walther Nernst. More recently, Traynor, an accomplished illustrator, put together on her computer side-by-side comparisons of Walther Nernst with some of Philip's male descendants. The hair loss pattern, facial profile, even the

---

[37] Author interview with Raymond Gill and Jane Gill. Brooklyn, New York. 18-19 January, 2017.

stance and the way the men folded their hands were suggestively similar.[38] Unfortunately, 'suggestive' is not on its own sufficient evidence, although, like other aspects of Maltby's history, an accumulation of historical oddities might eventually lead to a revelation.

Again the documentary evidence, if it ever existed, had been destroyed. Walther Nernst had burned all his personal papers just before his death. He, perhaps even more so than Maltby, would have been conscious that he had a great deal to lose were certain secrets to get out. A scientist of high national and international standing and 1920 winner of the Nobel Prize in Chemistry, Nernst was aware of his own importance—those who knew him occasionally remarked on his lack of humility. It is entirely possible that Nernst may have destroyed his papers in 1941 because he did not wish to give ammunition to the Nazi regime regarding contacts he had, or had had, with Jewish scientists. It is also possible that he did not want his wife, children, or future biographers coming across Philip's existence (Barkan 1999, pp. 21, 51, 219).

It should be noted that the exact nature of Maltby's and Nernst's relationship, if one existed, is not at all clear. Given the increasing awareness of sexual harassment and assault in academia in recent times, it is entirely possible for a contemporary person to read that type of situation backward onto Nernst and Maltby. However, there is not enough evidence to determine whether or not any sexual relationship was consensual. Still, it is true that at the time Maltby worked with Nernst, Nernst was married and had three children. One of his sons is even visible in the bottom right corner of the photograph of the Göttingen scientists. The discovery of an extra-marital child would have been scandalous to his professional and personal standing. And, because of his prominence, it still would today. Although the Maltby family has approached a few of Nernst's direct descendants, so far none has agreed to take a DNA test. The closest the Maltby family has come is a DNA comparison with a relative of Nernst's mother. There's a connection.[39]

The Role of the Historian

Most of the time, it is easy to accept as an incontrovertible good that a historian's role is to chronicle and tell stories of the past. However, it is also just as easy to see that there are many instances in which the role is complicated—which are the stories that need to be told? Telling some stories, or not telling them, is taking a political stance. Historians wield an incredible power over what is remembered, and how. Is this one of those stories in need of telling? The topic concerns the private life of an

---

[38] Author interview with Elizabeth Traynor, Guntersville, Alabama, 24–25 February 2018; Elizabeth Traynor email to author 26 November 2018.

[39] Author interview with Raymond Gill and Jane Gill. Brooklyn, New York. 18–19 January, 2017.

intensely private woman. Would Maltby have wanted this information revealed even decades after her death? What are the ethical implications of conducting and then publishing this research?

This story has similarities to that of Sally Hemings and Thomas Jefferson, particularly in the discomfort it gives to those invested in a rose-colored vision of Jefferson's (or Nernst's) life (Gordon-Reed 1997). However, history is frequently discomforting, and the historian should not be in the practice of whitewashing the truth. Moreover, the needs of the living outweigh the needs of the dead. Maltby, Philip, and Nernst, have all passed on. And the Maltby family's need for closure should not be outweighed by the Nernst family's desire for reputation, if that is their concern. The revelation of a biologically connected relationship between Maltby and Nernst's descendants would challenge the Nernst family's view of an important and admired ancestor. Fortunately or unfortunately, with the growth of genetic testing and family genealogy, such discomforting revelations will likely become more common (Zhang 2018).

Conclusion

Piecing together Maltby's life is a labor of historical archeology. With much documentary evidence lost or deliberately destroyed, reconstructing the narrative involves the examination of fragmentary evidence and a great deal of extrapolation from a very small base. And often, it is the silences in the archive that speak the most. This narrative was only possible through deliberately decentering what would be considered 'traditional' markers of a scientific biography: papers published, research milestones, prizes and awards. These markers were often absent from Maltby's life but were part of the standard to which she publicly held herself.

It was clear from what Maltby deliberately published, destroyed, or concealed, that she wanted her legacy to be primarily one of professional success. However, she was openly proud of her 'adopted' son. His impact on her life did not detract but rather enhanced her professional figure as the former researcher who became a caring, motherly, mentor to future generations of scientists. That is, as long as she was careful to disavow any biological relationship with Philip.

For Maltby, as for many women scientists, the private and personal was intimately intertwined with the public and professional, both during her life and even after her death. Reconstructing Maltby's life requires balancing and entangling the 'official' documents with the personal recollections of Maltby's descendants. Only when used in conjunction are the details of Maltby's life revealed. And yet, these two facets of her historical legacy still come into conflict. The desires of the Maltby and Nernst descendants still weigh against each other. It remains to be seen if Nernst's personal life will become publicly intertwined with his professional legacy in the same way that Maltby's always was.

Acknowledgements The author gratefully acknowledges the invaluable help of Margaret Maltby's descendants, including Raymond and Jane Gill and Elizabeth and Mary Traynor. Thanks are also due to the Barnard College Archives, Columbia University Archives, MIT Archives and Museum, Oberlin College Archives, Royal Swedish Academy of Science, Smith College Archives, Lake Erie College Archives, and the Wellesley College Archives, as well as Joris Mercelis and the Johns Hopkins Gender History Seminar. This work was made possible by a Research Fellowship of the Consortium for History of Science, Technology and Medicine.

References

Barkan, D. K. (1999). *Walther Nernst and the transition to modern physical science*. Cambridge: Cambridge University Press.
Barnard Bulletin. (1912 September 25). Alumnae parade, *16*(29), 3.
Barnard Bulletin. (1917 April 4). Dr. Welch gives the Stevens memorial lecture, *21*(23), 1.
Barnard Bulletin. (1917 November 9). Barnard at P. & S, *22*(6), 3.
Barnard Bulletin. (1929 December 12). Bulletin Concludes Department Survey, *34*(21), 4.
Barnard College Yearbook Class of 1923. (1922). The Xcitement of x: A moral melodrama in two acts. In *The Mortarboard* (pp. 40–41). New York: Barnard College.
Bartel, H.-G. (1989). *Walther Nernst*. Biographien hervorragender Naturwissenschaftler, Techniker und Mediziner 90. Leipzig: BSB B.G. Teubner Verlagsgesellschaft.
Bronner, S. (2012). *Campus traditions: Folklore from the old-time college to the modern mega-university*. Jackson: University of Mississippi Press.
Crawford, E. (1996). *Arrhenius: From ionic theory to the greenhouse effect*. Canton, Massachusetts: Science History Publications.
Ferris, H., & Moore, V. (1927). *Girls who did: Stories of real girls and their careers*. New York: E. P. Dutton & Co.
Gill, R. (2016). Genetics & genealogy—Miss Maltby and her ward: using DNA to investigate a family mystery. *American Ancestors, 17*(2), 49–52.
Gordon-Reed, A. (1997). *Thomas Jefferson and Sally Hemings: An American controversy*. Charlottesville: University Press of Virginia.
Hanisch, C. (1970). The personal is political. In S. Firestone & A. Koedt (Eds.), *Notes from the second year: Women's liberation*. http://www.carolhanisch.org/CHwritings/PIP.html.
Harrison, S. W. (1993). Margaret Eliza Maltby (1860–1944). In L. Grinstein, R. K. Rose, & M. Rafailovich (Eds.), *Women in chemistry and physics: A bibliographic sourcebook* (pp. 354–360). Westport, Connecticut: Greenwood Press.
Horowitz, H. L. (1984). *Alma mater: Design and experience in women's colleges from their nineteenth-century beginnings to the 1930s*. New York: Alfred A. Knopf.
Hunt, C. (1912). *The life of Ellen H. Richards*. Boston: M. Barrows and Company.
Jones, C. (2009). *Femininity, mathematics and science, 1880–1914*. New York: Palgrave Macmillan.
Kidwell, P. A. (2006). Margaret Eliza Maltby (1860–1944). In N. Byers & G. Williams (Eds.), *Out of the shadows: Contributions of twentieth-century women to physics* (pp. 26–35). Cambridge: Cambridge University Press.
Kremer, R. L. (2011). Reforming American physics pedagogy in the 1880s: Introducing 'learning by doing' via student laboratory exercises. In P. Heering & R. Wittje (Eds.), *Learning by doing* (pp. 243–280). Stuttgart: Franz Steiner Verlag.
Maltby, M. E. (1895). Methode zur Bestimmung grosser elektrolytischer Widerstande. *Zeitschrift für Physicalische Chemie, 22,* 133–158.
Maltby, M. E. (1896). A few points of comparison between German and American universities. Address given before the Association of Collegiate Alumnae. Nineteenth Century Collections Online. http://www.tinyurl.galegroup.com/tinyurl/49hrt7. Accessed 1 November 2018.

Maltby, M. E. (1897). Methode zur Bestimmung der Periode electrischer Schwingungen. *Annalen der Physik und Chemie, 61,* 553–577.

Maltby, M. E. (1920). The Physicist. In C. Filene (Ed.), *Careers for women* (pp. 430–433). Boston: Riverside Press.

Maltby, M. (Ed.). (1929). *History of the fellowships awarded by the American Association of University Women 1888–1929.* Washington, D.C.: American Association of University Women.

Oberlin Weekly News. (1882). The classical commencement, June 30: 2. http://dcollections.oberlin.edu/cdm/ref/collection/p15963coll38/id/6137.

Onion, R. (2016). *Innocent experiments: Childhood and the culture of popular science in the United States.* Chapel Hill: North Carolina Press.

Rayner-Canham, M. F., & Rayner-Canham, G. W. (1992). *Harriet Brooks: Pioneer nuclear scientist.* Montreal: McGill-Queen's University Press.

Rossiter, M. (1982). *Women scientists in America: Struggles and strategies to 1940.* Baltimore: Johns Hopkins University Press.

Rossiter, M. W. (1993). The Matthew Matilda effect in science. *Social Studies of Science, 23*(2), 325–341.

Science. (1912 February 2). Memorial to Mrs. Ellen H. Richards. *Science New Series, 35*(892), 176–177.

Stage, S. (1997). Ellen Richards and the social significance of the home economics movement. In S. Stage & V. Vincenti (Eds.), *Rethinking home economics* (pp. 17–33). Ithaca: Cornell University Press.

Shortland, M., & Yeo, R. (Eds.). (1996). *Telling lives in science: Essays on scientific biography.* Cambridge: Cambridge University Press.

Talbot, M. (1931). *The history of the American Association of University Women 1881–1931.* Boston: Houghton Mifflin.

Wellesley College. (1900). *Wellesley College record, 1875–1900.* Wellesley, Massachusetts: Wellesley College. Retrieved November 1, 2018 from, https://repository.wellesley.edu/wellesleyhistories/8.

Wellesley College. (1897). *Annual reports of the president and treasurer of Wellesley College 1897.* Boston: Frank Wood. http://repository.wellesley.edu/presidentsreports/2.

Whitman, B. (1987). Mary Frances Winston Newson. In L. Grinstein & Paul Campbell (Eds.), *Women of mathematics: A biobibliographic sourcebook* (pp. 161–164). Westport, Connecticut: Greenwood Press.

Wiebusch, A. (1971). Margaret Eliza Maltby. In E. T. James, J. W. James, & P. S. Boyer (Eds.), *Notable American women 1607–1950* (Vol. 2, pp. 487–488). Cambridge, Massachusetts: Harvard University Press.

Zhang, S. (2018 July 17). When a DNA test shatters your identity. *The Atlantic.* https://www.theatlantic.com/science/archive/2018/07/dna-test-misattributed-paternity/562928/.

Zott, R. (Ed.). (1996). *Wilhelm Ostwald und Walther Nernst in ihren Briefen sowie in denen einiger Zeitgenossen. Studien und Quellen zur Geschichte der Chemie 7.* Berlin: Verlag für Wissenschafts- und Regionalgeschichte.

Chapter 4
Erwin Schrödinger in the Second Spanish Republic, 1934–1935

Enric Pérez

> Si un hombre nunca se contradice, será porque nunca dice nada.
> *Miguel de Unamuno, quoted from a conversation*

This epigraph ("If a man never contradicts himself, the reason must be that he virtually never says anything at all") comes from the last chapter of Schrödinger's famous book "What is life?" (Schrödinger 1951). The authors of the other epigraphs in this book are Descartes, Spinoza and Goethe, and this I think illustrates quite well the admiration Schrödinger felt for the Spanish poet, philosopher and playwright Miguel de Unamuno, who was widely known in Germany in those years (Martín Gijón 2017). It also wonderfully symbolizes Schrödinger's own views on quantum mechanics and many other controversial philosophical issues; this was not the only time he quoted this sentence (Pérez 2018). Moreover Unamuno was not the only Spanish philosopher he met and quoted. In the first chapter of *Science and Humanism* we find praise for the work of another Spanish contemporary, José Ortega y Gasset (Schrödinger 1944).

Schrödinger's book on molecular biology was a compilation of the talks he had given the previous year in Dublin. It deals with biological life (the Greek *zoe*). Here I am of course going to deal with an episode of Schrödinger's life, but in a biographical sense (the Greek *bios*). I will attempt to portray its influence on the scientific development in Spain and its role in Schrödinger's career. In the early days of August 1934, Erwin Schrödinger—a Nobel laureate since the previous November—landed in Galicia, a north-western region of Spain. This was the first time he had been in the country of *El Quijote* (a book he also quoted many times). He liked Spain so much that in less than a year he returned to go on a lengthy tour by car together with Annie, his wife. Apparently, Schrödinger was planning to visit Spain for the third time in 1936, but in the end he cancelled his trip (Cabrera 1972, p. 71).

Why Spain? Were these trips just pleasure trips? Or was Schrödinger considering a position in Madrid, as some of his German colleagues did before? Since November

E. Pérez (✉)
Universitat de Barcelona, Barcelona, Spain
e-mail: enperez@ub.edu

© The Editor(s) (if applicable) and The Author(s), under exclusive licence to Springer Nature Switzerland AG 2020
C. Forstner and M. Walker (eds.), *Biographies in the History of Physics*,
https://doi.org/10.1007/978-3-030-48509-2_4

1934 Schrödinger had been in Oxford, in *voluntary* exile from Germany. It was there that his first daughter was born in May 1935 (Moore 1989, p. 296). It seems that he was not comfortable there, no less than the academic authorities were with his personal lifestyle. Thus, he had to think about where to go next. I would argue that, strange as it may seem, Spain was not a bad idea, at least in the early 1930s. Spanish academic authorities and scientists were doing their best to turn Spanish scientific institutions into a reasonable alternative.

Indeed, from a historiographical perspective, Schrödinger's trips can be seen as part of an interesting episode of the scientific community in Spain attempting to leave the periphery in order to become more central. But why Schrödinger? Why should he appear attractive to the Spanish physics community? Was he a good choice for the purpose of creating or fostering a new research group in physics? Apparently not, as it is known that Schrödinger was not prone to collaborations.

As far as I know, these visits have only been studied in detail in a paper by the Spanish science historian José Manuel Sánchez Ron (Sánchez Ron 1992), although the biography by Moore also contains some fleeting comments (Moore 1989). Sánchez Ron gives a quite good description of both trips and an interesting but brief review of the lectures in Santander and of the paper Schrödinger published in the *Anales de la Sociedad Española de Física y Química* in 1935 (Schrödinger 2001, 1935a). Since 1992 new research has revealed other actors in this story, and has increased our knowledge of the state of physics in Spain in those years (Sánchez Ron et al. 1993; Fernández Terán 2014; Gimeno 2015). Particularly, the papers by Clara Janés on Schrödinger's relationship with José Ortega y Gasset and Xavier Zubiri threw some light on this fascinating period of Spanish cultural and scientific life (Janés 2015a, b). These and other archival materials have enabled us to complete this account and to explore in greater depth the historiographical significance of the visits by the creator of wave mechanics to Spain. This paper is a preview of that account.

As an introduction I begin with a chronological description of the two trips, followed by a general view of the state of physics in Spain in the early 1930s, focusing on the two institutions that were most closely-related to these episodes: the *Instituto Nacional de Física y Química* (*National Institute of Physics and Chemistry*, henceforth INFQ) and the *Universidad Internacional de Verano de Santander* (*International University of Santander*, henceforth UIS). I will then go on to argue that Schrödinger's visits must be considered in the context of migration. Finally, after attempting to contextualize Schrödinger's conferences in Spain within his own philosophical and scientific evolution (even though the proximity of the Spanish Civil War left little time to appreciate the influence of the illustrious visitor), I will speculate on the impact of these trips, had not war broken out in July 1936. I have titled this last section "Schrödinger's cat in Spain" because he wrote his famous paper on the "present situation in quantum mechanics", his "general confession", some weeks after visiting Spain (Schrödinger 1935b). Therefore, we should look for traces of the cat in his talks in Spain.

The Trips: August 1934 and March–April 1935

Schrödinger made two visits to Spain in the consecutive years 1934 and 1935. The first was sponsored by the above-mentioned UIS, a summer university, and the second by INFQ, the main centre for research in physics in Spain.

From *Magdalene College* to the *Palacio de La Magdalena*

On 3 August 1934, Schrödinger gave a short talk at the 14th conference of the *Asociación Española para el Progreso de las Ciencias* (*Spanish Association for the Advancement of Science*, henceforth AEPC). As we can read in the manuscript of the talk (which is conserved in the *Archivo Xavier Zubiri* in Madrid), Schrödinger's contribution was not part of the original plan for the visit.[1] He had been formally invited in January to participate in the UIS (Sánchez Ron 1992, p. 10), where he gave three sessions on August 13th, 14th and 15th. Some days before Schrödinger arrived in Spain (probably by ship at La Coruña, as many ocean lines sailing to America, for instance, from Southampton made stops in the north of Spain) Blas Cabrera, his host, might have informed Schrödinger that a scientific conference was being held in Santiago de Compostela, and that a talk from the Nobel prize winner would be highly appreciated. He stayed at the *Hotel Compostela*, and fortunately for him, he did not join the trip to a traditional *pazo* (a typical country house), *o pazo de Oca*, on Sunday 5 August—when nearly 100 participants from the conference visited the *pazo*'s main room at the same time, the floor collapsed. Many people were seriously injured and a teacher even died.[2] The conference was cancelled.

Santander is 500 km from Santiago de Compostela. There Schrödinger spent his first days in Spain. He very likely travelled along the Cantabrian coast with Cabrera, who in addition to attending the conference in Galicia was the dean of the UIS. Although there are many charming places in that part of Spain, there is no detailed account of which of these spots the Austrian professor visited. Perhaps not many, as he later apologized to his Spanish friend and guide in Santander, the philosopher Xavier Zubiri, for not having been in a good mood in Santander (Janés 2015b). Apparently he was a bit depressed.

Be that as it may, according to the local press, Schrödinger and Cabrera arrived in Santander on 11 August, where Schrödinger gave his first lecture on Monday 13 August in the beautiful Palacio de la Magdalena.[3] The three lectures he gave in Santander were translated from French into Spanish by Zubiri, and became a nice short book titled *La nueva mecánica ondulatoria* (*The new wave mechanics*), which was published in Madrid in 1935 (Schrödinger 2001). Very few testimonies exist

---

[1] Documento 038_04_01, *Archivo Zubiri*, Madrid.
[2] Un hundimiento en el Palacio de Oca. Una maestra muerta y cuarenta heridos. *La Luz. Diario de la República*, Madrid, 6 August, 1934.
[3] En la Universidad Internacional. *El Cantábrico*, Santander, 11 August, 1935.

regarding Schrödinger's stay in Santander, nor do we know exactly when and how he went back to England. In a letter to Zubiri in November of that year, he says that he spent four weeks in Spain (Janés 2015b, p. 32), but we can only be certain that he was there during the first fortnight of August. We have not found any detailed account of those four weeks in his correspondence.

Tracing a *Big Eight* with a BMW

The following spring, 1935, Schrödinger traveled once again to Spain, this time not alone but with his wife Annie, in their BMW, "tracing a large 8 with the double point in Madrid".[4] This second time he was clearly in a very different mood, and accordingly planned visits to different places throughout Spain: Roncesvalles, Salamanca, Valencia, Alicante, Granada, and so on. He did not attend any conferences, but did give more technical talks than the previous year, at the INFQ on 27, 29 and 30 March, at the *Sociedad Española de Física y Química* (*Spanish Society of Physics and Chemistry*) on 1 April, and at the *Academia de Ciencias Exactas, Físicas y Naturales* (*Academy of Exact, Physical and Chemical Sciences*) on 10 April, all of them in Madrid. The organizer of these lectures was in all likelihood Cabrera, at that time the director of INFQ and president of the Academy. We also know, from a letter Schrödinger wrote to Einstein a few weeks later, that Erwin and Annie had a really good time in Spain (Sánchez Ron 1992, p. 17). Cabrera, and probably Enrique Moles, a chemist in charge of the journal of the *Academia*, asked Schrödinger for a paper, and as a consequence Schrödinger published a paper in Spanish: "¿Son lineales las verdaderas ecuaciones del campo electromagnético?" (Are the true equations of electromagnetic field linear?) (Schrödinger 1935a). Schrödinger gave his talks in the language of Cervantes, and his correspondence with Moles related to that paper demonstrates how fast he acquired competence in Spanish. Doubtless his knowledge of Romance languages such as Italian or French had been of great help.

The *Silver Age* of Spanish Physics

The full significance of Schrödinger's first visit to Spain only becomes clear if we consider its political and scientific context. The Second Spanish Republic was born on 14 April 1931. After seven years of the dictatorship of Primo de Rivera, in 1930 King Alfonso XIII tried to establish a parliamentary monarchy, without success. The proclamation of the new democratic regime gave a great boost to the eagerness for change and regeneration that had been growing in Spain for more than 20 years. As for

---

[4]Schrödinger to Einstein, 17 May 1935, Albert Einstein Archive. Also the document "Was ein kleiner BMW erzählen kann" by Annie Schrödinger, Nachlass Erwin Schrödinger, Universität Wien.

physics, in the early 1930s a so-called "First school" (Sánchez Ron and Roca-Rosell 1993) had already come into being in 1907.

Science (and particularly physics) as we know them today were greatly developed during the nineteenth century (Cunningham and Williams 1993), when the second industrial revolution took place. New technology and science were needed, and physics research experienced a massive increase in funding. Indeed, the most powerful countries began to consider it a key aspect of their policies. In Spain, due to causes that we are not going to discuss here, this process did not take place during the same decades as in other European countries, at least not at the same level and not at the same pace. But after Spain lost its last American colony in Cuba in 1898, the national debate on the backwardness of the country's industry, science and technology became a central issue and gave rise to *regenerationism*: many people from different areas of society called for a radical change in Spain's traditional policies and education, mainly focusing (but not exclusively) on the development of science and technology. Whether it was as a result of these efforts or for other reasons, in the first third of the twentieth century the cultural life in Spain flourished. It is called the *Silver Age*, a reference to the *Golden Age* of Spanish literature in the 16th and 17th centuries.

Based on these ideas, within a few years new institutions were created with the aim of changing this situation outside of the university, because inertia often led to strong resistance. The intention was not to deny the central role of universities, but to diminish their relevance by injecting fresh air infused with innovative views. The institution that plays a major role in this story, the *Junta para Ampliación de Estudios (Board for the Extension of Studies*, henceforth JAE) (Fernández Terán and González Redondo 2007, p. 71), was created in 1907. Its first president was the Nobel laureate for medicine Santiago Ramón y Cajal. One of the main purposes of the JAE, and certainly one of its most successful tasks, was to obtain funding for Spanish students to go abroad and work alongside international specialists in well-equipped laboratories around the world (Sánchez Ron 2000). Thanks to the JAE, in less than 10 years there were some research groups in physics working in Spain. They were located in Madrid, at the *Laboratorio de Investigaciones Físicas (Laboratory of Physical Research*), an institution created under the auspices of the JAE in 1910. These groups were led by the first generation of Spanish physicists (Fernández Terán 2014). From 1910 to 1937 (including its conversion into the later INFQ in 1932) the director of the *Laboratorio* was Cabrera, an authority on magnetism. Other notable figures included Enrique Moles, an international expert on precision measuring in chemistry, Julio Palacios, the man who introduced modern X-ray apparatus to Spain, and Miguel Catalán, a specialist in spectroscopy who developed a refined technique for analysing multiplets.

Thanks to the successful work of the JAE in producing international experts in such a short space of time, the *International Educational Board* of the *Rockefeller Foundation* responded to the requests by Spanish scientists and politicians by supplying funding for a new building with new laboratories and new equipment (Kohler 1991, pp. 188–198). It was built in the late 1920s and officially opened in 1932. This institute became the centre of the *Silver age* of Spanish physics and chemistry.

With the coming of the Republic in April 1931, new reforms and initiatives were proposed, and one of them was the creation of an international summer university in Santander, where foreign and national specialists could be invited in order to give Spanish students the opportunity to catch up with recent developments in all fields. The government decided to convert the king's summer residence, the *Palacio de la Magdalena*, into an academic facility. The first such course took place in 1933, but 1934 was the first year that the restructuring was complete, and Schrödinger could enjoy the new lodgings and a beautiful space where professors and students could attend lectures and have informal discussions, and where science and arts were mixed and taught by professors from different parts of the world. Weyl had also intended to go there in 1934, at the same time as Schrödinger, but in the end he did not. He wrote to Abraham Flexner that the UIS "gathers all the people who play a role in the intellectual life of Spain" (von Meyenn 2009, p. 534). Certainly, the UIS was very successful, and even when the political orientation of the government veered to the right in late 1933 (which entailed some cutbacks) it continued to be a meeting place which played both a symbolic and a real role in the young Republic. In its fourth year, in 1936, the Civil War broke out and professors and students (Cabrera among them) had to return to their respective homes, spread across a divided, war-torn country.

The invitation to Schrödinger to participate in the UIS, the event which motivated his first visit, was sent by its secretary, the Spanish poet Pedro Salinas (Sánchez Ron 1992, p. 10). That year, Cabrera—the Spanish physicist who probably had the closest connections with the European physics community—was appointed dean. He had been working in Zurich with Pierre Weiss in the 1910s thanks to a grant from the JAE, and his measurements of magnetic susceptibilities were widely cited, for instance, in the renowned book on magnetism by John H. van Vleck (van Vleck 1932). From 1928 onward, Cabrera was a member of the scientific committee of the Solvay conferences, and in 1930 he participated in the sixth conference, devoted to magnetism, with a presentation and by helping out with its organization. He was also deeply involved in the development of Spanish science at an institutional level. Therefore, it was only logical that in 1923 he was Einstein's host in Madrid. In many senses Cabrera could be considered the representative of physics in Spain in those years. He had written several books on relativity and modern physics, and was one of the main figures responsible for introducing modern physics into Spain (Fernéndez Terán and González Redondo 2007). When the Civil War broke out, Cabrera went to Paris. Despite having decided to resume his scientific activities after the defeat of the Republican government by General Franco's insurrectional forces, he was not permitted to return to Spain and travelled to Mexico, where he died in 1945.

Cabrera also attended the 7th Solvay meeting, focused on the atomic nucleus, though he did not contribute a talk (Stuewer 1995). The Solvay meeting was held in Brussels in October 1933, and from there Cabrera could have invited Schrödinger to go to Spain the following summer, since he was probably planning the international participation for the UIS. In November of that year, he sent Schrödinger a postcard congratulating him for the Nobel Prize.[5] Unfortunately Cabrera's correspondence is

---

[5]Cabrera to Schrödinger, 11 November 1933, *Nachlass Erwin Schrödinger*, Universität Wien.

almost completely lost, and only three letters from Schrödinger have survived, from 1937 to 1939 (Sánchez Ron 1992, p. 17). Reading these missives we can see that these two physicists enjoyed a true friendship. Schrödinger even proposed to Cabrera that they should both move to a Spanish-speaking country to begin a new life far from tumultuous Europe; both were in exile during those years.

But Schrödinger had another Spanish contact: Xavier Zubiri, a Basque philosopher who had been studying in Berlin in the academic year 1930/1931 (Corominas and Vicens 2006). Xavier Zubiri had been ordained in 1921, but resigned his position as a priest 14 years later. In 1929 he had travelled to Freiburg to study phenomenology with Edmund Husserl and Martin Heidegger, and then moved to Berlin in 1930 as a visiting professor (he lodged at the Kaiser Wilhelm Society's Harnack Haus) to study ancient philosophy with Werner Jaeger and modern physics with Schrödinger. In Berlin he attended lectures by Einstein and Schrödinger, and began a friendship with both physicists. In his archive in Madrid we can find the manuscript of Schrödinger's lectures for the academic year 1927/1928 (presumably in Schrödinger's own handwriting), and which may have been a gift from the Viennese physicist.[6] Zubiri was one of the few Spanish philosophers in those years who were aware of the latest developments in quantum mechanics, and in 1934 he even wrote an interesting paper in a Spanish philosophical journal—*Cruz y Raya*— on that subject, mainly based on Schrödinger's lectures on quantum mechanics in Berlin (Zubiri 1934). It is a long, wide-ranging paper which expounds a general but accurate vision of the new mechanics and its philosophical consequences for the conception of reality. In the Zubiri archive we find as well some letters related to Zubiri's translation of Schrödinger's lectures in Santander.[7] They spent many hours together there, walking along the Cantabrian coast (Janés 2015b).

Looking for a Position

The international connections of Spanish physicists led to the arrival of illustrious visitors to Spain in the 1920s and early 1930s. The best known was Einstein, who in 1923 went to Madrid, Barcelona and Zaragoza (Glick 1988). But Sommerfeld and Weyl also went to Spain, invited by the government in order to foster Spanish science and physics (Glick 1988, p. 71). It was on that visit to Madrid in 1922 that Sommerfeld learnt of the works of Catalán, who thanks to a grant from the JAE (Eckert 2013, p. 257) had previously been at the *Imperial College* of London, working with Alfred Fowler. Later on, Catalán went to Munich and published some important work on spectroscopy (Sánchez Ron 1994). In February 1932, Sommerfeld returned to Spain during the inauguration of the new building of the INFQ. Pierre Weiss, mentor to Cabrera in the physics community, was also among the invited people. Other visitors

---

[6]Documento 30_01_01, *Archivo Zubiri*, Madrid.
[7]Schrödinger to Zubiri, 5 August 1934, 11 September 1934, 8 November 1934, 15 December 1934. Carpeta 131_18 *Archivo Zubiri*, Madrid.

were Arthur Eddington in 1928, Paul Scherrer in 1929 and Marie Curie in 1919, 1931 and 1933.[8]

Though he was a Nobel laureate, Schrödinger's visits had little social impact on the Spanish Republic. If we compare them with Einstein's, the difference is huge. Certainly, his talks were advertised in the local and national press, but never in the headlines. Einstein rapidly became a celebrity after 1919, and as a matter of fact his trip to Spain was part of a plan to popularize science and physics. Einstein was invited to many social events, and he was a famous wise man many months before he arrived in Barcelona. Definitively, this did not happen in the case of Schrödinger: only experts and students were aware of his visits. As far as I know, he did not meet any political figures. Without any doubt, the relativity of time and space, along with the spectacular success of the deviation of solar rays were much more attractive for the press and the general public than wave mechanics. I would guess that only the uncertainty principle (and maybe the still non-existent cat) could be compared with those shocking discoveries. But Schrödinger was not its father, and in fact, he repeatedly used the uncertainty relations to cast doubt on the conceptual coherence of the theory. Thus a lack of interest in quantum mechanics might also have been responsible for the lack of media attention to Schrödinger's visits. The political situation may also have affected this difference in coverage, as turbulent times had commenced in Spain's political life at the end of 1933.

In any case, the short but intense affair that Schrödinger enjoyed with Spain is captivating. He felt attracted both by its climate and by the Spanish people. This is what he wrote to Zubiri some weeks after leaving Santander:

> Yours is a happy country. A little thanks to the external gifts that God has given her, but above all thanks to the internal divine gift of a temperament that is receptive to happiness, one which seeks joy and creates it. Four weeks are not enough to find one's feet there, when one has come from coarse Europe. Only later did I realized, and felt a bit ashamed, that I did such a bad job of adapting to it. There will be a next time. I am so glad that Spain exists, that it exists for me, now. Not only because I will come back. It is the knowing that a people exist who see the world in such a way, a people who shape and change, support and improve the very concept of the world.[9]

Since November 1933 Schrödinger had held a temporary position in Oxford, and was living there along with Annie, his wife, and Hilde March (Moore 1989, p. 279). He had left Berlin in May 1933, and in the same year became a Fellow of *Magdalen College*, mainly thanks to the good management of the physicist Frederick Lindemann (Moore 1989, p. 267). His position there was not as a professor, and it seems that he never really got used to this academic situation. When it was time to renew the grant, neither he nor his hosts were very enthusiastic, and the Schrödingers moved to Graz (Austria) in autumn 1936 for a couple of years. It is known that one of the aspects which made things difficult in Oxford was that Erwin had, so to speak, two wives—Annie and Hilde, who was married to the physicist Arthur March. In

---

[8]EPS Historic Sites—The Residencia de Estudiantes, Madrid, Spain. https://www.eps.org/page/distinction_sitesRE.

[9]Schrödinger to Zubiri, 11 September 1934, Carpeta 131_18, *Archivo Zubiri*, Madrid.

addition, Hilde was pregnant by Erwin. When the child was born on 30 May, it is easy to imagine how awkward that situation fitted into the life of Oxford University, especially for a Nobel laureate. Lindemann was extremely angry with Schrödinger, and this probably caused his reluctance to renew his grant (Moore 1989, p. 298). Of course, this personal situation also complicated the search for a new position. In the spring of 1934 Schrödinger had been at Princeton, at the *Institute for Advanced Studies*. But the possibility of moving there, as Einstein and Weyl had (Sánchez Ron and Glick 1983; Sigurdsson 1996, p. 49), did not come to fruition, and once again his non-traditional marital life could have affected his options negatively, as well as conditioned his preferences. It was in this context that he travelled to Spain, while Annie went to Zurich to be with Weyl, and Hilde, the mother of his daughter, probably remained in England.

However, we did not find any evidence to suggest that he had seriously considered taking up a position in Spain, as Einstein and Weyl had. Schrödinger could not understand why Einstein ultimately refused to go to Spain:

> I have just come back from Spain and I am very enthusiastic. When I think that you had the opportunity to go there and probably, as long as you want, to live outside Madrid...[10]

Be that as it may, the context of forced migration meant that Schrödinger was forced to consider different possibilities. Max von Laue stayed in Oxford with the Schrödingers for a few days in May–June 1934 (Moore 1989, p. 295). He wrote to Erwin Freundlich:

> Schrödinger also feels himself to be not at all happy, although he is the only one of the German scholars who has had the honour to be made a fellow of a college. He speaks of leaving and alternates in his mind the most distant lands of the earth.

Schrödinger may have thought that a trip to an emerging scientific nation was a good idea to help him mull over new options. Besides Einstein and Weyl, we also have evidence that Hertha Sponer considered the possibility of obtaining a position in Spain (Glick 1986, p. 282). The political situation in Germany forced many scientists to look for alternative places to develop their careers. Max Born was in Bangalore, India, in the winter semester of 1935–36, and even considered the option of settling there (Sigurdsson 1996).

Schrödinger's Cat in Spain

Scientifically and philosophically speaking, 1935 is usually considered the beginning of the end of the period of silence Schrödinger had kept since 1927 on his disagreement with the orthodox view of quantum mechanics (Bitbol 1996). In fact, two months after his return from Spain in the spring of 1935, he congratulated Einstein for the publication of the EPR paper (Fine 1996, p. 66). The proximity of

---

[10]Schrödinger to Einstein, 17 May 1934, *Albert Einstein Archive*.

Schrödinger's visits to Spain to his publication on the quantum theory, where his now famous cat was born (and died), suggests that we should look for some traces of the cat in Spain–and the result is definitively positive.

In his talks in Spain in 1934 we can fully appreciate his dissatisfaction with what was later called the Copenhagen interpretation. For instance, in Santiago de Compostela (the talk was titled "Quelques remarques sur l'interprétation de la méchanique quantique"[11]) he openly criticizes the idea of continuing to use classical variables in the new theory despite the strong restrictions that one must impose on them. Therefore, this critique points to the perturbation interpretation of the uncertainty relations, which justifies the use of physical images and classical magnitudes. The content is very similar to that of a short paper Schrödinger had sent to *Die Naturwissenschaften*, probably before going to Spain (Schrödinger 1934). It was an affirmative reaction to a previous paper by Max von Laue (von Laue 1934), who had been in Oxford some weeks before Schrödinger went to Spain. In Santiago, Schrödinger also presented the difficulties of defining a rigid body within the frame of quantum mechanics, basing his arguments again on uncertainty relations. It was a way of showing the problems of a proper treatment of time in the new mechanics.

In Santander he insisted on these points, but in the context of a general exposition of the fundamentals of the theory (Schrödinger 2001). In the published version of it, we can appreciate how masterly Schrödinger was in exposing the theory with a minimum of mathematical apparatus, and also find scattered comments that clearly show his disagreement with the shape the new theory was taking. For instance, referring to the uncertainty relations, he wrote:

> ... I am not sure that abandoning oneself to triumphant joy is justifiable here. You can judge for yourselves. It seems that, at least in the current state that the interpretation of this theory has reached, it should be content with a success that is a little more modest, namely: not having eliminated from the physics of the atom the questions that it was not forbidden to raise, but rather causing them to enter a precise, interesting system. It has, we might say, made this uncertainty into a fundamental law: Heisenberg's law of uncertainty or indeterminacy, which we will have to talk about right away. (Schrödinger 2001, p. 75)

Or as for the use of classical concepts:

> ... it would seem more prudent to confess frankly that the classical notion of energy (and many others that will be discussed immediately) are not really applicable to the new image of nature, since a well-defined state of a system does not generally contain a well-determined value of the variable in question. Having recognized it in that way, I consider a bit daring the attempt to interpret any experience as a determination of the current value of a variable that in general does not possess it. (Schrödinger 2001, p. 83)

In essence, although Schrödinger gave an overview of the fundamentals of wave mechanics, he transmitted it with a certain dose of scepticism.

Interestingly, the main character of the talks he gave on his next visit was slightly different in its nature. For instance, what we find in the manuscript of the three talks he gave in the INFQ written by Schrödinger himself (in Spanish),[12] and also

---

[11] Documento 038_04_01, *Archivo Zubiri*, Madrid.

[12] Vorlesungen für Madrid. *Nachlass Erwin Schrödinger*, Universität Wien.

the published notes by one of the members of the audience, Eduardo Gil Santiago (Gil Santiago 1941) is an elementary course for a specialized audience. These three conferences are the most technical and complete lectures he ever gave in Spain. Another Spanish physicist who also attended, Fernando Peña Serrano, published a paper two years later on the quantum treatment of the harmonic oscillator based on Schrödinger's lectures (Peña Serrano 1937). Nevertheless, he uses the formalism of creation and annihilation operators; that is, the Dirac approach.

In the *Spanish Society of Physics and Chemistry* Schrödinger dealt with a different, but related subject: the form of Maxwell equations (Schrödinger 1935a). It was titled "The equations of electromagnetic field," and we can assume that its content was the same as that which appeared in the form of a paper in the *Anales* the same year (Schrödinger 1935a). In it, Schrödinger tackles the issue of how the linear character of electromagnetic equations should be modified, appealing to an analogy with sound waves. Doubtless, Schrödinger was thinking about how to obtain a quantum theory for the electromagnetic field.

We have no manuscripts or papers related to the last talk, titled *The uncertainty principle and its influence in the geometry concepts of the world*. According to its title and the brief reviews we find in the press,[13] Schrödinger tackled once again the privileged status of time in quantum mechanics, and how this could be related with the lack of a good fit with the relativity theory.

Could Spanish students and professors of physics fully understand Schrödinger's lectures? Certainly, the state of quantum mechanical research in Spain was poor. As for the old quantum theory, there had been some lectures and some contributions in conferences: the first public exposition was made by Esteve Terradas, precisely at the first conference of the AEPC, held in Zaragoza in 1908 (Moreno 2000). However, this initial presentation had barely any continuation from its author or other colleagues. The translation of Reiche's book on quantum theory (Reiche 1922) was probably the most important text in Spanish on quantum theory in the early 1920s. Its translator, Julio Palacios, also published an original paper in *Annalen der Physik* on the emission theory in Bohr's atom, and its physical context was, again, the old quantum theory (Palacios 1926). The new quantum mechanics did not have followers in Spain. Not until the early 1930s did a few publications on the subject by Spanish physicists and engineers appear (Baig et al. 2012; Gimeno 2015).

In other words, the first school of physics in Spain was experimental: the groups of Cabrera, Catalán and Palacios did not focus their investigations on theoretical physics, but on experimental research (including magnetism, spectroscopy and X-rays). More than that it was probably this backwardness, this deficiency, that motivated the invitation to Schrödinger. And the success of the first lectures he gave in Santander helped to prompt the second invitation, where Schrödinger gave a more specialized overview of wave mechanics to a group of students who were eager to develop their theoretical skills.

---

[13]El professor austriaco Schrödinger, en la Academia de Ciencias. *La Nación*, Madrid, 11 April, 1935.

Final Remarks

The impending Civil War and the political atmosphere that preceded it make it very difficult to assess the relevance and consequences of Schrödinger's visits on the development of Spanish physics. Indeed, the few historiographical papers that deal with this question openly disagree. Sánchez Ron argues that "as far as Spanish physics is concerned, the visits of the creator of wave mechanics did not have too much practical importance", and goes on to compare them with Sommerfeld's visit in 1922 (Sánchez Ron 1992, p. 14; Sánchez Ron and Roca-Rosell 1993): in "the thirties, and even before, Schrödinger was, in many senses, more a philosopher of nature than a scientist along the lines of the competitive and dynamic quantum physicists". In contrast, Baig et al. show and document some of the implications of Schrödinger's visit in the papers by Gil Santiago and Peña Serrano (Baig et al. 2012, p. 164).

I do believe that choosing Schrödinger to attend the Summer school in Santander was a very good one. His proficiency as a teacher was well known, and particularly by Zubiri, who had attended his lectures on wave mechanics in Berlin. But what made the choice of Schrödinger a well-considered one was the deficiency in theoretical physics in Spain. None of the leading Spanish physicists mentioned above was a theorist. In the 1920s, some work had been done on the theory of relativity (Glick 1988), but practically nothing on quantum mechanics. Schrödinger was able to motivate the students that attended his lectures in Santander to start doing research into a theory that, he claimed, was not completely finished. The three lectures he gave in his visit to the INFQ the following year, where doubtless many of the attendees had also been at his talks in Santander, support this view. At least two students decided to publish his notes, or some related ones, in scientific journals (Gil Santiago 1941, Peña Serrano 1937). Both of these showed good understanding of the material.

Schrödinger certainly posed questions to a peculiar audience in Spain: they were receiving the orthodox view and a critique of it at the same time. Spanish students were eager to know about the new ideas, and presumably were still not ready to dwell on their contradictions or ambiguities. But this does not mean that Schrödinger's presentation was a failure. On the contrary, it consisted of a well-structured, well-argued exposition of the new mechanics which was followed, the next year, with a more technical exposition. Spanish physicists were ready to hear Schrödinger's talks. Among the numerous celebrities Nicolás Cabrera came to know thanks to the central role played by his father Blas, years later he especially remembered Schrödinger's visits:

> Perhaps the visit I remember most for several reasons is that of Erwin Schrödinger (…) During the year 1935 he visited Madrid, where he gave a seminar in Spanish on quantum mechanics which greatly benefited all the young people who were at that time playing with physics. (Cabrera 1972, p. 71)

Among the few publications on quantum mechanics in Spain in the early 1930s, almost all of them are directly or indirectly related to either Schrödinger or his visits. Firstly, the little book in Spanish which contains the talks he gave in Santander (Schrödinger 2001); secondly, the above-mentioned papers by Gil Santiago and Peña

Serrano (Gil Santiago 1941, Peña Serrano 1937). The paper "La nueva física" ("The new physics") by Zubiri must be considered, too (Zubiri 1934). This paper is probably the most ideal for undertaking a detailed study of the process of appropriation of the new ideas by a Spanish (that is, peripheral) philosopher (Gavroglu et al. 2008). In fact, in 1932 Schrödinger had published an interesting paper in the most important philosophy journal in Spain, the *Revista de Occidente* (Schrödinger 1932). It consists of a translation into Spanish of a talk he gave in Berlin that same year: "Are the natural sciences conditioned by the milieu?" In this talk, Schrödinger discussed cultural influences on physics theories, and used some of the most striking features of quantum mechanics as instances.

Given the almost non-existent bibliography on quantum mechanics in Spain in the early 1930s, these examples are significant. In other words, quantum mechanics was arriving in Spain via Schrödinger's views. Of course, we cannot tell what would have happened had the Civil War not broken out, but I believe that these few but existent signs show that in those years the Spanish community of physicists was ready to develop its own research on quantum physics. They struggled to make Spain a center of physics research in Europe. Nonetheless, the efforts and results they achieved in the last decades are impressive. They not only wanted to foster research, but to make Spain an appealing place to go for foreign and more experienced researchers. It is in this context that the UIS shows its true value, and helps us to understand why figures like Einstein and Weyl seriously considered working in Spain. Although I have not found any hard evidence, I believe that Schrödinger, at that time in a temporary position, also thought of Spain as an option. Maybe he even thought that the lack of theoretical research might allow him to transmit his own perspective on the new mechanics. In his first talk in Spain, he presented a paper in which he subscribed and developed critical statements on the new mechanics by von Laue.

In sum, Spain was not such a bad alternative. The great efforts the Spanish governments had made in the previous decades, together with the blooming of not only science, but also of the country's cultural, social, and political life, meant that Spain became a real alternative for some scientists to move to, once the situation in Germany had become unbearable. Unfortunately, at the beginning of 1934 the political situation in Spain also worsened, and it was no longer a place to begin a new career and a new life. After the war (1936–1939), the new regime completely decapitated the elite that led the INFQ, tried to eliminate the entire legacy of the Republican institutions, and plunged the country into an autarchic epoch that lasted, practically, until the 1960s (Herrán and Roqué 2012). The attempts to place Spain at the center of European scientific life were interrupted for decades.

Meanwhile, Schrödinger, unhappy with his situation in Oxford, decided to return to Austria in the autumn of 1936, where the government provided him with a position in Graz. Shortly before he had been about to move to Edinburgh, but the proposal did not bear fruit, apparently due to bureaucratic problems (Moore, p. 318). Austria was not a good choice: in September 1938 the Schrödingers had to leave the country to finally land in Dublin in October 1939.

None of this could have been foreseen in the spring of 1935, when the Schrödingers toured Spain in their BMW. However, the preamble to World War II had begun, not only in Germany, but also in Spain.

Acknowledgements I want to thank especially the organizers of the inspiring meeting on the biographical approach to the history of physics we attended in Bad Honneff in May 2018: Christian Forstner, Mark Walker and Dieter Hoffmann, as well as Arne Schirrmacher, the commentator on my paper. He offered me some valuable remarks and references to help improve the written version. I also want to thank Marià Baig, Michael Bunn, Tony Duncan, Gonzalo Gimeno and Mercedes Xipell for their help and assistance during the preparation of my talk.

References

Baig, M., Gonzalo, G., & Mercedes, X. (2012). La introducción de la mecánica cuántica en España: las primeras lecciones y los primeros textos. In N. Herrán & X. Roqué (Eds.), *La física en la dictadura. Físicos, cultura y poder en España, 1939–1975*, (pp. 161–176). Barcelona: Universitat Autònoma de Barcelona.
Bitbol, M. (1996). *Schrödinger's philosophy of quantum mechanics*. Dordrecht: Kluwer Academic.
Cabrera, N. (1972). Apuntes biográficos de mi padre D. Blas Cabrera y Felipe (1878–1945). In *En el centenario de Blas Cabrera*, (pp. 59–73). Las Palmas de Gran Canarias: Universidad Internacional de Canarias Pérez Galdós.
Corominas, J., & Vicens, J. A. (2006). *Xavier Zubiri: la soledad sonora*. Madrid: Taurus.
Cunningham, A., & Williams, P. (1993). De-centering the 'Big Picture': The origins of modern science and the modern origins of science. *British Journal for the Philosophy of Science, 26*, 407–432.
Eckert, M. (2013). *Arnold Sommerfeld: Science, life and turbulent times, 1868–1951*. New York: Springer.
Fernández Terán, R. E. (2014). El profesorado del "Instituto Nacional de Física y Química" ante la Guerra Civil, el proceso de depuración y el drama del exilio. Doctoral thesis. Universidad Complutense de Madrid.
Fernández Terán, R. E., & González Redondo, F. A. (2007). Blas Cabrera y la física en España durante la Segunda República. *Llull* 30, 65–103.
Fine, A. (1996). *The shaky game*. Chicago: University of Chicago Press.
Gavroglu, K., Patiniotis, M., Papanelopoulou, F., Simões, A., Carneiro, A., Diogo, M. P., et al. (2008). Science and technology in the European periphery: Some historiographical reflections, *46*, 153–175.
Gil Santiago, E. (1941). Nociones de la nueva mecánica cuántica. *Metalurgia y Electricidad* 47: 31–35, 48: 54–58, 51: 22–27.
Gimeno, G. (2015). *La matemática de los quanta en España. El andamiaje de la física teórica en el intervalo (1925,1955)*. Doctoral thesis. Universitat Autònoma de Barcelona.
Glick, T. F. (1986). *Einstein y los españoles. Ciencia y sociedad en la España de entreguerras*. Madrid: Alianza.
Glick, T. F. (1988). *Einstein in Spain: relativity and the recovery of science*. Princeton: Princeton University Press.
Herrán, N. & Roqué, X. (2012). Los físicos en el primer franquismo: conocimiento, poder y memoria. In N. Herrán & X. Roqué (Eds.), *La física en la dictadura. Físicos, cultura y poder en España, 1939–1975*, (pp. 85–104). Barcelona: Universitat Autònoma de Barcelona.
Janés, C. (2015a). Humanismo y ciencia. Erwin Schrödinger y José Ortega y Gasset. *Revista de Occidente, 408*, 63–81.

Janés, C. (2015b). El saber y el mar. Xavier Zubiri y Erwin Schrödinger. Eu-topías. *Revista de interculturalidad, comunicación y estudios europeos, 10,* 23–33.
Kohler, R. E. (1991). *Partners in science. Foundations and natural scientists: 1900–1945*. Chicago: University of Chicago Press.
Martín Gijón, M. (2017). *Excitator Germaniae*. La recepción de Miguel de Unamuno en la República de Weimar. *Revista de Literatura* LXXIX: 159–188.
Moore, W. J. (1989). *Schrödinger, life and thought*. Cambridge: Cambridge University Press.
Moreno, A. (2000). La teoría de los quanta en España. *Arbor, 167,* 603–620.
Palacios, J. (1926). Theorie der Lichtemission nach dem Modell von Rutherford-Bohr. *Annalen der Physik, 384,* 55–80.
Peña Serrano, F. (1937). Un método para determinar los niveles de energía del oscilador armónico. *Revista matemática hispano-americana* 12, 10–16.
Pérez, E. (2018). Schrödinger y Unamuno, un encuentro no casual. In M. D. Ruiz-Berdún (Ed.), *Ciencia y técnica en la universidad. Trabajos de historia de las ciencias y de las técnicas,* (pp 463–472). Alcalá de Henares: Universidad de Alcalá.
Reiche, F. (1922). *Teoría de los quanta. Su origen y desarrollo*. Trans. Julio Palacios. Madrid: Calpe.
Sánchez Ron, J. M. (1992). A man of many worlds: Schrödinger and Spain. In M. Bitbol & O. Darrigol (Eds.), *Erwin Schrödinger. Philosophy and the birth of Quantum Mechanics,* (pp 9–22). Paris: Frontières.
Sánchez Ron, J. M. (1994). *Miguel Catalán. Su obra y su mundo*. Madrid: Consejo Superior de Investigaciones Científicas.
Sánchez Ron, J. M. (2000). *Cincel, martillo y piedra: historia de la ciencia en España: siglos XIX y XX*. Madrid: Taurus.
Sánchez Ron, J. M., & Glick, T. F. (1983). *La España posible de la Segunda República. La oferta de Einstein de una cátedra extraordinaria en la Universidad Central (Madrid 1933)*. Madrid: Universidad Complutense.
Sánchez Ron, J. M., & Roca-Rosell, A. (1993). Spain's first school of physics: Blas Cabrera's Laboratorio de Investigaciones Físicas. *Osiris, 8,* 127–155.
Schrödinger, E. (1932). ¿Está la ciencia natural condicionada por el medio? *Revista de Occidente, 113,* 125–159.
Schrödinger, E. (1934). Über die Unanwendbarkeit der Geometrie im Kleinen. *Die Naturwissenschaften, 22,* 518–520.
Schrödinger, E. (1935a). ¿Son lineales las verdaderas ecuaciones del campo electromagnético? *Anales de la Sociedad Española de Física y Química, 33,* 511–517.
Schrödinger, E. (1935b). Die gegenwärtige Situation in der Quantenmechanik. *Die Naturwissenschaften* 23, 807–812, 823–828, 844–849.
Schrödinger, E. (1944). *Science and humanism. Physics in our time*. Cambridge: Cambridge University Press.
Schrödinger, E. (1951). *What is life?*. Cambridge: Cambridge University Press.
Schrödinger, E. (2001). *La nueva mecánica ondulatoria*. Madrid: Biblioteca nueva.
Sigurdsson, S. (1996). Physics, life, and contingency: Born, Schrödinger, and Weyl in exile. In M. G. Ash & A. Söllner (Eds.), *Forced migration and scientific change. Emigré German-speaking scientists and scholars after 1933,* (pp 48–70). Cambridge: Cambridge University Press.
Stuewer, R. (1995). The seventh Solvay conference: Nuclear physics at the crossroads. In A. Kox & D. M. Siegel (Eds.), *No truth except in details. Essays in honor of Martin J. Klein,* (pp. 336–362). Dordrecht: Kluwer Academic.
van Vleck, J. H. (1932). *The theory of electric and magnetic susceptibilities*. Oxford: Oxford University Press.
von Laue, M. (1934). Über Heisenbergs Ungenauigkeitsbeziehungen und ihre erkenntnistheoretische Bedeutung. *Die Naturwissenschaften, 22,* 439–441.
von Meyenn, K. (Ed.). (2009). *Eine Entdeckung von ganz ausserordentlicher Tragweite: Schrödingers Briefwechsel zur Wellenmechanik und zum Katzenparadoxon*. Berlin: Springer.
Zubiri, X. (1934). La nueva física (un problema de filosofía). *Cruz y Raya, 10,* 7–94.

Chapter 5
Ludwig Prandtl: Pioneer of Fluid Mechanics and Science Manager

Michael Eckert

Introduction: The Biographical Challenge

In general, the biographer of a scientist is already at pains to avoid the falling apart of the biographical narrative into chapters that describe the personal life on one side, and the scientific work on the other—with the former accessible to a broad readership and the latter to experts only. 'Scientific engineers' (a notion discussed in more detail below) like Ludwig Prandtl pose even more challenges: Their legacy is not confined to one academic discipline but extends to the realm of technology. At the same time they excelled as institution builders and in establishing research schools for novel specialties. Thus, the biographer has to account not only for their personal life and scientific work, but also for the emerging technologies such as aeronautics, naval architecture or rocketry that they have influenced. With such technologies, Prandtl and the likes of him also entered the political stage as advisers for research policy in fields of strategic importance.

In other words: As a biographer of a scientist like Prandtl one has to transgress the borders between the history of science, the history of technology and the history of institutions as well as the political and military history—without losing sight of neighboring fields such as Science and Technology Studies (STS) and the epistemology of science. The mastery of this challenge appears almost impossible, but other considerations render it less intimidating: biography is a genre of its own right. The biographer should not subordinate the narrative under the more analytic demands of STS, epistemology and other specialist studies. Biographical writing is not meant to explicate abstract features from other disciplines with concrete examples. In contrast, rather than exposing his or her expert knowledge in a variety of historical, sociological and political specialties, the biographer should focus on the narrative challenge (Nye 2006).

M. Eckert (✉)
Deutsches Museum, Munich, Germany
e-mail: m.eckert@deutsches-museum.de

From this vantage point, the available sources deserve particular attention. They should not merely be used as an argument for one or another thesis but deserve to be taken into account from the vantage point of the narrative. Original documents provide not only authenticity, but also add life to otherwise dreary passages. Hence, the availability of letters, notebooks or some other direct residues is mandatory for the biographer of a scientist or engineer. Another challenge poses the organization of the narrative: Although the biography may be divided into systematic chapters, a chronological order seems more desirable. The latter entails a more integrative approach, while the former tends to transform the biography into a treatise of more or less coherent accounts on particular subjects.

The subsequent section of this chapter provides a brief summary of Prandtl's life and career—for more details the reader is referred to the comprehensive biography (Eckert 2017a, 2019) and the personal biography by Prandtl's daughter (Vogel-Prandtl 2005). The following sections focus on aspects which deserve particular attention. Prandtl's most outstanding achievement, the boundary layer theory, is highlighted (Sect. 3) because the concept provides a link between theoretical fluid mechanics and practical applications—and thus serves as a key to understand Prandtl's role in both spheres. Prandtl's performance as an institution builder is portrayed in Sect. 4. Section 5 focuses on turbulence, which emerged as the major research field for both Prandtl and his school throughout the twentieth century. Prandtl's advisory activity for research policy is explained further in Sect. 6. This section prepares the ground for the concluding section that is dedicated to the notion—or rather ideology—of the unpolitical scientist that shines through many biographies of scientists who had assumed political responsibilities in Nazi Germany.

Prandtl's Career in a Nutshell

Prandtl studied mechanical engineering at the Technische Hochschule (TH) Munich in the late 1890s—at a time when THs were not yet allowed to award doctoral degrees. As the assistant of August Föppl in the mechanical laboratory of the TH, Prandtl analyzed the phenomenon of lateral buckling. The results of this study later became the subject of his doctoral dissertation. He addressed his dissertation to the physics and mathematics professors of the University of Munich who at that time belonged to the philosophical faculty. Thus he finished his study with a diploma from the TH as a mechanical engineer and a "Dr. phil." from the University of Munich.

Despite the additional degree Prandtl started his career as an engineer in a machine factory. But his dual orientation—practical engineering and academic science—foreshadowed his future professional life. Throughout his career, Prandtl was eager to bridge the gap between engineering and science. After a year in industry he became a professor of mechanics at the TH Hannover. In 1904 he was called to the University of Göttingen by the mathematician Felix Klein, who attempted to establish applied sciences under the umbrella of universities—facing fierce opposition from the THs. Together with Prandtl, Klein lured Carl Runge from Hannover to Göttingen. With

Prandtl and Runge as professors for applied mechanics and applied mathematics, the University of Göttingen became a breeding ground for these disciplines.

The rise of aeronautics in the early twentieth century added aerodynamics as another item to Klein's agenda. From a scientific perspective, the motion of air belonged to the mechanics of continuous media, like the flow of water or the elasticity of a solid body, and thus became a natural research subject of Prandtl's institute for applied mechanics. When airships were considered as a novel technology that needed more scientific underpinning, Klein established an extramural research laboratory for aerodynamics—with Prandtl as director. With the advent of airplanes and their use in World War I, the military funded the extension of this laboratory into a large aerodynamic research facility, the AVA. Thus began Prandtl's career as an institution builder.

In his own research, however, Prandtl was not satisfied with the focus on applications. He was eager to study more fundamental problems in fluid mechanics, in particular the boundary layer theory and turbulence. As early as in 1911, he had already made plans for the foundation of a Kaiser-Wilhelm-Institut (KWI) für Hydrodynamik und Aerodynamik (Kaiser Wilhelm Institute for Hydrodynamics and Aerodynamics), but lack of finances and the outbreak of the war had prevented their realization. After the war, he renewed his plans. In 1925 his wish came true with the foundation of the KWI für Strömungsforschung (KWI for Fluid Mechanics). His responsibilities now comprised the directorship of three establishments: the institute for applied mechanics at the university, the AVA and the KWI. At the age of fifty, Prandtl had reached the climax of his career. By this time he had also made himself a name as the head of a school of 'scientific engineers'—a label he had coined together with his master student Theodore von Kármán. Together with like-minded professors and engineers they had formed a new organization in 1922, the Gesellschaft für Angewandte Mathematik und Mechanik (Society for Applied Mathematics and Mechanics, GAMM), and established a series of International Congresses for Applied Mechanics. Prandtl was regarded as the father-figure of this group of academics who attempted to base applications in mechanics on solid scientific ground.

As a scientist of international reputation and director of the first aeronautical research facility in Germany, Prandtl also became an influential adviser for aeronautical research policy. He assumed this role first within Felix Klein's circle of influential industrialists and academics, the Göttinger Vereinigung zur Förderung der angewandten Physik und Mathematik (Göttingen Association for the Promotion of Applied Physics and Mathematics). In the late years of the Weimar Republic he directed the German Research Council for Aeronautics (Deutscher Forschungsrat für Luftfahrt) as an advisory board for the Ministry of Transport. In the Nazi era he served as an adviser for the Air Ministry under Hermann Göring. In this capacity he assumed responsibilities for two organizations aimed at fostering the aeronautical research effort under the umbrella of the Air Ministry, the Lilienthal Gesellschaft für Luftfahrtforschung (Lilienthal Society for Aeronautical Research) and the Deutsche Akademie für Luftfahrtforschung (German Academy for Aeronautical Research). In 1942 he served as chairman of the so-called "Forschungsführung des Reichsministers der Luftfahrt und Oberbefehlshaber der Luftwaffe" (Research Leadership of the

Reich's Air Minister and Commander in Chief of the Air Force), a four-men group charged with the coordination of the entire German aeronautical research.

At the end of the Second World War Prandtl was 70 years old—and no longer able to perform a similar advisory role in the Federal Republic of Germany. He dedicated the final years of his life to his book *Führer durch die Strömungslehre*, literally translated as "Guide through Fluid Mechanics", which became an authoritative textbook of this discipline. In 1948 Prandtl published its third edition. Actually the *Führer* had originated from his *Abriss der Strömungslehre* (Outline of Fluid Mechanics) in 1931. It is perhaps one of the most influential textbooks for fluid mechanics, with a 14th edition published in 2017, and translations in many languages.

Prandtl died in 1953, at the age of 78. Despite his dubious responsibilities for Göring's air ministry, he is widely revered for his scientific merits. In memory of Prandtl, the Deutsche Gesellschaft für Luft- und Raumfahrt (German Society for Aeronautics and Space Flight) honors outstanding achievements in aeronautical research since 1957 annually with a "Ludwig-Prandtl-Ring".

Early Success in Science: The Boundary Layer Concept

Prandtl's own scientific achievements are reflected by the concepts and specialties that are named after him. At least a dozen of technical items in fluid mechanics bear his name: Prandtl number, Prandtl-Meyer expansion, Prandtl-Glauert rule, to provide only some examples (Zierep 2000, p. 3). Among these Prandtl's boundary layer concept deserves a top rank. It emerged from practical problems he had to master in his first job as an engineer in a machine factory, where he had to improve an exhaust system for sucking off shavings from workbenches. The arrangement of exhaust tubes involved wider and narrower tubes connected to one another in such a way that the flow separated from the walls.

> The question why a flow, instead of flowing along the wall becomes detached from it, did not go out of my mind until three years later the boundary layer theory brought the solution, (Prandtl 1948, p. 90).

This was the way how Prandtl recalled the origins of the boundary layer concept many years later. Prandtl's contemporary notes offer a glimpse on his effort to grasp the process of flow separation. Although he attempted to analyze this process also by extending earlier theoretical approaches, his primary goal was a qualitative understanding. He built a small water channel in order to take photos of the vortex shedding from the curved surfaces. In August 1904 he presented his results at the Third International Congress of Mathematicians in Heidelberg (Prandtl 1905). Prandtl's presentation must have appeared strange to the assembled mathematicians as he displayed many photographs from his water channel while keeping the mathematical analysis rather short. He left it to his first doctoral student, Heinrich Blasius, to elaborate the mathematical details for the boundary layer along a flat plate. Only with this and other elaborations by Prandtl's doctoral students did it become clear that his Heidelberg paper marked the beginning of a new era in fluid mechanics. According to Prandtl the motion of fluids with small viscosity (such as water and air) along a solid surface may be divided into two regimes: close to the surface as a viscous flow subject to simplified 'boundary layer equations'; and at some distance from the surface as an inviscid flow for which the ideal flow theory offered an arsenal of methods such as the potential theory.

Although the mathematical analysis of the flow within the boundary layer remained elusive in many practical cases, Prandtl's concept involved a discrimination of the causes of fluid resistance into 'skin friction' inside the boundary layer and 'form drag' due to the separation of the boundary layer and vortex shedding. In particular it justified the use of the ideal flow theory outside the boundary layer such as for the air flow around airships and wings—a consequence which enormously contributed to bridging the gap between theory and practice in aeronautics.

Institution Builder: AVA and KWI

In May 1906 representatives from industry, politics and the military formed a "Study Group for Motorized Airships" ("Motorluftschiff-Studiengesellschaft") in order to improve the technology of airships. Klein regarded this as another opportunity to attract application oriented research to Göttingen. He suggested the establishment of a small facility for model tests under Prandtl's supervision. In 1908, the Motorluftschiffmodell-Versuchsanstalt (MVA) was installed in a small hut at the outskirts of Göttingen (Rotta 1990).

The facility was dedicated to measurements of airship models in a wind tunnel. Prandtl and his assistant, Georg Fuhrmann, attempted to find a shape with minimal air resistance. Prandtl reported:

> The shape of the models was determined by a mathematical procedure that allowed us to compute by a special approach the flow and particularly the distribution of pressure under the assumption of a frictionless fluid in order to arrive at a comparison between these hydrodynamic methods and the measurements. It may be added that the agreement between theory and experiment is very satisfactory. (Prandtl 1911, pp. 43–44).

During the First World War the MVA was moved to a new site where it developed into a facility for a broad range of aerodynamic investigations. By now the major part of wind tunnel measurements was concerned with airplanes. The war ministry financed the construction of a larger wind tunnel housed in a new building. As in the first wind tunnel (which was modernized and kept in use in an attached building) the air was propelled in a closed circuit; compared to the 30 horsepower fan of the first wind tunnel, the new wind tunnel had a ten times more powerful motor which blew the air at a maximal speed of 200 km/h through a test section of a circular cross section with a diameter of 2 m. The Aerodynamische Versuchsanstalt (AVA), as the MVA was renamed in 1919, was the "most powerful and largest facility of its kind", Prandtl noted proudly, which "we owe, of course, to the generosity of our military administration" (Prandtl 1920), p. 87.

As the head of the world largest aerodynamic research laboratory, Prandtl had become a leading expert for aeronautical research. Despite the outcome of the war, which resulted in a reduction of manpower and a struggle to keep the facility in operation, the AVA became a role model for modern aerodynamic model testing. Furthermore, when the theoretical results that accompanied the airfoil tests were published in the early 1920s, Prandtl and his collaborators became sought-after experts, not only for the design of wind tunnels and model testing, but also for the theory of the lift and drag of airfoils.

But Prandtl was not satisfied with a reputation in applied aerodynamics only. When he was offered a chair for mechanics at the TH Munich in 1920, he made the fulfillment of his old plans to found a KWI for aerodynamics and hydrodynamics the condition for him to stay in Göttingen. The Munich offer and his international reputation became assets in a tug-of-war between Göttingen and Munich. In view of the dire financial situation the plan of a new KWI was doomed to failure—until a wealthy industrialist appeared on the scene. He offered a considerable donation if the University of Göttingen presented him with an honorary doctoral degree. The university agreed with this deal, and in July 1925, shortly after his fiftieth birthday, Prandtl's dream came true. On paper, the new institute was considered as an extension of the AVA. Its official name read in full: "Kaiser-Wilhelm-Institut für Strömungsforschung, verbunden mit der Aerodynamischen Versuchsanstalt in Göttingen". But in practice Prandtl regarded the new KWI as his true institutional home. He entrusted his deputy, Albert Betz, with the directorship of the AVA.

The Great Challenge: Turbulence

The KWI afforded Prandtl the opportunity to reorient his research towards more fundamental areas. He chose turbulence as the most challenging phenomenon in fluid mechanics where a breakthrough would be regarded as a major achievement (Bodenschatz and Eckert 2011). Already in 1916 he had sketched a "working program for a theory of turbulence", but postponed it in view of more urgent war work. One part of this research program concerned the onset of turbulence, the other part would deal with a theory of fully developed turbulence. The onset of turbulence was perceived as the result of a hydrodynamic instability which turns a laminar motion into an eddying flow. Fully developed turbulence became the subject of statistical approaches.

By the early 1920s—after there had been persistent failures to solve the onset of turbulence with the help of the hydrodynamic instability theory, which had been developed independently by the Irish mathematician William McFadden Orr in 1907 and the German theoretical physicist Arnold Sommerfeld in 1908—the onset of turbulence was regarded as the major challenge (Eckert 2010). Even in the simple case of a laminar flow between two plates moving at constant speed in opposite directions (plane Couette flow) it proved impossible to derive from this theory a critical speed beyond which the flow would become turbulent, in contrast to empirical experience. The Orr-Sommerfeld approach was doomed to failure—until Werner Heisenberg and

Walter Tollmien derived limits of stability for other special flows in their doctoral dissertations under Sommerfeld and Prandtl in 1924 and 1929. Tollmien's solution was further studied in the 1930s by another student of Prandtl, Hermann Schlichting. The limit of stability in this flow (close to Blasius' boundary layer flow) became known as "Tollmien-Schlichting instability". But it took another decade until this theory was corroborated experimentally in the course of secret war research in low-turbulence wind tunnels at the National Bureau of Standards in Washington and the Guggenheim Aeronautical Laboratory of the California Institute of Technology (GALCIT) in Pasadena (Eckert 2017b). However, it was only after World War II that the Orr-Sommerfeld-approach was rehabilitated as a viable method to account for hydrodynamic instability. Nevertheless, the transition from laminar to turbulent flow could hardly be explained in terms of instabilities alone. In the second half of the twentieth century the onset of turbulence was recognized as a much more complex process. As one expert of this research area observed a few decades later, "the development of a general theory of transition [to turbulence] is yet a utopia" (Herbert 1988).

In the 1920s Prandtl also launched research on the second topic of his "working program" from 1916: fully developed turbulence. In contrast to the theoretical efforts about the onset of turbulence, Prandtl hoped to gain "a deeper understanding of the inner processes of turbulent fluid motions" from experimental observations of vortices with a moving camera on top of a water channel. However, he did not expect an early success in this venture. The "big problem of fully developed turbulence", he supposed in 1926 at the Second International Congress of Technical Mechanics in Zurich, "will not so soon be solved" (Prandtl 1927), p. 62. What he presented at this occasion was "far from accomplished" and represented only "the first steps of a new approach". It became known as Prandtl's mixing-length approach. The 'great problem' of fully developed turbulence was of enormous practical importance, in particular with regard to the skin friction due to turbulent flow in the boundary layer along a solid wall. The quest for a turbulent 'wall law' became subject of a fierce rivalry between Prandtl and Theodore von Kármán. In 1938, turbulence stood at the core of a symposium at the Fifth International Congress for Applied Mechanics in Cambridge, Massachusetts, chaired by Prandtl (for a more detailed account on turbulence research between the two world wars, see Eckert 2018).

To what extent Prandtl was regarded internationally as the representative of this field is illustrated by his correspondence with his British counterpart, Geoffrey Ingram Taylor. They exchanged their views quite frankly despite rival research efforts, particularly on the turbulence problem. Taylor felt "very strongly", as he wrote in a letter to Prandtl in 1935,

> that if the Nobel Prize is open to non-atomic physicists it is definitely insulting to us that our chief—and I think that in England and USA at any rate means you—should never have been rewarded in this way. (Eckert 2017a, Chap. 7.6; see also Sreenivasan 2011, pp. 169–170).

Actually Prandtl had been already nominated for the physics Nobel prize in 1928 by German colleagues, and again in 1936 by the British Nobel prize laureate William Lawrence Bragg, apparently on the initiative of Taylor. However, to

Taylor's and Prandtl's chagrin, the Nobel prize was not awarded for achievements in fluid mechanics.

Adviser for Research Policy

Scientific engineers like Prandtl received their accolades from other circles than the Nobel committee. Furthermore, Prandtl's research facilities at Göttingen (which could be categorized as big science many years before this label was used for the nuclear physics laboratories) were regarded since World War I as assets of national importance. Therefore, it is no accident that Prandtl assumed the advisory roles for research policy such as in the German Research Council for Aeronautics (Deutscher Forschungsrat für Luftfahrt) for the Ministry of Transport at an early stage. But the dire economic situation of the Weimar Republic prevented the expansion of German aeronautical research as Prandtl had hoped.

When the Nazis came to power in 1933, the buildup of an air force became a major priority from which aeronautical research benefitted to an heretofore unimagined extent. The expansion of the AVA with a new wind tunnel for which Prandtl and his deputy, Albert Betz, had applied in vain a few years before, "now found willing ears", as Prandtl commented the approval of the expansion in his speech at the ground-breaking ceremony on May, 7th, 1934. "Our wishes were not at all modest, but we could realize to our satisfaction that for the new Ministry only the best was good enough" (Eckert 2017a, p. 223).

Prandtl was not a member of the Nazi party, but he evinced his loyalty to the regime in other ways, like financial support of the SS (Eckert 2017a, p. 223, note 54). In the early years of the "Third Reich" he had trouble with Nazi fanatics among the employees of his institute, but this did not prevent him from defending Nazi politics in his correspondence with foreign colleagues. Since its foundation in 1935, he served Göring's Air Ministry as an advisor for aeronautical research. Under the umbrella of this Ministry, he also assumed representative roles in the Lilienthal Gesellschaft für Luftfahrtforschung and the Deutsche Akademie für Luftfahrtforschung, newly founded organizations which aimed at a broad (self)mobilization of the aeronautical research community for the purposes of the regime (Trischler 1994; Schmaltz 2018).

By 1938, participation in conferences abroad had also become a matter of political meddling. When the Fifth International Congress for Applied Mechanics took place in September 1938 in Cambridge, Massachusetts, Prandtl assumed the responsibility as the leader of the German participants. With the Sudeten crisis in full swing, politics became a natural subject of debate among the conference participants—and Prandtl alienated his colleagues by his defense of Hitler's aggression against Czechoslovakia. "I realized that you know nothing of what the criminal lunatic who rules your country has been doing", Taylor reproached Prandtl after the Congress, "so you will not be able to understand the hatred of Germany which has been growing for some years in every nation which has a free press" (For this and the subsequent quotes in this section see Eckert 2017a, Chap. 7.11).

Prandtl responded with a five-page letter to which he included propaganda material. "American news agencies are entirely in Jewish hands", he argued with respect to the reports about the Sudeten crises during the Congress, and

> the fight which Germany unfortunately had to struggle against the Jews was necessary for its self-preservation. It is only regrettable that very many Jewish scientists who had no part in these intrigues had to suffer, and there are many people in Germany who wish that one had not pursued these actions in this regard so hard.

After the "crystal night" on November 9, 1938, Taylor hoped once more that Prandtl would finally see reason:

> I don't suppose you can have any idea of the horror which the latest pogroms have inspired in civilized countries. People of all shades of political opinion in England, America, Holland, Scandinavia and France are utterly disgusted at the revolting savagery of a regime which continues its attacks on a defense-less minority which is in its power.

But Prandtl persisted in his views. "If there will be war", he wrote in a four-page letter to Taylor's wife in August 1939, "this time clearly England is at fault due to its political arrangements… Germany attempts only to remove the last remains of the treaty of Versailles…" Around the same time he justified in a letter to a French colleague "the readjustment of affairs in Czechoslovakia by the German government" and denied "that Germany is going to subordinate the neighboring states under its rule". At the bottom of this "lie of warmongers" he identified

> of course the big moneyed men, the arms dealers etc. They have the press in their hands and create artificially a temper of war in countries where they are in power.

When a few weeks later German troops invaded Poland, Prandtl participated in a propaganda campaign in which the invasion was justified. It was launched by the Deutsche Kongress-Zentrale (German Congress Center), a part of the Propaganda Ministry, and aimed at foreigners in neutral countries, in particular "such men and women who dispose in their own country of prestige and influence", as Prandtl was informed. Prandtl submitted a list with the names of 45 colleagues in 11 countries whom he regarded as receptive for this propaganda. The recipients would receive a copy of Hitler's speech from October 6th, 1939, which outlined the aims for the reorganization of Europe after the invasion of Poland. In some cases, Prandtl sent such letters himself, accompanying Hitler's speech with his personal views:

> Every good German is concerned to make known abroad the ideas of the German government, which have been often distorted in foreign newspapers, in their true form. I therefore seize the opportunity to send you an English translation of the Führer's speech in which he reports about the result of the Polish war and his plans for the reorganization of Europe after this campaign.

Beyond his role as a kind of goodwill ambassador for the Nazi regime Prandtl remained an active advisor for aeronautical research. In 1942 the Air Ministry assigned this task to a four-men advisory board with the pompous title "Research Leadership of the Reich's Air Minister and Commander in Chief of the Air Force", FoFü for short. Prandtl acted as chairman of this group until the end of the war. The FoFü had in particular the following tasks: 1) planning and monitoring the execution of the aeronautical research, 2) control of the uses of the available means, facilities and installations for aeronautical research as well as its research personal, 3) exchange of experiences with science, industry and front. (Eckert 2017a, Chap. 8.5)

Thus Prandtl assumed far-reaching responsibilities. The expenditures of the FoFü exceeded by far those of other research organizations. They were used to build new research facilities like a giant wind tunnel in the Ötztal Alps and other constructions that involved forced labor from prisoners of war or concentration camp inmates. In March 1945 Göring awarded Prandtl with the War Merit Cross ("Ritterkreuz des Kriegsverdienstkreuzes mit Schwertern") usually reserved for the military. Prandtl responded that he was "not overly satisfied with my achievements in this war", but he accepted it in his capacity as chairman of the Fofü

> as a sign of recognition for the achievements of the entire German aeronautical research. May the many beautiful results that it could accomplish just recently become effective for the defense of the German fatherland and thus contribute its share to the hoped for final victory. (Eckert 2017a, pp. 292–293)

A few weeks later the war was over. Allied intelligence teams inspected what was left of Germany's aeronautical research effort—and were astonished about its extent. In his account on German Research in World War II. An Analysis of the Conduct of Research from August 1945, Leslie E. Simon, director of the United States Ordnance Ballistic Research Laboratory, regarded the FoFü as "the most powerful scientific organization of the world" because it was in control of eight big science facilities and many other institutes. He admired in particular the Göttingen AVA.

> Its former leader was the illustrious Prandtl, who was also a member of FoFü. At the time of the surrender, the leader was Professor Betz. Before the war and in the early part of the war, almost all the aerodynamics research on the ballistics of projectiles was done at AVA (Simon 1947, pp. viii, 73–75, 107).

Conclusion: The Self-image as "Unpolitical" Scientist

In the absence of a comparative historical study of aeronautical research in Germany and the U.S.A. in World War II it is difficult to arrive at a sober evaluation, but postwar operations like "Paperclip" and the relocation of wind tunnels to Great Britain and the U.S.A. attest to the high esteem of the Allies for the German research effort in aeronautics. A scientific administrator from the British Ministry of Supply commented in 1946 at a session of the Royal Aeronautical Society the interest in German achievements on high-speed aerodynamics:

> As a result there was quite a pilgrimage after the war from this country—from the Royal Aircraft Establishment, from the universities and from the industry by people who talked to the scientists and learned much about the work they had done (Smelt 1946, p. 899)

In view of such appraisals Prandtl's advisory activity for aeronautical research raises once more the question of his political responsibility. Beyond his role as a goodwill ambassador Prandtl had authorized in his capacity as chairman of the FoFü research efforts at a large scale. There can be little doubt that he sincerely wished that his services for Göring's Air Ministry would result in "the hoped for final victory" of Germany—which meant Nazi Germany. Nevertheless he regarded himself as unpolitical. In an attempt to prevent the dismissal of colleagues who had joined the Nazi party not as a result of political conviction but in order to pursue their career, he addressed a memorandum to the British military government in Göttingen in 1946 titled "Thoughts of an unpolitical German regarding de-nazification". He demanded to keep the "valuable men among the party members" in office "for the undisturbed pursuance of their activity" and exonerate them "from the odium of being a Nazi" (Eckert 2017a, p. 322). What Prandtl enounced in this memorandum could not agree more with the thoughts of white-wash certificates ("Persilscheine") for scientists and engineers that aimed to separate actual activities from party affiliation. Only the latter was perceived as unpolitical.

To some extent these thoughts prevail to this day. Prandtl's legacy as a role model is preserved at the German Aerospace Center (Deutsches Zentrum für Luft-und Raumfahrt, DLR) which emerged in the 1950s from the dismantled AVA in Göttingen. Significant birthday celebrations of this "cradle of aeronautical research" serve as reminders of Prandtl as "the father of modern aerodynamics" (DLR 2017). Prandtl is also celebrated for the establishment of the KWI for Fluid Mechanics and its successor, the Max Planck Institute for Dynamics and Self-Organization in Göttingen. Last, but not least, Prandtl's name provides symbolic capital (to speak with Pierre Bourdieu) for outstanding achievements in fluid mechanics. In addition to the Ludwig-Prandtl-Ring for excellent work in the sciences of flight (DGLR 2018), each year a scholar with outstanding achievements in this field is invited to present an annual Ludwig Prandtl Memorial Lecture (GAMM 2017). Scientists who are rewarded in this manner enjoy the symbolic capital that comes along with Prandtl's role as pioneer of modern fluid mechanics—without aligning themselves to Prandtl's political engagement for the Nazi regime.

The myth of the unpolitical scientist relies on the separation of scientific merits from political attitude. Dividing the life of a scientist like Prandtl into different realms of action either results from a lack of biographical knowledge or a conscious exclusion of what is regarded as irrelevant from a 'scientific' vantage point. Here we are back at the biographical challenge as articulated in the introductory section— to strive for an integrative chronological narrative. A glimpse into the literature on scientific biography as a genre of its own right (e.g. Daston and Sibum 2003; Nye 2006) shows that the same challenge arises in quite different disciplines and periods. From this perspective Prandtl's self-image as an unpolitical scientist serves as another lesson for the biographer to tell a scientist's life story in a dense narrative that avoids the compartmentalization of specialties and spheres of action into different chapters on personal life and scientific work.

References

Bodenschatz, E., & Eckert, M. (2011). Prandtl and the Göttingen school. In P. A. Davidson, Yokio Kaneda, K. Moffatt, & K. Sreenivasan (Eds.), *A voyage through turbulence* (pp. 40–100). Cambridge: Cambridge University Press.

Daston, L., & Sibum, O. (2003). Introduction: Scientific personae and their histories. *Science in Context, 16*(1–2), 1–8.

DGLR. (2018). Auszeichnung—Ludwig-Prandtl-Ring. http://www.dglr.de/?id=2643. Accessed 25 Oct 2018.

DLR. (2017). DLR's precursor was founded in November 1907 in Göttingen. The cradle of aeronautical research celebrates its 110 anniversary. https://www.dlr.de/content/en/articles/news/2017/20171109_the-cradle-of-aeronautical-research-celebrates-its-110-anniversary_24902.html. Accessed 21 May 2020.

Eckert, M. (2010). The troublesome birth of hydrodynamic stability theory: Sommerfeld and the turbulence problem. *European Physical Journal History, 35*(1), 29–51.

Eckert, M. (2017a). *Ludwig Prandtl—Strömungsforscher und Wissenschaftsmanager*. Berlin, Heidelberg: Springer.

Eckert, M. (2017b). A Kind of Boundary-Layer 'Flutter': The Turbulent History of a Fluid Mechanical Instability. https://arxiv.org/abs/1706.00334. Accessed 21 May 2020.

Eckert, M. (2018). Turbulence research in the 1920s and 1930s between mathematics, physics, and engineering. *Science in Context, 31*(3), 381–404.

Eckert, M. (2019). *Ludwig Prandtl: A life for fluid mechanics and aeronautical research* (Translation of Eckert 2017a, translated by D. A. Tigwell, Trans.).Berlin, Heidelberg: Springer.

GAMM. (2017). Ludwig Prandtl Memorial Lecture. http://jahrestagung.gamm-ev.de/index.php/2017/programme/prandtl-lecture. Accessed 21 May 2020.

Herbert, T. (1988). Secondary instability of boundary layers. *Annual Review of Fluid Mechanics, 20,* 487–526.

Nye, M. J. (2006). Scientific biography: History of science by another means? *Isis, 97,* 322–329.

Prandtl, L. (1905). Über Flüssigkeitsbewegung bei sehr kleiner Reibung. In A. Kratzer (Ed.), *Verhandlungen des III. Internationalen Mathematiker-Kongresses in Heidelberg vom 8. bis 13. August 1904*, (pp. 484–491). Leipzig: Teubner.

Prandtl, L. (1911). Bericht über die Tätigkeit der Göttinger Modellversuchsanstalt. *Jahrbuch der Motorluftschiff-Studiengesellschaft* 43–50.

Prandtl, L. (1920). Die Modellversuchsanstalt in Göttingen. *Zeitschrift für Flugtechnik und Motorluftschiffahrt, 11,* 84–87.

Prandtl, L. (1927). Über die ausgebildete Turbulenz. In S. Ernst Meissner (Ed.), *Verhandlungen des II. Internationalen Kongresses für Technische Mechanik in Zürich vom 12. bis 17. September 1926*, (pp. 62–75). Zürich: Füßli, Zürich.

Prandtl, L. (1948). Mein Weg zu Hydrodynamischen Theorien. *Physikalische Blätter, 4,* 89–92.

Rotta, J. C. (1990). *Die Aerodynamische Versuchsanstalt in Göttingen, ein Werk Ludwig Prandtls. Ihre Geschichte von den Anfängen bis 1925*. Göttingen: Vandenhoeck und Ruprecht.

Schmaltz, F. (2018). Die Deutsche Akademie der Luftfahrtforschung 1936–1945: Hermann Görings nationalsozialistische Muster-Akademie? In J. Balcar & N. Balcar (Eds.), *Das Andere und das Selbst. Perspektiven diesseits und jenseits der Kulturgeschichte. Doris Kaufmann zum 65. Geburtstag*, (pp. 69–92). Bremen: Edition Themmen.

Simon, L. E. (1947). *German research in World War II. An analysis of the conduct of research*. New York: Wiley.

Smelt, R. (1946). A critical review of German research on high-speed airflow. *The Royal Aeronautical Society, 50,* 899–934.

Sreenivasan, K. R. (2011). G. I. Taylor: The inspiration behind the Cambridge school. In P. A. Davidson, Y. Kaneda, K. Moffatt, & K. Sreenivasan (Eds.), *A voyage through turbulence*, (pp. 127–186). Cambridge: Cambridge University Press.

Trischler, H. (1994). Selfmobilization or resistance? Aeronautical research and national socialism. In M. Renneberg & M. Walker (Eds.), *Science, technology and national socialism* (pp. 72–87). Cambridge: Cambridge University Press.

Vogel-Prandtl, J. (2005). *Ludwig Prandtl. Ein Lebensbild. Erinnerungen, Dokumente.* Göttinger Klassiker der Strömungsmechanik, Bd. 1. Göttingen: Universitätsverlag Göttingen.

Zierep, J. (2000). Ludwig Prandtl, Leben und Werk. In G. E. A. Meier (Ed.), *Ludwig Prandtl, ein Führer in der Strömungslehre. Biographische Artikel zum Werk Ludwig Prandtls*, (pp. 1–16). Braunschweig/Wiesbaden: Vieweg.

Chapter 6
Rudolf Tomaschek—An Exponent of the *"Deutsche Physik"* Movement

Vanessa Osganian

Introduction

"Aryan Physics" or *"Deutsche Physik"* is considered a prime example of ideologically influenced science (Beyerchen 1982, pp. 113–114; Eckert 2007, p. 139; Richter 1980, p. 116–117). This movement is strongly connected with Philipp Lenard (1862–1947) and Johannes Stark (1874–1957), two Nobel laureates who tried to invigorate their reactionary beliefs by accommodating their scientific work to National Socialist ideology (e.g. Beyerchen 1982, p. 114).

However, the categorization of "Aryan physics" as some kind of political physics separate and different from purely scientifically working physics was created in retrospective. The representatives of the German Physical Society (*Deutsche Physikalische Gesellschaft, DPG*) especially promoted this version in order to conceal their own collaboration with the Nazi regime (Eckert 2007, pp. 159–163). The relationship between politics and science was by no means as one-sided and unbalanced as it might appear at first glance (Ash 2002, 2010).

"Aryan Physics" was not a two-man-show, albeit historical research has so far mostly concentrated on the two Nobel Prize winners. Various younger scientists committed themselves to the idea of ideologically influenced physics as well. Among them was Rudolf Tomaschek (1895–1966), who was Lenard's assistant between 1919 and 1927 and qualified to become a professor during his time in Heidelberg. While working for the Nobel laureate Rudolf Tomaschek mainly focused on experiments proving Lenard's ether theory, but he also researched phosphorescence, fluorescence, and gravimetric analysis (Hoffmann 2016, p. 343).[1]

---

[1] TUMA PA Prof. Tomaschek, Rudolf (Personalakt Dresden): Saxon Ministry of Education to R. Tomaschek, 10 April 1934.

V. Osganian (✉)
Deutsches Museum/Ludwig Maximilian University, Munich, Germany
e-mail: v.osganian@deutsches-museum.de

© The Editor(s) (if applicable) and The Author(s), under exclusive licence to Springer Nature Switzerland AG 2020
C. Forstner and M. Walker (eds.), *Biographies in the History of Physics*,
https://doi.org/10.1007/978-3-030-48509-2_6

In 1927, Tomaschek became a non-tenured professor in Marburg (Auerbach 1979, p. 917) and seven years later he was appointed a full professor in Dresden.[1] By 1936, it was obvious that the physicist would not stay in Dresden because the head of the Technical University (TH) in Munich had voiced his desire to appoint Tomaschek. The decisive factor for the TH to propose him as the successor of Jonathan Zenneck (1871–1959) was that Tomaschek was "one of Lenard's best pupils",[2] and thus part of the so-called "*Deutsche Physik*".

However, different factors and actors slowed down the process. One explanation for this can be found in the workings of the Bavarian Ministry of Education. The responsible state councillor, Ernst Boepple (1887–1950), opted for extending Zenneck's tenure on his own initiative several times.[3] However, the Technical University was still pushing to appoint Tomaschek—they even appealed directly to the Reich Ministry of Education (*Reichserziehungsministerium*, REM). This in turn induced Boepple to try to discredit the university's candidate of choice: He reported to the REM that Tomaschek had previously been married to a "non-Aryan woman".[4] Nevertheless, this attempt failed, as the deciding authorities already knew about this "blemish".[5]

Another factor for the delay was Johannes Stark, who had apparently objected directly at the REM in Berlin to Tomaschek's appointment.[6] Stark's behaviour in this case is quite surprising because he and Tomaschek were both known as supporters of "Aryan Physics". Consequently, one would have expected Stark to support Tomaschek's case in order to strengthen this movement. However, this story shows that "Aryan Physics" was by no means a coherent group, but instead was characterised by constantly changing alliances of convenience (Litten 2000, p. 380–384).

Subsequently, it required an intervention of the NSDAP in order to finally appoint Tomaschek in 1939. The party emphasised—similar to the university's reasoning—the fact that Tomaschek had once been a pupil of Lenard as one of the main reasons to appoint him. They stated that it would be of great importance not only for the TH but also for the Ludwig-Maximilians-University (LMU) and thus for the city as a whole to have "a different school of physics" in Munich.[7] The correspondence between

---

[2]TUMA RA C 143, Berufung Tomaschek: Faculty director F. Boas to the rector of the TH Munich, 20 July1936; TUMA RA C 143, Berufung Tomaschek: Head of Lecturers to the rector of the TH Munich, 4 March 1938.

[3]BayHStA MK 67439, Professur für Experimentalphysik THM: Bavarian Ministry of Education to Reich Ministry of Education (REM), 30 January 1937; BayHStA MK 67439, Professur für Experimentalphysik THM: Bavarian Ministry of Education to REM, 25 March 1938.

[4]BayHStA MK 67439, Professur für Experimentalphysik THM: Bavarian Ministry of Education to REM, 24 November 1937.

[5]TUMA RA C 143, Berufung Tomaschek: Faculty director F. Boas to the rector of the TH Munich 14 February 1938.

[6]BayHStA MK 43325, PA Tomaschek, Rudolf (Personalakt Dresden): Informational letter by W. Studentkowski, 22 February 1938.

[7]TUMA RA C 143, Berufung Tomaschek. NSDAP to REM, 21 July 1938.

Werner Heisenberg (1901–1976) and his former teacher Arnold Sommerfeld (1868–1951) demonstrates that contemporaries regarded Tomaschek's appointment on the 1st of April 1939 as a victory or at least as a reinforcement for *"Deutsche Physik"*.[8]

Several historians have adopted this viewpoint and often ascribe to Tomaschek a central role regarding "Aryan Physics" (e.g. Beyerchen 1982, pp. 196, 265; Richter 1980, p. 123). It may therefore come as a surprise that Rudolf Tomaschek has not yet been the object of a detailed study and only been mentioned occasionally, e.g. in histories of the TH Munich (Fleckenstein 1968, pp. 61–133; Pabst 2006, p. 229–352; Wengenroth 1993) or Bavarian Academy of Sciences and Humanities (Berg 2011; Krauss 2009). Moreover, in studies focussing on "Aryan Physics" as such, his name only appears marginally (Beyerchen 1982; Hentschel 1996; Walker 1995). When it comes to interpreting Tomaschek's role in this movement, historians usually agree that he gradually abandoned his former beliefs after the so-called Munich synod/debate in 1940. The fact that he supported the appointment of Fritz Sauter (1906–1983) a theoretical physicist at the TH Munich is widely considered a confirmation of this interpretation (Beyerchen 1982, pp. 244–245; Pabst 2006, p. 281; Wengenroth 1993, pp. 247–250).

By approaching a person like Tomaschek biographically, it is possible to analyse not only his public statements but also to show the complex and diverse network in which he acted. This approach also sheds light on the interdependency between the individual and his social and political environment from a historical point of view (Hashagen 2003, pp. 6–11; Szöllösi-Janze 2000, pp. 17–22, 29–32; Szöllösi-Janze 2015, pp. 9–15). Furthermore, a biographical study can help to understand—or at least illustrate—the complexity of science and the scientific community itself (Litten 2000; Lindner 2014). The networks in which Tomaschek was operating were much more complex than the term *"Deutsche Physik"* indicates (See Fig. 1).

Tomaschek as Professor of Experimental Physics at the TH in Munich

Munich, which the National Socialists called their "Capital City of the Movement" (*"Hauptstadt der Bewegung"*), played a special role in the context of the "Aryan Physics" movement. The city was among the three centres of modern physics in Germany during the Weimar Republic (Beyerchen 1982, pp. 25–29). Therefore, when Arnold Sommerfeld, the tenured professor of Theoretical Physics at the LMU, had to retire in 1935 his succession became highly political (Beyerchen 1982, pp. 207–227; Eckert 2013, pp. 459–479; Eckert 2007, p. 139; Eckert and Märker 2004, pp. 360–377; Richter 1980, pp. 126–128; Walker 1995, pp. 28–40; Wengenroth 1993, pp. 242–245). The supporters of "Aryan Physics" were eager to transform Munich into a centre for their type of science, so that Tomaschek's appointment at the TU Munich on the 1st of April 1939 was seen as a first important step towards their aim.

---

[8]DMA HS 1977-28/A, 136: W. Heisenberg to A. Sommerfeld, 13 May 1939.

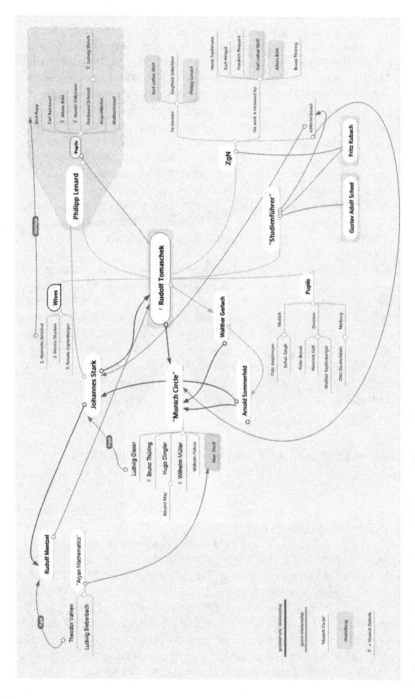

Fig. 1 Illustration of the networks in which Tomaschek operated

"Aryan Physics" in the Lecture Catalogue?

A look into the lectures given by Tomaschek during his six years in Munich reveals some surprising findings: His teaching clearly resembled that of his predecessor. Each semester, he lectured on experimental physics and delivered an additional one-hour course in which he focused on selected topics. Furthermore, he gave a class on how to work scientifically. Tomaschek also supervised different practical courses.[9] Finally, starting in the winter term of 1942/43 he organised an annual physics colloquium, which he held in cooperation with the other professors of physics at the TH Munich, Walther Meißner (1882–1974) and Fritz Sauter (1906–1983).[10] The available sources, which are mainly the lecture catalogues, do not reveal what Tomaschek actually taught. What stood behind titles like *"Experimentalphysik I"* (Experimental Physics I) or *"Ausgewählte Kapitel aus der Experimentalphysik"* (Selected Chapters in Experimental Physics) can therefore only be speculated about. It is quite possible that Tomaschek criticised the theory of relativity in his courses and defended the idea of the ether, as he had already done in his textbook.

Nonetheless, it is striking that he did not choose titles that were more obvious.[11] In contrast to his previous lectures in Marburg and Dresden, one cannot find any courses that clearly deal with topics of "Aryan Physics" in Munich. In view of the fact that his affiliation with "another school of physics (Lenard)"[12] was the decisive factor for his appointment, why did Tomaschek not affiliate himself more openly with "Aryan Physics" in Munich?

Doctoral Dissertations at the Physical Institute

Another important task of a professor, apart from giving lectures, is the supervision of doctoral dissertations. Between the summer term of 1939 and 1945, 24 students did their doctorate in the department of physics. In ten of these cases, Zenneck supervised the doctoral dissertations, although he had already retired. However, what is most surprising is that Zenneck not only examined his own students in the first year after

---

[9] See the lecture catalogues of the TH Munich: *Vorlesungsverzeichnis* (*VV*) Munich WiSe 1939/40 and SoSe 1940, pp. 58–59, *VV* Munich 2nd and 3rd Trimester 1940, p. 24, *VV* Munich 1st Trimester and SoSe 1941, pp. 8–9, *VV* Munich WiSe 1941/42 and SoSe 1942, pp. 72–73, *VV* Munich WiSe 1942/43 and SoSe 1943, pp. 74–75, VV Munich WiSe 1943/44 and SoSe 1944, pp. 3–4, *VV* Munich WiSe 1944/45 and SoSe 1945, pp. 4–5.

[10] See the lecture catalogues of the TH Munich: *VV* Munich WiSe 1942/43 and SoSe 1943, pp. 74–75, *VV* Munich WiSe 1943/44 and SoSe 1944, pp. 3–4, *VV* Munich WiSe 1944/45 and SoSe 1945, pp. 4–5.

[11] In Marburg he delivered a lecture on *"Grundlagen der Relativitätstheorie"* and in Dresden he lectured for example on *"Relativitätstheorie und Ätherfrage"*. See: *VV* Marburg WiSe 1928/29, p. 31; *VV* Marburg WiSe 1930/31, pp. 33–34; *VV* Dresden SoSe 1935, p. 31.

[12] TUMA RA C 143, Berufung Tomaschek: NSDAP to REM, 21 July 1938.

Tomaschek had taken over his chair, but that he continued to do so at least until 1942.[13]

In six cases, Tomaschek was the second supervisor, but the reports he wrote were quite short and superficial: They seldom consisted of more than two sentences and these often appear as simple handwritten notes on the sheet of Zenneck's first report.[14] The assessments Tomaschek wrote for students of other professors were mostly about half a page long, written on a separate piece of paper and a bit more detailed.[15] Since the reports other professors wrote for Zenneck's doctoral candidates were also at least half a page long, one can assume that Tomaschek's behaviour was by no means common practice at the TH Munich.[16] It seems more plausible that he acted with special reserve in cases in which long-time students of his predecessor were involved.

Furthermore, Tomaschek only supervised two doctoral dissertations: surprisingly, one of them was in chemistry by Sohan Singh, an Indian student who had investigated rare earth elements.[17] The other doctoral dissertation Tomaschek supervised was in physics in 1941 by Fritz Asselmeyer (1911–2005), who had moved to Munich only one year before. In both cases, the reports have not been preserved, so nothing can be said about Tomascheks's evaluation of these two students.[18]

Nevertheless, it is striking that Tomaschek obviously did not seek to attract younger scientists, since the research of scientific institutes at universities was to a large degree conducted by doctoral candidates and graduate students (Lindner 2014, pp. 88–124). He thereby probably failed the high expectations that might have been placed on him as a former student of Philipp Lenard, who had supervised numerous doctoral candidates (Becker 1942). As a professor, however, Tomaschek decided not to follow in his teacher's footsteps. While he took a stand for the scientific reorientation of the institute during the negotiations preceding his appointment, he did not promote this plan in all matters.[19] Instead, his predecessor continued to supervise the majority of the doctorates done in physics and thus still had some influence on the institute. One reason for this might be Zenneck's excellent connections to the industry, which may well have made him the more attractive supervisor. Hence, in comparison to his predecessor Tomaschek remained rather insignificant.

---

[13]TUMA Promotionsbücher: List of Doctorates 1938–1939; TUMA Promotionsbücher: List of Doctorates 1940–1943; TUMA Promotionsbücher: List of Doctorates 1943–1950.

[14]E.g. TUMA Prom. A. Burger, Bernhard: First and Second Report, 09 May 1940; TUMA Prom. A. Hechtel, Richard: First and Second Report, 19/20 March 1940; TUMA Prom. A. Schmidtmann, Hans Joachim: First and Second Report 25 October 1939.

[15]E.g. TUMA Prom. A. Hilz, Robert: Second Report, 07 July 1939; TUMA Prom. A. Hitzelsberger, Anton: Second Report, 29 February 1940; TUMA Prom. A. Meyer, Erich: Second Report, 12 July 1939.

[16]E.g. TUMA Prom. A. Ochmann, Wilhelm: Second Report, 02 December 1937; TUMA Prom. A. Zbinden, Kurt: Second Report, 18 April 1939.

[17]TUMA Prom. A. Singh, Sohan: Doctoral Certificate, 04 September 1943.

[18]TUMA Promotionsbücher: List of Doctorates 1940–1943.

[19]TUMA RA C 143, Berufung Tomaschek: Bavarian Ministry of Education to the rector of the TH Munich, 17 July 1939.

Fritz Sauter's Appointment—a Complete Turnaround?

Historians have often assumed that Tomaschek gradually abandoned his former commitment to "Aryan Physics" from 1940 onwards. The appointment of Fritz Sauter to the TH Munich in 1942 is generally cited as an evidence of this. As a theoretical physicist, Sauter represented the branch of physics that had been fiercely attacked by the supporters of "Aryan Physics" for a long time. However, when Tomaschek and his colleague Walther Meißner restarted the attempt to appoint Sauter in December 1940, the Munich synod/debate had taken place and thus the "Aryan physicists" had already publicly recognised the theory of relativity and quantum mechanics as acceptable science.

In fact Tomaschek's behaviour was hardly a total "break with his [...] colleagues" (Beyerchen 1982): First of all, Sauter was an active member of different National Socialist groups (e.g. the NSDAP, the NSKK and the SA), which might have enabled the cooperation between him and Tomaschek. Indeed Sauter stuck to his political beliefs even after 1945, as letters from his former colleagues Walther Meißner[20] and Walther Gerlach reveal (Rammer 2004, pp. 131–135).

Second, the negotiations for creating a professorship for theoretical physics at the TH Munich date further back to Tomaschek's predecessor Jonathan Zenneck. In cooperation with his colleague Walther Meißner, Zenneck had been trying since the mid-1930s to set up such a professorship with Sauter as his candidate.[21] However, the REM saw no reason for another chair on theoretical physics besides the one at the LMU.[22] Nonetheless, Zenneck and Meißner did not give up. In 1938, shortly before Tomaschek's appointment, the Ministry finally relented and allowed the TH Munich to create a professorship for theoretical physics. However, the question of who should be appointed had to be postponed until the succession of Zenneck had been decided.[23]

The case had thus rested for about two and a half years before Meißner and Tomaschek undertook a new attempt in 1940. The former appears to have been the driving force, which is indicated e.g. by his signature coming first in the proposal (Lindner 2014, pp. 86–88). Furthermore, the rector, Lutz Pistor (1898–1952), instructed Tomaschek to approach the REM and to promote Sauter's appointment.[24] But even two months after Tomaschek's visit in Berlin, nothing had happened. Nevertheless, the persistent efforts of the TH finally showed success when Sauter was appointed on the 1st of January 1942.[25] Tomaschek did not play the leading role in

---

[20]DMA NL 045, 031 (1: W. Meißner to W. Hanle, 2 September 1948.
[21]TUMA RA C 142, Berufung Sauter: J. Zenneck to the rector of the TH Munich, 20 December 1934.
[22]TUMA RA C 142, Berufung Sauter: REM to Bavarian Ministry of Education, 25 May 1936.
[23]TUMA RA C 142, Berufung Sauter: J. Zenneck and W. Meißner to faculty director F. Boas, 16 July 1936; TUMA RA C 142, Berufung Sauter: Bavarian Ministry of Education to the rector of the TH Munich, 29 June 1938.
[24]TUMA RA C 143, Berufung Sauter: Rector's report, 13 September 1941.
[25]TUMA RA C 143, Berufung Sauter: REM to F. Sauter, 24 January 1942.

this whole process, but instead followed his colleagues' efforts and often still had to be explicitly requested to intervene actively.

Furthermore, when it came to promoting Sauter to a higher rank, the head of the lecturers revealed that he and his colleagues had expected Sauter to teach theoretical physics

> in a way that would fit the views of our colleague Tomaschek, one of Ph. Lenard's foremost disciples. However, this hope has not been fulfilled, so we have experienced a disappointment in this regard.[26]

This short statement indicates that in 1943 Tomaschek—at least from the viewpoint of the head of the local lecturers organization, who was closely aligned to NS party organizations—was still counted as part of the circle around Philipp Lenard and thus connected to "Aryan Physics".

Tomaschek as a Part of *"Deutsche Physik"*

As already mentioned above, Rudolf Tomaschek's appointment to the TH Munich can be tied to him being a pupil of Philipp Lenard, someone who had tried to establish an "Aryan Physics" by accommodating science to National Socialistic ideology. However, as shown in the previous chapter, once in Munich Tomaschek appears to have restrained himself in this setting, in contrast to the expectations that others might have had for him. This begs the question as to how he acted in a somewhat broader framework. The following chapter will therefore analyse his public statements about "Aryan Physics" and his former teacher. Furthermore, his behaviour at the Munich synod/debate will be examined in order to find out whether or not he had been trying to detach himself from the *"Deutsche Physik"* movement.

The Munich Synod/Debate (1940) and the Conference in Seefeld (1942)

As a consequence of the different campaigns the supporters of "Aryan Physics" had launched against Werner Heisenberg and other modern physicists, a counter-attack was started (Walker 1995, pp. 50–52). A meeting organised by National Socialist officials between the two sides took place on the 15th November 1940 in Munich (Beyerchen 1982, pp. 240–241). Scientists from both sides were invited to discuss their divergent schools of physics under the supervision of two supposedly unbiased physicists and the National Socialist German Lecturers League (*Nationalsozialistischer Deutscher Dozentenbund,* NSDDB). This debate is (and was) generally considered a victory for modern physics because the advocates of the *"Deutsche Physik"*

---

[26]TUMA RA C 146, Beförderung Sauter: Head of Lecturers to the rector of the TH Munich, 09 July1943.

had to sign a compromise according to which they officially recognized the theory of relativity and quantum mechanics as acceptable science (Beyerchen 1982, pp. 240–242; Eckert 2013, pp. 485–490; Richter 1980, pp. 127–128; Walker 1989, pp. 72–73; Wengenroth 1993, pp. 247–248). Although two supporters of "Aryan Physics"—Wilhelm Müller and Bruno Thüring (1905–1989)—had left the meeting before the end, most of them stayed and Tomaschek and Alfons Bühl (1900–1988) were even involved in crafting the final compromise (Beyerchen 1982, pp. 240–241, see especially Footnote 3; Scherzer 1965, pp. 56–57).

It is interesting that Wolfgang Finkelnburg (1905–1967) and Otto Scherzer (1909–1982), who both attended the meeting on the side of the modern physics and who wrote reports on this debate in retrospect, described Tomaschek as the only representative of "Aryan Physics" with a certain amount of technical competence (Finkelnburg 1996, p. 342; Scherzer 1965, p. 57).

Finkelnburg interpreted Tomaschek's behaviour as follows: "To begin with, when he felt our resistance, Mr Tomaschek carefully tried to distance himself from his original friends" (Finkelnburg 1996, p. 343). Historical research regards the Munich synod/debate as the trigger for Tomaschek's separation from the ideas of the "*Deutsche Physik*" (Beyerchen 1982, p. 242; Pabst 2006, p. 281; Wengenroth 1993, pp. 247–248). It seems to be correct that Tomaschek's behaviour, at least in the context of the Munich synod/debate, was less fanatical than that of many of his allies. However, it is questionable to what extent Tomaschek really tried to detach himself from the ideas of "Aryan Physics". Finkelnburg's phrasing suggests that Tomaschek's behaviour may have been based on tactical considerations. Faced with the fierce resistance of the theoretical physicists to "Aryan Physics", Tomaschek made a careful effort to distance himself from his colleagues. Perhaps Tomaschek had realized that he and his associates would not be able to overcome the theoretical physicists at this point. Furthermore, he may as well have taken into consideration that the two guiding figures of "Aryan Physics" had substantially lost ground in matters of science itself as well as of science policy since the mid-1930s. The Munich synod/debate and the willingness to compromise that Tomaschek and Bühl had displayed certainly led to tensions within the "*Deutsche Physik*" movement (Beyerchen 1982, pp. 241–242; Litten 2000, pp. 150–152).

Despite the final declaration, the phenomenon of ideologically influenced physics had not yet been laid to rest. As a result, a second meeting among physicists took place from the 1st till the 3rd November 1942 in Seefeld (Beyerchen 1982, pp. 258–259; Eckert 2007, pp. 155–158; Hoffmann 2007, pp. 190–194). However, the representatives of "Aryan Physics" were clearly outnumbered: Only three out of 30 participants (Bühl, Tomaschek and Thüring) belonged to this circle. Once again, Tomaschek and Bühl were quite active in the discussion, as the report written by Fritz Sauter and Carl Friedrich von Weizsäcker (1912–2007) indicates.[27] Furthermore, Tomaschek—unlike his colleagues—was apparently involved in the preparation of this second

---

[27] AIP Samuel A. Goudsmit Papers, Alsos Mission, Box 25, Folder 12: Vorläufiger Bericht über das Physiker-Lager in Seefeld (Tirol) im November 1942 [Digital resource]. https://repository.aip.org/islandora/object/nbla:253079#page/1/mode/2up. Accessed 12 April 2018.

meeting. This involvement—possibly due to his connections to the NSDDB—might have been another reason for the tension between Tomaschek and some other supporters of "Aryan Physics", especially the so-called "Munich Circle" led by Thüring and Hugo Dingler (1881–1954) (Eckert 2007, pp. 155–156). In a letter to his close friend Dingler, the astronomer Thüring accused Tomaschek of being insensitive. Apparently, the latter shared this opinion, because he feared that Tomaschek "might be ranked among the opposing party".[28] Tomaschek's behaviour during the Munich synod/debate in 1940 and the conference in Seefeld 1942, as well as the perception of Thüring and Dingler, suggest that Tomaschek carefully avoided presenting himself as a fanatical supporter of the *"Deutsche Physik"*.

Ties to the *"Deutsche Physik"* Movement

Tomaschek by no means completely renounced his old views in the following years, even though he kept his distance from the "Munich Circle": Otherwise, his public statements on the *"Deutsche Physik"* can hardly be explained.

When examining Tomaschek's connection to the "Aryan Physics" movement, one can observe that he did not publish any books or articles with obvious anti-Semitic titles as some of his colleagues did (e.g. Müller 1939, 1941). His major publication was the republication of Ernst Grimsehl's (1861–1914) *Textbook of Physics*, which he revised in large parts (e.g. Tomaschek 1929). He added another volume to the already existing two, which he dedicated to the topics of matter and ether (e.g. Tomaschek 1934). Although he did not use clear anti-Semitic language in the preface—compared to Lenard's textbook *Deutsche Physik* (Lenard 1936, pp. IX–XV)—it is obvious that he was very influenced by his former teacher's ideas (Tomaschek 1929, pp. III–IV).

Apart from this textbook, Tomaschek only published a few articles and gave occasional speeches during the National Socialist period. In 1936, he clearly portrayed himself as being part of the *"Deutsche Physik"* movement when he delivered a lecture in celebration of the opening of the *Forschungsabteilung Judenfrage* (Research Section for the Jewish Question), which was part of Walther Frank's *Reichsinstitut für Geschichte des neuen Deutschlands* (Reich Institute for History of the New Germany). Its advisory board consisted of many conservative or even National Socialist scientists with Lenard among them. It is not clear why Tomaschek gave a talk on this occasion. However, since Lenard was not able to participate, Tomaschek might have been invited as his substitute. Unfortunately, unlike the other speeches, his was not published afterwards. Therefore, when investigating the content of Tomaschek's talk, one is dependent on newspaper articles: According to these reports, Tomaschek's speech – which he had entitled "Natural Science and Ideology" – aimed to show the difference between an "Aryan" and a "Jewish" attitude and thus clearly promoted the ideas of an "Aryan Physics" (N. N. 1936a, b).

---

[28] HBA Dingler papers: H. Dingler to B. Thüring, 23 August 1942.

Only one month later, Tomaschek took up a similar position in a talk he gave at a meeting of the National Socialist German Students' League (*Nationalsozialistischer Deutscher Studentenbund,* NSDStB). Once again, he equated a "Jewish" way of thinking – using Einstein's theory of relativity as an example—with being something foreign to the species ("*artfremd*") (F. R. 1936). One year later, Tomaschek reviewed Lenard's textbook *Deutsche Physik* in the *Zeitschrift für die gesamte Naturwissenschaft* (ZgN, see below). Similar to other former pupils, he was full of praise, especially for the "*völkisch*" ideas (Tomaschek 1937/38, p. 95). This review also shows that he strongly appreciated Lenard's romantic way of investigating and understanding nature's secrets, which is often regarded as one of the characteristics of "Aryan Physics".

In the following years, Tomaschek did not publicly promote ideologically influenced physics. Speeches such as the one he delivered in 1939 at a meeting of the NSDDB, in which he explained the difference between the "German" and the "Jewish" way of thinking, were rare exceptions (Ba. 1939). This is quite surprising, as one could have expected the new professor of experimental physics to clearly promote his way of understanding physics, especially since his affiliation to "Aryan Physics" seemed to have been the decisive factor in his appointment. In Munich, however, he instead abstained from clearly committing himself to ideologically influenced physics.

This situation did not change until Philipp Lenard celebrated his 80th birthday in 1942. In honour of this occasion, Tomaschek came to the fore again and published some articles in different journals, e.g. the *Deutsche Technik,* and in newspapers, e.g. the *Völkischer Beobachter.* What distinguishes the publications and statements in both formats is the strong emphasis on the hard times which Lenard had to cope with, especially before 1933. Tomaschek especially highlighted the claim that other scientists often used Lenard's research results or the devices he had developed without giving him credit. Thus, Tomaschek concluded that his former teacher had never gotten the scientific and public acknowledgement he deserved. However, as Tomaschek stated, Lenard had not withdrawn but rather continued his fight against a "dogmatic formalism" in physics (Tomaschek 1942a, p. 270, 1942d).

In the same year, ZgN dedicated one issue to Lenard with various former students and other exponents of the "*Deutsche Physik*" contributing articles. It is not surprising to find Rudolf Tomaschek among them: In his essay, he explained Lenard's theory of the ether, which seems to be—at first glance—a non-judgmental presentation of his former teacher's ideas. Nevertheless, at second glance one can notice that Tomaschek still connected himself to Lenard: e.g. when criticising the unwillingness of most physicists to accept the existence of the ether, Tomaschek described this as a "strange state" since this theory would lead to a deeper understanding of nature itself (Tomaschek 1942b, p. 117). The celebration of Lenard's birthday at the TH Munich was another occasion on which Tomaschek publicly expressed his close connection to his former teacher (F. 1941).

In the following year, Tomaschek reviewed Lenard's *Wissenschaftliche Abhandlungen* (Scientific Papers), which he described as a "swan song of an epoch, which ends [...] for science as well" (Tomaschek 1943b, p. 131). Historians usually cite this short expression in order to prove that Tomaschek did distance himself from "Aryan Physics" (Pabst 2006, p. 281; Richter 1980, p. 131; Wengenroth 1993, p. 249), overlooking that he also referred to the publication as a "gift to the German physicists", which makes the "whole greatness of Lenard's achievements" visible (Tomaschek 1943b, p. 130). Moreover, as in many of his earlier statements, Tomaschek pointed out that his former teacher had never received the recognition he deserved. Therefore, he once again adopted Lenard's self-perception.

Moreover, it seems that Tomaschek did not refer to Lenard's theories per se when talking of the "swan song of an epoch", (Tomaschek 1943b, p. 131) but rather to the nature of his scientific work. He went on to describe Lenard as a lone fighter, but added that in the mean time the importance of collaboration and exchange of scientific results had greatly increased. Immediately afterwards, Rudolf Tomaschek clearly expressed his aversion to this change when he wrote that "the real progress [...] is coupled with the ingenious thoughts of a gifted man." Therefore, he considered Lenard's publication as a "cornerstone of conscious and constructive contribution to building our new world" (Tomaschek 1943b, p. 131). The use of and the special emphasis on the possessive pronoun make it clear that in 1943 Tomaschek had not put the ideas of his former teacher behind him. Although Tomaschek admitted that new methods were finding their way into physics—which can to some degree be regarded as insight—he was by no means open-minded about this change. Instead, he obviously longed for the "good old days" (Tomaschek 1943b, p. 131).

What is striking about all of Tomaschek's remarks on "Aryan Physics" is that most of them were directly related to his former teacher. In addition, the content of his comments on "Aryan Physics" is very consistent. Although in some of his later articles he refrained from clearly voicing anti-Semitism, he still adhered to Lenard's rather romantic way of understanding science and shared his scepticism towards technology. These findings indicate that even after 1940 Tomaschek did not dismiss his former beliefs, but rather stuck to many of them.

Networks Between "*Völkisch*" Sciences and Party Organisations

Historians have labelled Rudolf Tomaschek as a supporter of "Aryan Physics". This consideration is mainly based on his close connection to his former teacher Philipp Lenard. However, focussing on the category of "Aryan Physics" can be highly problematic because this movement lacked a consensus concerning its contents. Even Lenard and Stark—the two main actors—did not agree on what should be the characteristics of "Aryan Physics" (Schneider 2016, pp. 130–131). The two physicists scientifically pursued their own agendas, sometimes without acting in their colleague's interests. The Nobel laureates mainly shared a common interest regarding personnel policy and organisational matters, apart from their anti-Semitic political attitude (Beyerchen 1982, pp. 188–194; Hermann 2008, pp. 614–615; Hoffmann 2013, pp. 71–72; Kleinert 2000, pp. 252–260, 1980, pp. 35–38).

Based on this incoherence with regard to the content and inconsistency by its supporters, Freddy Litten has criticized the use of the category of "Aryan Physics" as such (Litten 2000, pp. 377–384). Even though Litten has not been able to fully assert his claim so far, he decisively sharpened the historians' awareness of the difficulties associated with the use of this term (e.g. Eckert 2007, pp. 140–141). Furthermore, a potential pitfall of focussing too much on the category of "Aryan Physics" is that connections scientists established in a different context can be easily overlooked. Tomaschek not only maintained contacts with representatives of "Aryan Physics" but was also involved in a much more complex network between *"völkisch"* scientists and party organisations.

The *Zeitschrift Für Die Gesamte Naturwissenschaft* (ZgN)

In 1935, a new scientific journal was established: *Zeitschrift für die gesamte Naturwissenschaft einschließlich Naturphilosophie und Geschichte der Naturwissenschaften* (The Journal for All Science including Natural Philosophy and the History of Science, ZgN). As the title suggests, this new journal originally aimed to be interdisciplinary. This orientation is also visible in the journal's editorial board, which in the beginning consisted of Karl Beurlen, a palaeontologist, Alfred Benninghoff, an anatomist, the chemist Karl Lothar Wolf and the philosopher Kurt Hildebrandt. Supporters of the *"Deutsche Physik"* movement did not come into particular prominence during the first years. The *"völkisch"* thought in science must nevertheless have played an important role from the very beginning because some of the editors eagerly tried to fuse the National Socialist ideology with their branches of science: Karl Beurlen, for example, tried to establish some kind of "Aryan chemistry". Furthermore, Ludwig Bieberbach the protagonist of *"Deutsche Mathematik"* can be found among the contributors of this journal (Bechstedt 1980; Lindner 1980; Rieppel 2012).

In the third year of its publication, the responsibility for editing the ZgN was transferred to the *Reichsstudentenführung* (RSF) and thus fell under the jurisdiction of Gustav Adolf Scheel (1907–1979). In the course of this transfer, the editorship changed as well: The new editorial board consisted of the biologist Ernst Bergdolt, the astronomer Bruno Thüring and Fritz Kubach (1912—disappeared 1945, pronounced dead in 1957), who was in a leading position in the NSDStB. In addition, the magazine was published in 1939 by the newly founded *Ahnenerbe* Foundation Publishing House in Berlin, leaving no doubt about its connection to the SS-*Ahnenerbe* and thereby the SS (Jagemann 2005, pp. 65–87; Simonsohn 2007, pp. 269–271).

Moreover, the number of articles written by supporters of "Aryan Physics", especially by the so-called "Munich Circle", increased in the following years (Litten 2000, pp. 377–384). Some of them, such as Hugo Dingler or Wilhelm Müller even joined the extended editorial board. Therefore, the ZgN was not only strongly connected to the different party organisations (the NSDStB, the RSF and the SS-*Ahnenerbe*). It was also transforming itself into a formal platform of the *"Deutsche Physik"* movement (Richter 1980, pp. 126; Simonsohn 2007, pp. 269–271). Furthermore the journal seems to have been dominated by the "Munich Circle" around Thüring and Dingler during this time (Litten 2000, pp. 380–384). This impression is at least conveyed when looking at the members of the editorial board and the table of contents of the ZgN. The members of the extended editorial board did—similar to Thüring—publish many articles in the journal. Apart from these, some other scientists, e.g. Ludwig Glaser, Eduard May and Max Steck—who can as well be identified as members of the "Munich Circle"—can be found among the contributors.

It might therefore be surprising that Tomaschek participated in the journal's work. In light of the growing tension between Tomaschek on the one and Thüring and Dingler on the other hand, it is at least remarkable. Why did Thüring, who played an important role in this journal, let Tomaschek, whom he was extremely suspicious of, join the editorial board? Even though Thüring might already have doubted Tomaschek's loyalty before November 1940, it would have been most likely in the interest of the different representatives of "Aryan Physics" to present themselves as a cohesive group—especially at a time when the external pressure was increasing noticeably.

The dedication of one volume in honour of Lenard in 1942 is probably also related to these efforts. Although the Nobel laureate and Dingler were not particularly close, the latter also contributed an essay to this volume. Thus, the editorial staff was probably trying to maintain contact with Lenard and his students, and thus with the broader network of "Aryan physicists".

In 1937, Tomaschek's first article in the ZgN, a review of Lenard's textbook *Deutsche Physik*, remained his only contribution during the following four years. Nevertheless, his scientific work, especially the new editions of Grimsehl's textbook, was reviewed frequently and positively by many of his colleagues (Bühl 1938/39; Requard 1939; Teichmann 1939, 1940; Wolf 1935).

In 1940, Tomaschek's connection to the journal became closer when he joined its editorial board. In this position, he may have been actively involved in selecting authors and topics for the respective volumes. However, he still rarely appeared as an author in this journal. Some reviews and one article in honour of Lenard's birthday in 1942 constitute some exceptions. The majority of his remarks are linked to his former teacher. However, after 1940 he also reviewed other scientists' publications for the journal, which could be interpreted as a sign of his deeper involvement. His review of Karl Lothar Wolf's *Theoretische Chemie* was quite detailed. Wolf himself had earlier reviewed Tomaschek's textbook (Wolf 1935) and honoured the physicists in the foreword as one of the most important "teachers [...] of today's chemistry", (Wolf 1941, p. 2) which indicates that both scientists maintained a rather close contact. Tomaschek returned the favour with his review, which was obviously influenced by ideological matters: For example, he emphasized the "constant use of the experiment while avoiding relativity-oriented flourishes" in a particularly positive way (Tomaschek 1942c, pp. 94–95). The physicist also praised the romantic manner of depicting nature, a characteristic sign of *"Deutsche Chemie"*. Apparently, he felt very close to the chemist in their common support for a *"völkisch"* science. Finally, by the frequent use of the personal pronoun "we", he clearly expressed solidarity with his colleague against the backdrop of possible criticism of "exact" scientists (Tomaschek 1942c, p. 94). The praise with which he described Wolf's textbook indicates that Tomaschek still cherished the idea of an 'Aryan' way in science.

Tomascheks Contribution to the Student Guide (*Studienführer*)

Because of Tomaschek's involvement in the ZgN, his contact with Fritz Kubach and Gustav Adolf Scheel, who were both affiliated with the NSDStB, the NSDDB, and the SS, appears to have intensified. This is interesting because Tomaschek thereby became involved in a major project of the SS-*Ahnenerbe*: The Student Guide—a series of study guides which was planned to contain about 300 volumes on different subjects, but was never completed (Jagemann 2005, pp. 25–36).

Fritz Kubach was in charge of organising the entire series, although formally the publication was under the control of the RSF, and thus its director Scheel. Despite the fact that the series was—due to the personal involvement of Kubach, Scheel and many of the volume's authors—closely tied to the SS-*Ahnenerbe*, this connection was never officially promoted. A reason for that might be the universities' sceptical attitude towards the SS. Moreover, the RSF may have been considered the more suitable sponsor for the study guide, which aimed to reach all German students. (Jagemann 2005, pp. 63–88; Kater 2006, pp. 353–360) However, this did not change the entanglement with the SS-*Ahnenerbe* in the background.

Tomaschek benefited from his ties in this context: After Scheel had taken over the leadership of the NSDDB in July, Tomaschek, according to his own statement, started to intensify his relationship to the same organization.[29] However, the physicist seems to have already had contacts with the NSDDB, since he gave at least one talk at one of their meetings in 1939 (Ba. 1939). His personal acquaintance with Scheel helped Tomaschek once again in 1944: due to the increased risk of air strikes, the TH started to evacuate its institutes to surrounding and more rural regions. Tomaschek's Institute of Physics was also affected, but he experienced some trouble in finding an appropriate place. Finally, Gustav Scheel helped him by providing him with a suitable space in Gmund near Lake Tegernsee.[30]

The project was divided into eight groups, with Tomaschek taking part in Fritz Kubach's group III on the natural sciences. It is nearly impossible to evaluate the entire project because the different groups were very heterogeneous. However, group III is considered to be highly anti-Semitic and influenced by "*völkisch*" thought because Kubach had recruited many authors from the ZgN. According to an advertising folder, 16 out of 23 authors had already been under contract in 1945. Nine out of these can be identified as authors of this journal. This impression is already

---

[29]DMA NL 089, 013: R. Tomaschek to A. Sommerfeld, 25 July 1945.
[30]DMA NL 089, 013: R. Tomaschek to A. Sommerfeld, 25 July 1945.

conveyed by the introduction Kubach himself had written: he stated that some special preconditions, which are typical of Germans and passed on by heritage, are required in order to successfully finish one's studies (Kubach 1943b, pp. 3–6).

Rudolf Tomaschek also participated in the work on the aforementioned Student Guide. In 1943, the introductory volume to the guide on natural sciences and mathematics was published (Kubach 1943a), in which Tomaschek contributed one article on physics. Being an overview, his essay was kept quite superficial, but he gave a general view of the issues regarding physics. At first glance, no obvious influence of "Aryan Physics" in terms of their racist theories attracts the reader's attention. However, it is noteworthy that Tomaschek tried to explain his discipline in Lenard's sense and did not mention theoretical physics at all. Furthermore, he subordinated technique and practicality to "pure physics" (Tomaschek 1943a, p. 27) and dedicated a much shorter chapter to the former than to the latter. He went on by highlighting that the immediate experience with physical reality and thus an intimate contact with nature itself was the basis for every creative activity (Tomaschek 1943a, p. 33). Tomaschek's statements in the Student Guide strongly resemble the review he wrote in 1937 on the textbook *Deutsche Physik*. This, in turn, indicates that Tomaschek stuck to the ideas and beliefs of his former teacher even in 1943.

Moreover, Tomaschek was chosen for editing the special volume on physics as part of the series.[31] Thus, Tomaschek apparently was not only involved superficially in this project by writing one short essay but he can rather be considered an active participant. In turn, this testifies both to his close contacts with the *ZgN*, which can be regarded as a collection point for various nationalist scholars, and his attachment to various party organisations.

Conclusion

The aim of the biographical approach pursued in this study—even though it is not a comprehensive biography—was to reconstruct and analyse Tomaschek's diverse connections to a very complex network of conservative and even "*völkisch*" scientists. One important advantage of a microscopic and biographical study is that one can combine various methodical approaches and can thus answer a variety of questions, even macroscopic ones. As a result, this essay did not only focus on Tomaschek himself but also on the networks in which he operated.

Tomaschek's connections to "Aryan Physics" were largely based on his commitment to Philipp Lenard. It was this connection that became the decisive factor for the TH Munich to advance his appointment as Professor of Experimental Physics. In this context, even the NSDAP expressed its desire to establish "another school of physics" in Munich with the help of Tomaschek. However, Tomaschek

---

[31] TUMA PA Prof. Tomaschek, Rudolf: R. Tomaschek to the rector of the TH Munich, 29 August 1944.

remained rather insignificant in Munich with regard to "Aryan Physics". Moreover, this observation shows—in combination with his behaviour at the Munich synod/debate in 1940—that Tomaschek did not position himself as a fanatical supporter of "Aryan Physics". Tomaschek's willingness to cooperate with the theoretical physicists alarmed the members of the "Munich Circle", especially Dingler and Thüring doubted his loyalty. One must admit, however, that although he was also based in Munich, Tomaschek never maintained a particularly close relationship with this group.

Nevertheless, by no means did he dissociate himself from the ideas of "Aryan Physics" as such. It can instead be assumed that his attempt to distance himself from the "Munich Circle" was more likely based on tactical considerations with regard to his career than on real professional insight. This is at least what can be suggested when analysing his public statements during the same time and his involvement in projects which were clearly influenced by "*völkisch*" thoughts, like the ZgN and the Student Guide.

The popular assumption that Tomaschek had dismissed his former beliefs after the Munich synod/debate in 1940 seems to stem from his own version of dealing with the past. Shortly after the Allied Military Government had prohibited him from entering the TH, he wrote a letter to Arnold Sommerfeld in which he tried to explain his actions during National Socialism. He obviously feared for his future career, as he frankly voiced his desire "to be able to continue the scientific work which was interrupted by the war, in the sense of pure physics".[32] Tomaschek tried to portray himself as a physicist committed only to scientific work itself, which has been a common motive for apologising one's own actions. Nevertheless, it is remarkable that even an exponent of "Aryan Physics" attempted to use this strategy (Eckert 2007, pp. 159–169; Hoffmann 2005, pp. 313–319; Hoffmann and Walker 2004, pp. 57–58).

Tomaschek began by explaining his relationship to the "Aryan Physics", which he called "Müller Group". He accused them of not working scientifically and strictly demarcated his own scientific endeavours. However, he acknowledged that he needed some "time of orientation" until he was able to distance himself from this group. Interestingly enough, he did not mention Philipp Lenard at all, maybe because it would have been much more complicated for Tomaschek to dissociate himself plausibly from his former teacher. Instead, he ascribed the leading role regarding "Aryan Physics" to Wilhelm Müller, who was, for Sommerfeld, the main person in Munich to blame.[33]

In order to prove his dissociation from the ideas of "Aryan Physics", he cited his efforts in favour of Fritz Sauter's appointment to the TH. Although Tomaschek was not able to convince his contemporaries and thus to rescue his scientific career, he at least succeeded in influencing the retrospective perception of his person. However, his self-portrayal did not coincide with his actions.

Approaching an exponent of "Aryan Physics" biographically reveals some pitfalls of this category: First, by simply labelling Tomaschek as "Aryan physicist" in

---

[32]DMA NL 089, 013: R. Tomaschek to A. Sommerfeld, 25 July 1945.
[33]DMA NL 089, 013: R. Tomaschek to A. Sommerfeld, 25 July 1945.

the common sense, his scientific work does not fit. For instance, he was entrusted with numerous application-oriented projects for war purposes, although historical research has so far mostly denied the representatives of "Aryan Physics" such an involvement in utilitarian research (Beyerchen 1982, pp. 253–260; Schönbeck 2000, pp. 38–39).

Second, as shown in this paper, the common concept of what it meant to be an "Aryan physicist" also falls short in terms of his personality in a broader sense. His involvement in this network around Lenard, Stark and other supporters of the "*Deutsche Physik*" made up only one part of his professional connections. At the beginning of his career, the connection to Lenard was certainly of great importance. Nevertheless, the contacts he made with a much wider network of "*völkisch*" scientists, which was also closely tied to party organisations, increasingly gained importance. Tomaschek's complex interrelations reveal the weaknesses of "Aryan Physics" as a terminological category, as Freddy Litten has already pointed out.

Nonetheless, completely rejecting this concept is not a good idea either, because it is not as useless as Litten has postulated: How else could one explain Tomaschek's appointment to Munich? How could one describe the ZgN's aspirations to appear as a cohesive group on the occasion of Lenard's birthday in 1942?

"Aryan Physics" should be seen as an extremely complex network integrated into a broader network of "*völkisch*" sciences, rather than focussing on its possible subject-related contents. However, when using the category of "Aryan Physics", historians are confronted with the undeniable difficulty of adequately presenting its heterogeneous structure, which was shaped by constantly changing alliances. Indeed, exactly this complexity might have enabled Tomaschek to act successfully—both within and outside this "Aryan Physics" movement—without taking a distinctive stand or clearly marking his position.

References

Ash, M. G. (2002). Wissenschaft und Politik als Ressourcen für einander. In R. vom Bruch & B. Kaderas (Eds.), *Wissenschaften und Wissenschaftspolitik* (pp. 32–51). Stuttgart: Steiner.

Ash, M. G. (2010). Wissenschaft und Politik: Eine Beziehungsgeschichte im 20. *Jahrhundert. Archiv für Sozialgeschichte, 50,* 11–46.

Auerbach, I. (1979). *Catalogus Professorum Academiae Marburgiensis.* (Vol. 2, pp. 1911–1971). Marburg: Elwert.

Ba. 1939. Im Lager der Wissenschaften. *Münchner Neueste Nachrichten,* August 4.

Bechstedt, M. (1980). "Gestalthafte Atomlehre"—Zur "Deutschen Chemie" im NS-Staat. In H. Mehrtens & S. Richter (Eds.), *Naturwissenschaft, Technik und NS-Ideologie: Beiträge zur Wissenschaftsgeschichte des Dritten Reichs* (pp. 142–165). Frankfurt: Suhrkamp.

Becker, A. (1942). Philipp Lenard und seine Schule. *Zeitschrift für die gesamte Naturwissenschaft, 8*(5–6), 143–152.

Berg, M. (2011). Nationalsozialistische Akademie oder Akademie im Nationalsozialismus? Die Bayerische Akademie der Wissenschaften und ihr Präsident Karl Alexander von Müller. In W. der Akademiegeschichte (Ed.), *Friedrich Wilhelm Graf* (pp. 173–202). Regensburg: Pustet.

Beyerchen, A. D. (1982). *Wissenschaftler unter Hitler: Physiker im Dritten Reich.* Frankfurt: Ullstein.
Bühl, A. (1938/39). Rezension von Grimsehl/Tomaschek "Lehrbuch der Physik". *Zeitschrift für die gesamte Naturwissenschaft, 2, 4,* 246–247.
Eckert, M. (2007). Die Deutsche Physikalische Gesellschaft und die "Deutsche Physik". In D. Hoffmann & M. Walker (Eds.), *Physiker zwischen Autonomie und Anpassung* (pp. 139–172). Weinheim: Wiley-VCH.
Eckert, M. (2013). *Arnold Sommerfeld: Atomphysiker und Kulturbote 1868–1951. Eine Biografie.* Göttingen: Wallstein.
Eckert, M., & Karl M. (Eds.) (2004). *Arnold Sommerfeld: Wissenschaftlicher Briefwechsel: Band 2: 1919–1951.* Berlin, Diepholz, München: Verlag für Geschichte der Naturwissenschaften und der Technik.
F. 1941. Lenard-Feier der Technischen Hochschule. *Münchner Neueste Nachrichten,* June 9.
F. R. 1936. Artgemäße Wissenschaft: Vortrag in der Berliner Universität. *Berliner Tageblatt,* December 4.
Finkelnburg, W. (1996). The Fight against Party Physics [1946]. In K. Hentschel (Ed.), *Physics and national socialism: An anthology of primary sources* (pp. 339–345). Basel: Springer.
Fleckenstein, J. O. (1968). Hundert Jahre Lehre und Forschung. In *Technische Hochschule München: 1868–1968*, ed. Technische Hochschule München, 61–133. Munich: Oldenbourg.
Hashagen, U. (2003). *Walther von Dyck: Mathematik, Technik und Wissenschaftsorganisation an der TH München.* Stuttgart: Steiner.
Hentschel, K. (Ed.). (1996). *Physics and national socialism: An anthology of primary sources.* Basel: Springer.
Hermann, A. (2008). Stark, Johannes. In *Complete Dictionary of Scientific Biography* (Vol. 12, pp. 613–616) [Online]. Detroit: Charles Scribner's Sons. http://go.galegroup.com.scientificbiography.emedia1.bsb-muenchen.de/ps/i.do?id=GALE%7CCX2830904121&v=2.1&u=bayern&it=r&p=GVRL&sw=w&asid=2f8de98d4628376e5b6d1e4edad60977. Accessed 3 August 2017.
Hoffmann, D. (2005). Between autonomy and accommodation: The German physical society during the third reich. *Physics in Perspective, 7*(3), 293–329. https://doi.org/10.1007/s00016-004-0235-x.
Hoffmann, D. (2007). Die Ramsauer-Ära und die Selbstmobilisierung der Deutschen Physikalischen Gesellschaft. In D. Hoffmann & M. Walker (Eds.), *Physiker zwischen Autonomie und Anpassung* (pp. 173–216). Weinheim: Wiley-VCH.
Hoffmann, D. (2013). Stark, Johannes. In *Neue Deutsche Biographie,* (Vol. 25, pp. 71–72) [Online]. Berlin: Duncker & Humblot. https://www.deutsche-biographie.de/gnd118798499.html#ndbcontent. Accessed 14 December 2017.
Hoffmann, D. (2016). Tomaschek, Rudolf Karl Anton. In *Neue Deutsche Biographie* (Vol. 26, pp. 343–344). Berlin: Duncker & Humblot.
Hoffmann, D., & Walker, M. (2004). The German physical society under national socialism. *Physics Today, 57*(12), 52–58. https://doi.org/10.1063/1.1878335.
Jagemann, N. (2005). *"Der Studienführer" zur Wissenschaftspolitik der SS.* Hamburg: Kovač.
Kater, M. H. (2006). *Das "Ahnenerbe" der SS. 1935–1945: Ein Beitrag zur Kulturpolitik des Dritten Reiches* (4th ed.). Munich: Oldenbourg.
Kleinert, A. (1980). Lenard, Stark und die Kaiser-Wilhelm-Gesellschaft: Auszüge aus der Korrespondenz der beiden Physiker zwischen 1933 und 1936. *Physikalische Blätter, 36*(2), 35–43.
Kleinert, A. (2000). Der Briefwechsel zwischen Philipp Lenard (1862–1947) und Johannes Stark (1874–1957). *Jahrbuch Reihe 3/ Deutsche Akademie der Naturforscher Leopoldina, 46,* 243–261.
Krauss, S. (2009). Die Akademie im Dritten Reich. In *Helle Köpfe: Die Geschichte der Bayerischen Akademie der Wissenschaften 1759–2009,* ed. Generaldirektion der Staatlichen Archive Bayerns, pp. 211–232. Regensburg: Pustet.
Kubach, F. (Ed.). (1943a). *Das Studium der Naturwissenschaft und der Mathematik.* Heidelberg: Carl Winter Universitätsverlag.

Kubach, F. (1943b). Einführung. In F. Kubach (Ed.), *Das Studium der Naturwissenschaft und der Mathematik* (pp. 1–6). Heidelberg: Carl Winter Universitätsverlag.

Lenard, P. (1936). *Deutsche Physik. Vol. 1: Einleitung und Mechanik.* München: Lehmann.

Lindner, H. (1980). "Deutsche" und "gegentypische" Mathematik: Zur Begründung einer "arteigenen" Mathematik im "Dritten Reich" durch Ludwig Bieberbach. In H. Mehrtens & S. Richter (Eds.), *Naturwissenschaft, Technik und NS-Ideologie: Beiträge zur Wissenschaftsgeschichte des Dritten Reichs* (pp. 88–115). Frankfurt: Suhrkamp.

Lindner, S. A. (2014). *Walther Meißner: Physiker und Institutsgründer. Ressourcenmobilisierung in drei politischen Systemen.* Augsburg: Rauner.

Litten, F. (2000). *Mechanik und Antisemitismus: Wilhelm Müller (1880–1968).* Munich: Institut für Geschichte der Naturwissenschaften.

Müller, W. (1939). Jüdischer Geist in der Physik. *Zeitschrift für die gesamte Naturwissenschaft, 5,* 162–175.

Müller, W. (Ed.). (1941). *Jüdische und deutsche Physik: Vorträge zur Eröffnung des Kolloquiums für theoretische Physik an der Universität München.* Leipzig: Heling.

N. N. (1936a). Das Judentum in Naturwissenschaft und Recht: Die Tagung der "Forschungsabteilung Judenfrage". *Deutsche Allgemeine Zeitung,* November 22.

N. N. (1936b). Judentum und Wissenschaft: Die erste Tagung der neuerrichteten Forschungsabteilung Judenfrage. *Münchner Neueste Nachrichten,* November 24.

Pabst, M. (2006). Zwischen Vereinnahmung und Autonomie: Die Hochschule im Dritten Reich. In W. A. Herrmann (Ed.), *Technische Universität München: Die Geschichte eines Wissenschaftsunternehmens* (Vol. 1, pp. 229–352). Berlin, Munich: Metropol.

Rammer, G. (2004). *Die Nazifizierung und Entnazifizierung der Physik an der Universität Göttingen.* Göttingen.

Requard, F. (1939). Rezension von Grimsehl/Tomaschek "Lehrbuch der Physik" vol. 1. *Zeitschrift für die gesamte Naturwissenschaft, 5,* 75–76.

Richter, S. (1980). Die "Deutsche Physik". In H. Mehrtens & S. Richter (Eds.), *Naturwissenschaft, Technik und NS-Ideologie: Beiträge zur Wissenschaftsgeschichte des Dritten Reichs* (pp. 116–141). Frankfurt: Suhrkamp.

Rieppel, O. (2012). Karl Beurlen (1901–1985): Nature, Mysticism, and Aryan Paleontology. *Journal of the History of Biology, 45*(2), 253–299. https://doi.org/10.1007/s10739-011-9283-7.

Scherzer, O. (1965). Physik im totalitären Staat. In A. Flitner (Ed.), *Deutsches Geistesleben und Nationalsozialismus: Eine Vortragsreihe der Universität Tübingen mit einem Nachwort von Hermann Diem* (pp. 47–58). Tübingen: Wunderlich.

Schneider, M. (2016). Die "Deutsche Physik"—Wissenschaft im Dienst von Ideologie und Macht. *Naturwissenschaftliche Rundschau, 69*(3), 125–134.

Schönbeck, C. (2000). *Albert Einstein und Philipp Lenard.* Berlin, Heidelberg, New York et al.: Springer.

Simonsohn, G. (2007). Die Deutsche Physikalische Gesellschaft und die Forschung. In D. Hoffmann & M. Walker (Eds.), *Physiker zwischen Autonomie und Anpassung* (pp. 237–299). Weinheim: Wiley-VCH.

Szöllösi-Janze, M. (2000). Lebens-Geschichte - Wissenschafts-Geschichte: Vom Nutzen der Biographie für Geschichtswissenschaft und Wissenschaftsgeschichte. *Berichte zur Wissenschaftsgeschichte, 23*(1), 17–35. https://doi.org/10.1002/bewi.20000230104.

Szöllösi-Janze, M. (2015). *Fritz Haber: 1868–1934. Eine Biographie* (2nd ed.). München: Beck.

Teichmann, H. (1939). Rezension von Grimsehl/Tomaschek "Lehrbuch der Physik". *Zeitschrift für die gesamte Naturwissenschaft, 2, 1, 5,* 244.

Teichmann, H. (1940). Rezension von Grimsehl/Tomaschek "Lehrbuch der Physik". *Zeitschrift für die gesamte Naturwissenschaft, 3, 6,* 96.

Tomaschek, R. (1929). *Lehrbuch der Physik. Mechanik, Wärmelehre, Akustik: Begr. von Ernst Grimsehl* (7th edn., Vol. 1). Leipzig, Berlin: Teubner.

Tomaschek, R. (1934). *Lehrbuch der Physik. Materie und Äther: Begr. von Ernst Grimsehl* (6th edn., Vol. 2) Leipzig, Berlin: Teubner.

Tomaschek, R. (1937/38). Rezension von Philipp Lenard "Deutsche Physik in vier Bänden". *Zeitschrift für die gesamte Naturwissenschaft, 1–3*, 95–99.

Tomaschek, R. (1942a). Ein deutsches Forscherleben: Zum 80. Geburtstag von Philipp Lenard. *Deutsche Technik, 10,* 268–271.

Tomaschek, R. (1942b). Lenards Äthervorstellung. *Zeitschrift für die gesamte Naturwissenschaft, 8*(5–6), 117–136.

Tomaschek, R. (1942c). Rezension von K. L. Wolf "Theoretische Chemie". *Zeitschrift für die gesamte Naturwissenschaft, 8,* 93–95.

Tomaschek, R. (1942d). Das weltanschauliche Gewissen der deutschen Physik: Philipp Lenard. Zu seinem 80. Geburtstag am 07. Juni 1942. *Völkischer Beobachter,* June 6.

Tomaschek, R. (1943a). Physik. In F. Kubach (Ed.), *Das Studium der Naturwissenschaft und der Mathematik* (pp. 26–33). Heidelberg: Carl Winter Universitätsverlag.

Tomaschek, R. (1943b). Rezension von Philipp Lenard "Wissenschaftliche Abhandlungen aus den Jahren 1886–1932". *Zeitschrift für die gesamte Naturwissenschaft, 9,* 130–132.

Walker, M. (1989). National socialism and german physics. *Journal of Contemporary History, 24,* 63–89.

Walker, M. (1995). *Nazi Science: Myth, Truth, and the German Atomic Bomb.* Cambridge (Mass.): Perseus.

Wengenroth, U. (1993). Zwischen Aufruhr und Diktatur: Die Technische Hochschule 1918–1945. In U. Wengenroth (Ed.), *Technische Universität München: Annäherungen an ihre Geschichte* (pp. 215–260). Munich: Faktum.

Wolf, K. L. (1935). Rezension von Grimsehl/Tomaschek "Lehrbuch der Physik". *Zeitschrift für die gesamte Naturwissenschaft, 1–2,* 345–346.

Wolf, K. L. (1941). *Theoretische Chemie* (Vol. 1). Leipzig: Das Atom.

Chapter 7
The 'Better' Nazi: Pascual Jordan and the Third Reich

Dieter Hoffmann and Mark Walker

Introduction

Because Hitler's regime strove for a totalitarian mobilization of German society for the National Socialist "People's Community" (*Volksgemeinschaft*) and both rearmament and modern warfare depended on science-based technology and industry, it was difficult, if not impossible for most scientists to ignore or avoid National Socialist policies and ideology. Yet one of the most enduring and influential stereotypes about science during the Third Reich suggests that professional competence or proficiency was incompatible with National Socialism, or in other words, that real scientists were not Nazis.

The converse stereotype suggests in turn that any scientist politicized in the National Socialist sense was incompetent. Although the so-called "Aryan Sciences", Nazi movements within the different scientific disciplines, are usually presented as proof, these had few followers and their activists were usually scientific outsiders (Walker 2003). The two pillars of "Aryan Physics", the Nobel laureates Philipp Lenard and Johannes Stark, had increasingly become outsiders with regard to both science and politics after the First World War. By the start of the Third Reich they were already well past their prime (Beyerchen 1977).

This stereotypical distinction between competent and apolitical on one hand, and incompetent and politically active on the other, allowed scientists and their communities to portray themselves after the war as victims of Nazism (Walker 1990). In fact there was a great deal of scientific progress under National Socialist rule, in part

---

D. Hoffmann
Max Planck Institute for the History of Science, Boltzmannstr. 22, 14195 Berlin, Germany

M. Walker (✉)
Department of History, Union College, Schenectady, NY 12308, USA
e-mail: walkerm@union.edu

© The Editor(s) (if applicable) and The Author(s), under exclusive licence to Springer Nature Switzerland AG 2020
C. Forstner and M. Walker (eds.), *Biographies in the History of Physics*, https://doi.org/10.1007/978-3-030-48509-2_7

because many competent and talented scientists entered into productive cooperative and collaborative relationships with the National Socialist regime.

A prominent example of such a scientist is the physicist Pascual Jordan, an early and dedicated supporter of National Socialism. A cofounder of the matrix mechanics and pioneer of quantum field theory, Jordan was certainly competent. Firmly anchored in the network of contemporary modern physics and accepted by his physicist colleagues, he was no scientific outsider or eccentric. Politically savvy, he also was not innocent or naïve. This essay will complement and extend work by Richard Beyler (Beyler 1994, 1996, 2003, 2011b) and Norton Wise (Wise 1994) by focusing on Jordan's advocacy of National Socialism during the Third Reich and how he took the lead in reinterpreting it after the war. An article by Ryan Dahn (Dahn 2019) appeared while our chapter was being edited and has also been incorporated into our argument. In a sense, Jordan was able to reinvent his biography. As a result, obituaries and other biographical studies, if they even discuss the Third Reich, have often portrayed Jordan as the 'prodigal son', if not the 'Good Nazi' (Schücking 1999; Ehlers and Schücking 2002).

Weimar

Pascual Jordan was born on October 18, 1902 in Hannover, finished his Ph.D. in 1924 with Max Born at the University of Göttingen, and completed his *Habilitation* (similar to a second doctorate) the very next year. Like Werner Heisenberg and Wolfgang Pauli, Jordan was one of the 'boy wonders' in physics during the 1920s, scientists who attracted attention because of their outstanding contributions to quantum theory while still students. By the summer of 1929, after a short stint teaching at the University of Hamburg, Jordan became a second-level (*außerordentlich*) Professor for Theoretical Physics at the University of Rostock. Although together with Pauli and Heisenberg the 26-year-old Jordan belonged to the youngest German university professors, neither the rank of the University of Rostock, nor his position there was comparable to what Heisenberg and Pauli respectively enjoyed at the University of Leipzig and ETH Zürich. Jordan recognized that Rostock was in the scientific provinces and tried through negotiations with the university officials in Hamburg and Rostock to delay accepting the position. After his hopes of staying in Hamburg quickly faded, he accepted the call to Rostock in the autumn of 1929. Jordan was not even the first choice of the faculty, mainly because "when excited a speech defect (stutter) becomes apparent".[1]

---

[1] UK Nr. 69, Personnel File (PA) Pascual Jordan (HUA), 9. Archives of the Humboldt University in Berlin (HUA).

According to his own account,[2] Rostock was neither scientifically nor intellectually interesting or inspiring. However, the political conservatism of provincial Mecklenburg corresponded to Jordan's own nationalist and conservative worldview. Writing under the pseudonym Ernst Domeier at the beginning of the 1930s in the *völkisch* (racist-conservative-nationalist) newspaper *German Folk* (*Deutsches Volkstum*), he articulated the general mistrust of conservative and nationalist educated middle class Germans towards the unloved Weimar Republic. These articles "expressed a distrust of democratization and, conversely, an unapologetically elitist conception of culture and science; they resented the decline of German national power; they bemoaned the erosion of traditional social values and artistic tastes" after the First World War (Beyler 2011b). The war, the subsequent Treaty of Versailles, and the fear of Bolshevism were politically traumatic for the educated German elite and significantly influenced Jordan's writings.

National Socialism

It was therefore no accident in January of 1933 when Jordan became a member of Alfred Hugenberg's German National Peoples Party (*Deutschnationale Volkspartei*, DNVP), apart from the NSDAP the strongest rightwing party of the Weimar Republic, a receptacle for *völkisch* and anti-Semitic groups, and the representative of extreme conservative and nationalistic circles. The DNVP had sought an alliance with the National Socialists early on and belonged to Hitler's cabinet up until Hugenberg's resignation in June of 1933. Immediately after the DNVP dissolved itself on May 1, 1933, Jordan joined the National Socialist German Workers Party (NSDAP).

This was not opportunism. In 1936 the National Socialist Student Organization in Rostock testified that Jordan had been in "comradely contact... even before the takeover of power".[3] Jordan himself recognized that the NSDAP and especially Hitler had unified the hitherto splintered nationalist conservative camp. In the introduction to his book, *Physics and the Mystery of Organic Life*, he asserted that "National Socialism not only found a synthesis between nationalism and socialism, two concepts, which for the postwar imagination were completely contradictory". After conflicts had "ripped and divided Germany", National Socialism "found a new, surprising solution at a higher level" (Jordan 1941a, p. 6).

Jordan repeatedly eulogized National Socialism. In the autumn of 1933 he also joined the SA and functioned there as a section leader (*Rottenführer*), who "because of his eagerness to serve and his comradely essence is generally beloved".[4] Such strong and active political commitment was very unusual for a German professor

---

[2]Interview of Pascual Jordan by Thomas S. Kuhn, Hamburg 20 June 1963, 16. Archive for the History of Quantum Physics.

[3]Student Body of Rostock University to the Rector 29 February 1936, PA Jordan 69/11, 27. HUA.

[4]See Footnote 3.

and internationally respected scholar. Jordan's writings also contributed to the propagation of National Socialist ideas. In May of 1933 he argued in the newspaper of Rostock University that it was no longer a question of "whether or not a reformed university should be integrated into the new National Socialist state", rather "determining the forms and paths in which such an integration will occur". Although the many resulting problems would be difficult and diverse, one thing was clearly needed: "the emphatic reorientation of the entire work of the University towards today's military and political tasks" (Jordan 1933, p. 3).

During the winter semester of 1933–1934 Jordan taught an "Introduction to Ballistics", which was one of the first military science courses at the University. Jordan also taught a "Mathematical Seminar with Special Consideration Given towards Military and Aviation Science" in 1939 (Vorlesungsverzeichnis der Universität Rostock 1938–1939). The foreword to his 1935 book, *Physical Thought in the New Age*, shows that the military and political significance of modern physics was central for Jordan's thinking at this time. Although the "quiet scholarliness" of the mathematician or physicist appeared "far removed from the hustle and bustle of our eventful age", the technical applications of their work "in aircraft, radio technology, and weapons of all kinds" had provided the "strongest and fiercest means for political power for the tremendous struggles of our century". Indeed the "peak scientific performance of a nation" was an "indispensable foundation of its technical capabilities" and "ability to assert itself economically and politically" (Jordan 1935, p. 7f.).

He also urged his colleagues to join him. The "academics of today" had the task of "reconceiving" their scientific research in the "context of the new state" and to "embrace the substance and tasks" of National Socialism (Jordan 1933, p. 4) like the "Four Year Plan for the German Economy and Technology". This program, inaugurated in 1936, was supposed to secure autarky for raw and basic materials and within four years make the German economy and armed forces "operational", that is, capable of going to war. For Jordan this could "only be solved by leading scientific nation" (Jordan 1936a, 9). Jordan argued that all economic activity should be incorporated into the "effort to achieve national predominance" and economic life should be "organized according to political—and in the end military—considerations". This would mean a "further amplification of the political significance and urgency of scientific research" (Jordan 1935, 47). Several years later, at the height of German military victories, Jordan was even more explicit. The "commanding events in the struggle of the young greater German Reich", the "Four Year Plan, rearmament, and war", had given "scientific work its secure place" in National Socialist life. Alongside the German worker, the "world power of German science" stood behind the "wonderful victories of the National Socialist armed forces" and the "superiority of German weapons" (Jordan 1941b, 452).

According to Jordan, modern research laboratories now had decisive military significance. Because of the connection between science and armaments, the nations and political leaders of Europe would appreciate scientific research more and more, both in the present and the future (Jordan 1935, p. 46). After Germany had unleashed the Second World War and the strategy of lightning war appeared victorious, Jordan became a propagandist for aggression: "… now that military power has demonstrated its compelling reconstructive power in the creation of a New Europe, we are not willing to see the connection of science with military power as misuse..." (Jordan 1941a, p. 9). Jordan's vision for science was also in some respects very similar to what became known as "Big Science" after the war (Dahn 2019).

As these examples demonstrate, Jordan was no mere 'technocratic fellow traveller' or political opportunist. Through his strongly-worded publications Jordan actively contributed to the National Socialist system of domination. Almost all the political and ideological essentials of National Socialism can be found in Jordan's writings: the rejection of democracy; the revision of the Treaty of Versailles, if necessary through military means; and the reestablishment of German military and economic power and greatness. Only anti-Semitism is missing: Jordan did not use the well-known anti-Semitic stereotypes in his publications and public statements. This may have been the result of his academic socialization in Göttingen, where Jewish scientists like Max Born belonged to his teachers, because of familial influences, or his own ethical-moral convictions. According to Jordan's own postwar statements, he not only opposed the dismissal of his Jewish teachers and colleagues in 1933, rather also considered emigration at that time.[5] One could certainly be a National Socialist without being anti-Semitic, but it was not possible for a follower of Hitler actively and publicly to oppose anti-Semitism, and indeed Jordan never did.

The Struggle Against "Aryan Physics"

Jordan was undoubtedly fascinated by National Socialism and especially Adolf Hitler. Of course, Jordan was not alone in this. What instead distinguished Jordan was his propagandistic advocacy of the political goals of National Socialism, which brought him into conflict with a few NSDAP offices and especially with the conservative and radical currents in the party. This was not a conflict with the Party itself and its fundamental goals, rather Jordan wanted to contribute to the evolution and de-radicalization of the NSDAP. But Jordan did more than try to exert a moderating influence on party politics; he also wanted the potential inherent in modernity, and especially in modern physics, to serve National Socialism.

When his *Physical Thought in the New Age* received a positive review in *Völkischer Beobachter*, the leading National Socialist newspaper, Jordan sent the review along "with friendly greetings" to NSDAP comrade Franz Bacher, the director of the universities section in the Reich Ministry of Science, Education, and Culture (*Reichsministerium für Wissenschaft, Erziehung und Volksbildung*, REM).[6] Jordan sought the attention and support of people like Bacher because his book had been criticized by representatives of the so-called "Aryan Physics" and "Aryan Science".

---

[5]Pascual Jordan to Niels Bohr, Göttingen May 1945. Bohr Papers, folder Pascual Jordan. Niels Bohr Archive, Copenhagen (NBA).
[6]REM J 107 (PA Pascual Jordan), 9356. Federal German Archives, Berlin-Lichterfelde.

The *Journal for the Whole of Science* (*Zeitschrift für die gesamte Naturwissenschaft*) was founded in 1935 and became the main journal for "Aryan Physics". A review of Jordan's book in this journal criticized its programmatic title, since it gave "the impression that there is a connection between the contents and the [National Socialist] revolution in Germany" (Ramsauer 1935, 342–343). An article by the journal editor Kurt Hildebrandt attacked Jordan even stronger. Although Hildebrandt's critique remained philosophical and within the bounds of academic discourse, it also included ideological statements like: "behind the positivism of the present stands a great weariness with the world, a sinking life instinct" and "science does not suffer, rather only unfolds more fruitfully, if it lines up with the creative spiritual life" of National Socialism (Hildebrandt 1935, 18–19).

After further attacks in this journal, Jordan became one of the preferred targets of this ideologically and politically influential group. He saw in this criticism "the attempt to politically defame the entire mathematical-physical research" (Jordan 1935, 9). Because the political implications of Hildebrandt's criticism were clear to Jordan, he took the gloves off and polemicized in the Mecklenburg Student Newspaper:

> But in the sciences it is today still possible for obsolete theories and moldy hypotheses from the past century to be dished up and declared to be specifically suitable for our times—without laughing heartily, which one would have to do as the only suitable reply, and shaving such great grandfatherly beards. (Jordan 1936a, 8)

In the preface to *Physical Thought in the New Age* Jordan confronted these critics by presenting himself, not merely as a good National Socialist, but instead as the better one. Responding to those who were trying to "defame mathematical and physical research" and to make this defamation both "amusing" and to appear as "the consequence of a National Socialist stance", Jordan argued that: "We are living in the age of technical war: any attempt to sabotage Germany's leading position in the areas of mathematical, physical, and chemical research" was equivalent to trying to "disrupt the military strength of the National Socialist state" (Jordan 1935, p. 9).

Jordan's harsh polemics made the followers of "Aryan Physics" into irreconcilable enemies. Up until their political fall at the beginning of the 1940s, Jordan was the main target for their attacks on modern physics. When Jordan's 1936 book *Physics of the Twentieth Century* argued that "modern physics, and its characteristic revolutionary transformation of centuries old scientific concepts, is an integrative element of the new world of the twentieth century that is unfolding …" (Jordan 1936b, p. VII), the philosopher Hugo Dingler (Wolters 1992) responded in a book review by criticizing modern physics and its uncertain epistemological basis (Dingler 1937). This dispute is interesting because it was actually Jordan, more than Dingler, who introduced ideological and political arguments into the conflict. This included denouncing Dingler to REM and NSDAP offices in what Viktor Klemperer called the *Lingua Tertii Imperii* (LTI), or language of the Third Reich (Klemperer 1975).

For example, in letters written in February of 1938, he described "a certain metamorphosis in worldview" for Dingler, who "before the seizure of power had emphasized the 'familiar drive', the 'industriousness and acumen' of the Jewish race in his philosophical writings".⁷ Dingler's most important work, *The Collapse of Science* from 1926, also contained "countless bows and attempts to curry favor with Jewry". Indeed Dingler was "proud that the chief Rabbi of Vienna, the 'most competent' judge", had recognized his work. Furthermore Jordan accused Dingler of "pro-Jewish propaganda" and having chosen for his book a "downright Jewish title for sensation and publicity",⁸ which could in no way "serve the positive constructive work in the sense of National Socialist cultural policy".⁹ As a "party member and SA man", Jordan refused to accept "worldview instruction and censorship" from such an opportunist and "fanatical philosophical propagandist of Jewry".¹⁰

Jordan was even more explicit in a letter to Fritz Kubach, at the time the head of the Office for Science and Technical Education of the Reich Student Leadership and subsequent Reich Student Leader. (All of the quotations in this and the following paragraph come from this letter.)¹¹ Once again, Jordan used the conflict with Dingler and anti-Semitism to portray himself as the better National Socialist. A philosopher, he argued, who before the National Socialists had come to power and had "characterized his philosophical life's work as closely paralleling Jewish legal doctrine and felt flattered by the applause of a Rabbi" was "unsuitable as a contributor to, or advisor of the National Socialist reconstruction of science". Furthermore the

> former husband of a Jew, a coworker of Georg Bernhard [a critic during Weimar of National Socialism], a man who admitted his financial dependence on Zionist circles, who up until 1932 had worked with fanatical consistency for the jewification of the German universities, who himself is suspected of having Jewish ancestry, who after the seizure of power met secretly with an agent of the Jewish world leadership for several hours, etc., etc., is also unsuitable.

Such people, Jordan concluded, should not be allowed to "influence National Socialist cultural work".

However, Jordan emphasized that the "positive tasks of the National Socialist leadership of science, a spiritual penetration of scientific work by National Socialism, and in turn, a comprehensive incorporation of German science and research into the National Socialist cultural mission" were far more important than these negative things. Besides the "very significant connection between scientific work and National Socialism in the context of the Four Year Plan", the major vehicle for rearming Germany and making it autarkic, Jordan argued that "today there is no area in which we have fallen further behind than in incorporating scientific and mathematical work

---

⁷Pascual Jordan to the Rector of Rostock University, 23 February 1938, PA Pascual Jordan, PA 69,/II, 54. HUA.

⁸Pascual Jordan to the NS Regional Faculty Leader Gißel, 10 February 1938, PA Pascual Jordan, PA 69,/II, 56. HUA.

⁹Pascual Jordan to the NS Regional Faculty Leader Gißel, 10 February 1938, PA Pascual Jordan, PA 69,/II, 55. HUA.

¹⁰Pascual Jordan to the Rector of Rostock University, 23 February 1938, PA Pascual Jordan, PA 69,/II, 54. HUA; Pascual Jordan to the NS Regional Faculty Leader Gißel, 10 February 1938, PA Pascual Jordan, PA 69,/II, 56. HUA.

¹¹Pascual Jordan to F. Kubach, Rostock 28 March 1938 (transcription). Universität Konstanz, Philosophische Archiv, Nachlass Hugo Dingler. PA Pascual Jordan, PA 69,/II, 57–59. HUA.

into National Socialist cultural politics". This was not because of the work itself, rather was a consequence of the "sabotage by certain camouflaged friends of Jews, who we in particular can thank for the fact that the Einstein legend invented by the Jewish cultural propaganda was not destroyed and overcome, rather on the contrary was hammered even harder into people's heads". Solving the problems inherent in the tasks facing the National Socialist state "cannot be separated from the necessity of eliminating the confusion that people like Dingler have deliberately created". Three years later, Jordan also tried to denounce Kubach (Dahn 2019, 77–79).

Since anti-Semitic remarks cannot be found in Jordan's publications or personal behavior, his vicious words were probably the result of the tactical considerations at the time, rather than his own convictions. Jordan's conflict with Dingler and the advocates of "Aryan Physics" allied with him explains why Jordan could claim after the war that he was "alongside Planck and Heisenberg the physicist attacked the hardest" in influential Party publications[12] and definitely earned Jordan recognition and respect from his physicist colleagues. But these attacks had hardly anything to do with Jordan keeping political distance from, let alone opposing National Socialism.

These disputes created personal enemies in the National Socialist hierarchy, not least because of Jordan's aggressive manner. This certainly was part of the reason why Jordan was bypassed for appointments. For example, in 1937 a Rostock official wrote that: "… his work, in particular his ideological statements on the foundation and significance of physics have caused considerable criticism."[13] Such political assessments were made in the context of appointments, but also for applications for travel abroad. In both cases they were supposed to ensure the political reliability of the respective candidates. Jordan travelled to Copenhagen in 1936. The circumstances surrounding this trip also illustrate Jordan's compliant and opportunistic conduct at the time (Hoffmann 1988).

In 1936 Niels Bohr invited Jordan to give a talk in one of a series of now legendary physics conferences, which since the end of the 1920s had brought together a select circle of Bohr students as well as interested colleagues to debate current developments in quantum physics. The 1936 physics conference was connected to the "Second Congress for the Unity of Science" because the theme of this congress, "the problem of causality with special consideration of physics and biology", was of special interest to Bohr's circle. The double invitation presented a delicate problem for the German participants. Although it was possible to act relatively 'neutral' when discussing physics, at least as far as an external observer was concerned, this would be difficult to do at the congress. Not only did the neo-positivist philosophical orientation of the congress clash with National Socialism, because of their Jewish origins the leading representatives of this school (Rudolf Carnap, Philipp Frank, Otto Neurath, Hans Reichenbach, and Moritz Schlick) had recently been forced to emigrate from Germany and opposed both the ideology and politics of the Third Reich.

---

[12] Pascual Jordan to Max Born, Hamburg 23 July 1948. Preußischer Kulturbesitz, Handschriftenabteilung, Max Born Papers, Folder 353. Berlin State Library.

[13] Student Body of Rostock University to the Rector, Rostock 23 August 1937. 74. HUA.

7 The 'Better' Nazi: Pascual Jordan and the Third Reich

Moreover, the theme of the congress dealt with precisely those areas of philosophy where Jordan and the representatives of "Aryan Physics" were in conflict.

After some back and forth, Jordan received ministerial approval for participating in both meetings. Before and after the conference, Jordan demonstrated a great willingness to work with National Socialist officials. This is clear from the report he prepared for REM, which was also written in the language of the Third Reich. The report embraced Nazi policies and reported on the conduct and ideological positions of other participants (Hoffmann 1988). For this reason Jordan argued that it was "urgently desirable, that the facts and content of this report not become known outside of Germany". He had actually "with great care always avoided discussing this report with comrades or colleagues" and requested that the Ministry "treat the matter confidentially".[14]

The ideological attacks on Jordan by Nazi ideologues were concentrated during the second half of the 1930s. At the beginning of 1936 the University of Rostock requested a promotion to full professor for Jordan, arguing that this: "only corresponds to the outstanding status he has had for a long time within our university".[15] This request was approved without further commentary or questions.[16] There were also no political or ideological concerns about Jordan at this time. The Student Organization testified that Jordan "stands in honest conviction with the National Socialist movement",[17] while the National Socialist University Lecturers League even claimed that "politically J. is in every way reliable. He was already very interested in politics".[18]

Only after the representatives of "Aryan Physics" had opened a broad offensive in the summer of 1937 against the representatives of modern physics (Beyerchen 1977, pp. 141–167; Litten 2000) and defamed Jordan and others did the same official now write that: "a definitive judgment of Jordan cannot be made at this time".[19] During this period the conflicts with Dingler and the representatives of "Aryan Physics" hindered Jordan's career. When discussions began in the autumn of 1936 about the successor to Erwin Schrödinger at the University of Berlin, the search commission argued for Jordan,[20] but the Ministry ignored the recommendation and instead used a series of younger theoretical physicists to cover the teaching duties of the professorships (Hoffmann 2010, p. 570).

---

[14]Reichserziehungsministerium Nr. 2744, Atomphysikkongresse Kopenhagen 1936/37/38, 27. Federal German Archives, Berlin-Lichterfelde.

[15]Dean to the State Ministry of Mecklenburg, Abteilung Unterricht, Rostock 15 April 1936 PA j, 46. HUA.

[16]MfS-HA 1X711, PA 2697, 15. BStU MfS-HA Archives, Berlin.

[17]Studentenschaft der Universität Rostock to the Rector, Rostock 29 February 1936, PA j, 27. HUA.

[18]Gutachten des Leiters der Dozentenschaft der Universität Rostock (Gißel), Rostock 20 March 1936, PA j, 26ff. HUA.

[19]Student Body of Rostock University to the Rector, Rostock 23 August 1937, PA j, 74. HUA.

[20]Report on the Commission Meeting Regarding Schrödinger's Successor of, Berlin 19 August 1936. PA Schrödinger Nr. 248/III, 12. HUA.

Rehabilitation

By the beginning of the 1940s the influence of "Aryan Physics" had weakened. Their ideologues had been marginalized and modern physics rehabilitated (Hoffmann 2012), so that the attacks on Jordan and the other representatives of modern physics no longer had a public forum and faded away. Although a National Socialist revolution could be carried out very well with politicized physicists like Johannes Stark (Hoffmann 1982; Walker 1995, 5–63), physicists like Jordan were needed for the autarkic military state and its development of modern technology and innovative weapons systems. As rearmament and the politics of war were pushed forward, the ideologically motivated disrespect and discrimination towards modern theoretical physics steadily declined and the advocates of this physics enjoyed increasing influence and acceptance among the politically powerful (Heisenberg and Heisenberg 2011, 201–202).

The Max Planck Medal for Theoretical Physics was a symptom of this change. After five years of enforced pause, in 1943 the German Physical Society (DPG) awarded Jordan this medal, the highest honor in German physics. This was no coincidence. By giving this medal to someone who was both a pioneer of modern physics and a National Socialist, the DPG asserted some limited scientific autonomy while simultaneously demonstrating its political loyalty (Beyler et al. 2012, p. 182ff.). When a successor for Max von Laue in Berlin was also needed that same year, the Ministry now turned to Jordan. Beginning in the winter semester of 1944/1945, Jordan became full professor for theoretical physics at the University of Berlin and the personal ideological attacks on him came to an end. Jordan now had one of the most prestigious physics professorship in Germany and finally the professional recognition he deserved. However, because of the catastrophic conditions during the last terrible days of the Third Reich, Jordan's time in Berlin was ineffective.

Jordan had already left unloved Rostock and the scientific provinces by volunteering for the military at the start of the Second World War. Jordan's decision was probably influenced by the role the Armed Forces played in the Third Reich. Many nationalist and conservative members of the bourgeoisie saw the military as an antipole to the Nazi regime and potential alternative for a better future for Germany. Jordan certainly also hoped that military service would take him out of the political firing line. At first Jordan served in the meteorological units of the Air Force and held the rank of a major. In the autumn of 1942 he was posted to Peenemünde, where he carried out calculations for wind tunnel experiments. Peter Wegner, who shared an office with him, remembered that Jordan sat in front of his typewriter all day composing a textbook on algebra "without ever consulting notes". Usually toward the end of the day he would suddenly recall the original purpose of his assignment at Peenemünde. "He had never worked in fluid dynamics, but rather than read books as I did, he proceeded to derive the equations of motion of supersonic flows. He delighted in the discovery of such phenomena as shock waves" (Wegener 1996, p. 28).

Paul Rosbaud, the editor of the journal *The Sciences* (*Die Naturwissenschaften*) and British spy, knew Jordan and gathered information from him about Peenemünde for the western Allies (Kramish 1986). According to Rosbaud, in 1943 Jordan tried to get transferred from Peenemünde by sending "a cry for help" to a group of scientists working for the Naval Research Division under Helmut Hasse. Rosbaud claimed that Jordan was "very happy when he was transferred, against the will of a very stubborn colonel, to the navy, and there he did nothing else but develop his cosmology along with [Albrecht] Unsöld".[21]

In fact, along with his cosmological hobby, Jordan researched high pressure physics for weapons development. After the war, Jordan claimed that: "During the war, after year-long difficult arguments with military superiors, I withdrew from working on 'revenge weapons' or atomic energy" (Hoffmann 2003, 35). Since the Naval Research Division was in Berlin, Jordan could begin teaching at the University. He also reestablished contact to the group of researchers at the Kaiser Wilhelm Institute for Brain Research in Berlin-Buch under Nikolai Timoféef-Ressovsky, with whom Jordan had been working intensively since 1935 on biophysical questions and problems of applying the quantum theory to biology (Beyler 2011a).

Although Jordan's war work was certainly more engaged and valuable for weapons production than he implied after the war, it does beg a question: how does Jordan compare to other German scientists in this regard? The list of prominent researchers who provided their expertise to the National Socialist state at war is very long, including Walther Gerlach, Otto Hahn, and Werner Heisenberg, even if they, like Jordan, self-servingly systematically obscured and minimized this after the war (Walker 1989, 2006, 2009). Jordan stands out not because of his contribution to the war, for others like Gerlach made much more of an impact, rather because of the resonance between his previous public advocacy of science in the service of war and his actual military service.

Mastering the Past

Jordan experienced the end of the war in Göttingen, where Hasse's institute had been evacuated during the last weeks of the war. It was clear to Jordan that the collapse of the Third Reich now threatened his academic career, for despite his conflicts with "Aryan Physics", he had been closely involved with National Socialism and had propagated its goals. Now he had to reinterpret this involvement and adapt his biography to the new political circumstances. Jordan took up this task very quickly and directly, drawing upon his scientific expertise, which everyone recognized, and networking within the scientific community.

As early as May of 1945, just a few weeks after the occupation of Göttingen by British troops and a few days after the capitulation of Hitler's Germany, Jordan wrote a letter to Niels Bohr in order to: "to give a short coherent account of what I did during these black 12 years". (All the quotations in this and the following

---

[21] Niels Bohr Library, Samuel Goudsmit Papers, Box 27, Folder 4L. American Institute of Physics. https://repository.aip.org/islandora/object/nbla%3AAR2000-0092, accessed April 24, 2019.

four paragraphs come from this letter.)[22] It was "inevitable" that some of Jordan's "old friends" had misunderstood his actions. First of all, he explained why he had not emigrated: (1) he could not leave with his aged mother; (2) his speech defect, which caused Jordan "many difficulties in professional and in daily life" would have multiplied abroad; (3) as a "voluntary emigrant", Jordan could not justify taking funds from an "organization created for helping those who were inevitably *forced* to emigrate"; and finally (4) Jordan had falsely believed that the Nazi Party would probably evolve after 1933, that the "radicalism shown at the beginning" would fade with time and a "tolerable situation" would return step-by-step after a few years. Indeed Jordan had "hoped to be able to accelerate this evolution to a certain little extent. When I had to convince myself that on the contrary the tendencies grew more and more radical, there remained no possibility to emigrate".

Jordan next described the attacks he suffered on behalf of modern physics during the Third Reich as the student of "famous emigrated teachers and a notorious defender" of Einstein's theories. Jordan was the "most visible exponent" left in Germany of theoretical physics, a discipline that set "a record in attracting the animosities of Nazis". Jordan claimed that he had had only two options. He could have acted like Max von Laue by consistently opposing National Socialist policies, "an attitude which, though much more admirable than mine, would have meant for me, being much less prominent a figure than he, certainly [the] ruin of my existence, and most probably would have landed me in a concentration camp". Jordan's other alternative was to: "acquiesce in matters which seemed to me of minor importance, to look out for connections with people in the Nazi party which did not belong to the dangerous clique of Stark and Lenard, and to endeavor to concentrate my efforts in fighting those tendencies..."

Jordan told Bohr that he had written his book, *Physics of the Twentieth Century*, including its appreciation of relativity and quantum theory, as an answer to the influential Nazi ideologue Alfred Rosenberg's *Myth of the Twentieth Century* (*Der Mythus des zwanzigsten Jahrhunderts*, Rosenberg 1930) and claimed that it had been understood as such by many students and colleagues. Once again, Jordan brought up the specter of the concentration camps: "the number of persons imprisoned or murdered by the SS is so great that there seems to be no necessity to say it was really a little hazardous to attack in such a manner the *standard book* of nazistic *Weltanschauung* [world view]".

Perhaps, Jordan suggested, he had made a difference. It was a "fact, that theoretical physicists in Germany from 1933 till 1939 have *not* been persecuted in a similar manner as for instance theologists and clergymen who in great numbers went through (or ended in) the concentration camps". This was not trivial, Jordan argued, for certainly "Lenard, Stark, Thüring, Kubach would have been very satisfied if they had been able to induce similar persecutions against all believers in relativity- and quantum-theory, and they did their best to start them". Now affecting modesty, Jordan continued: "Perhaps I contributed a little to prevent such an evolution; I prefer to

---

[22] Pascual Jordan to Niels Bohr, Göttingen May 1945. Bohr Papers, folder Pascual Jordan. NBA.

7 The 'Better' Nazi: Pascual Jordan and the Third Reich   123

leave this question open. Let me only say that the mentioned antiscientists were awfully sorry when they learnt that I was a member of the party".

Jordan also attempted to explain the difference between militarism and Nazism in Germany and to emphasize the differences between the NSDAP and the Armed Forces. Things supposedly had changed considerably after 1939. While Jordan recognized that outside of Germany people were probably "inclined to regard militarism and Nazism as *identical* perversities". In reality, Jordan argued, the attempt to assassinate Hitler demonstrated that the Armed Forces and NSDAP were "sharply opposite". According to Jordan, during the first phase of the war many Germans were "certain (and I believe they were right) that in the case of a positive or tolerable result of the war the successful Wehrmacht would liquidate the NSDAP".

Finally while Jordan admitted that it might "sound a little cheap to give these explanations now, after all is over", he claimed that he was not trying to clear himself of all possible charges, or to argue that everything he had done was "necessary" or in "good taste". Instead Jordan wrote: "I did what I did, but I hope that you, who know me for long time, will understand that my intentions at least were good…".[23] A 1957 letter from Max Born to Jordan mentioned that, immediately after the end of the war, Born and his wife had been disturbed by a letter from Jordan. Born now reminded his former student that: "As an answer I sent you a list of my relatives and friends who were murdered by the Nazis and you answered, that you had known nothing about this and were shocked".[24]

Jordan's immediate postwar letters to colleagues were typical attempts to justify someone's conduct during the Third Reich and can hardly be taken as accurate or objective descriptions of actions or motives. Yet in any case a strategy was established: Jordan gave the impression that he had been the target of continual political attacks by the National Socialist rulers during the Third Reich in the hope that his commitment towards and service for National Socialism would be forgotten or pushed into the background. The new editions of his books naturally eliminated all references and language from the Third Reich, which was criticized by a few contemporaries. Ursula Martius, a physics student who had been discriminated against during the Third Reich, noted in a contribution to the newspaper *German Review* (*Deutsche Rundschau*) that in the new editions passages like Jordan's celebration of the connection between science and military power (see above) had been removed (Rammer 2012, 393). However, such voices were in the minority, so that Jordan was able to reintegrate himself relatively quickly and successfully into the German scientific community. He was thereby supported by influential colleagues and friends in both the west and east who were not willing to do without his scientific competence. As early as the autumn of 1946, Werner Heisenberg gave Jordan 'political absolution' by writing that: "During the entire National Socialist period I often discussed all

---

[23] See Footnote 22.

[24] Max Born to Pascual Jordan, Bad Pyrmont 30 October 1957. Preußischer Kulturbesitz, Born Papers. Berlin State Library.

sorts of political problems openly with Jordan and never considered the possibility that he could have been a National Socialist".[25]

Robert Rompe, who was a communist of long standing and had actively resisted the Nazis, now enjoyed great influence as a science policymaker in the Soviet occupation zone (Hoffmann 2005). During the war Jordan and Rompe had met in Berlin-Buch. Now Rompe wanted to bring Jordan back to East Berlin. In the autumn of 1946 the two physicists had worked out a concept for a large research institute for biophysics and medical biophysics that would be affiliated with the newly founded German Academy of Sciences, which was then under construction.[26] Jordan's National Socialist past naturally came up, since this should have meant that Jordan could not be given a leadership position.[27] Nevertheless in the summer of 1948 the Soviet Military Administration (SMAD) requested a "political assessment of Prof. Jordan", since his service in the Soviet occupation zone was being considered and his "disciplinary assessment by the SMAD had already been completed in a positive sense".[28]

The assessment by Rompe provided a very euphemistic portrayal of Jordan's conduct during the Third Reich. Although the two physicists were friendly and admired each other's work, Rompe's generous assessment was more the result of the Cold War and the competition between political systems it embodied. Thus Rompe tried to explain away "Jordan's political conduct during the Nazi-period" by the "characteristics of his personality" and his origins. Jordan came from the "bourgeois intelligentsia", which Communists argued had collaborated with the Nazis. Yet Jordan had "remained faithful to his anti-fascist and Jewish friends". Despite being a member of the NSDAP, Jordan "never belonged to it".[29]

As early as 1946 the University of Hamburg in the British occupation zone was interested in Jordan.[30] After he gave a talk in the summer of 1946, the director of the Institute for Theoretical Physics in Hamburg, Wilhelm Lenz, recommended that the University offer Jordan a three-year visiting professorship. Since the University officials knew that they were in competition with other universities and institutions, they quickly made an offer. After the negotiations were successfully completed in April, Jordan began his guest professorship on May 1, 1947. Jordan's political liabilities do not appear to have been particularly important in Hamburg. Lenz's recommendation noted that Jordan had joined the NSDAP and SA "relatively early", but added that

---

[25] Werner Heisenberg, 1946. Persilschein for Pascual Jordan, Nachlass Werner Heisenberg. Archives of the Max Planck Society (MPGA).

[26] Akademieleitung, Naturwissenschaftliche Einrichtungen, Bd. 5 Archives of the Berlin-Brandenburg Academy of Sciences (ABBAW) and Dahn 2019.

[27] Protocol of the Mathematical-Scientific Class, 20 February 1947. ABBAW.

[28] Memo for Director Naas, Berlin 24 August 1948. ABBAW.

[29] Robert Rompe: Regarding the Appointment of Prof. Dr. Pascual Jordan. Berlin 4 October 1948. ABBAW.

[30] Pascual Jordan to H. Schimank, Göttingen 1 June 1946. Hans Schimank- Gedächtnis-Stiftung, NL Schimank, Correspondence Jordan. Archives of the University of Hamburg.

he had not served in any higher Party position. The "Lenard group" had caused difficulty for Jordan because of his clear "advocacy of Einstein's great achievement."[31] Sometime in 1947/48 Jordan successfully traversed his denazification proceeding. Jordan's Nazi past played no role in the early Federal Republic and was irrelevant for his further academic career. Soon he could once again travel outside of Germany and accept speaking invitations.

When in 1950 the University of Hamburg wanted to transform his visiting position to a full professorship, the recommendation of the faculty contained no mention of Jordan's Nazi involvement. Instead his scientific merits were honored in great detail.[32] Wolfgang Pauli was also asked for a recommendation. In his response Pauli noted that everyone agreed that Jordan's work was valuable and should continue, but added that "his contradictory personality" made it difficult to judge him. Indeed many found it difficult "to follow his path up until 1945". Jordan's incorporation of political viewpoints into popular scientific books, for example the "comparison of a cell with the state, the cell nucleus with Hitler, etc.", as well as "discussions of weapons that sometimes rose to bellicose fanfares" into scientific and philosophical discussions, had been "unjustifiable". However, for Pauli two things were decisive. First of all, Jordan's early work on quantum mechanics and quantification of fields had earned him international recognition. Indeed the latter subject was an important field of research at that time. Second, Germany had "very few theoretical physicists of such rank". Pauli believed that "the part of Germany belonging to western Europe" could not afford to "pass over a man" like Pascual Jordan.[33]

Yet within the international physics community many people had reservations about Jordan. The fact that, despite repeated nominations, Jordan never received the Nobel Prize was probably due in part to these resentments. In a 1949 letter to Arnold Sommerfeld, Erwin Schrödinger criticized Jordan's cosmological ideas and saw in them "a curious (because only partial) impairment of the intellect through years of imbibing Nazi philosophy" (Eckert and Märker 2004, p. 639). Philipp Frank described Jordan as the "former Nazi" in a letter to Helen Dukas.[34] For some scientists, for example the Dutch physicist Jacob Clay, Jordan's Nazi past was so serious that Clay would not participate in a congress if anyone like Jordan would be there.[35] Lise Meitner also had not forgotten Jordan's Nazi past. In a 1958 letter to Max von Laue she remarked that it was difficult for her to imagine Pascual Jordan "as a celebratory speaker for Einstein".[36] This émigré nightmare came true two decades later

---

[31] Wilhelm Lenz to the Rector of the University of Hamburg, Hamburg 6 August 1946. Hochschulwesen, Dozenten- und Personalakten, IV 2076, PA Pascual Jordan, 2. Hamburg State Archives (HSA).

[32] W. Blaschke to the Rector of Hamburg University Paul Harteck, Hamburg 3.1.1953. Hochschulwesen, Dozenten- und Personalakten, IV 2076, Personalakte Pascual Jordan, 25–27. HSA.

[33] Wolfgang Pauli to P. Willer, Zurich 8 May 1952, from Karl von Meyenn.

[34] Philipp Frank to Helen Dukas, 1950. Albert Einstein Archives. No. 16410. The Jewish National und University Library, Hebrew University Jerusalem.

[35] Jacob Clay to Wolfgang Gentner, 7 February 1952. Ulrich Schmidt-Rohr Papers. Max Planck Institute for Nuclear Physics, Heidelberg.

[36] Lise Meitner to Max von Laue, Cambridge 17 January 1958. Max von Laue Papers. MPGA.

in 1979, when Jordan was the honorary guest of the West Berlin Senate for the Berlin celebration of Einstein's hundredth birthday.

Jordan's enthusiasm for the intersection of the military, politics, and science resurfaced in the late 1950s in response to the Göttingen Declaration, a public manifesto signed by sixteen prominent scientists, including Max Born and Werner Heisenberg, who promised that they would not participate in the manufacture, testing, or use of atomic weapons. These scientists were reacting to the fact that chancellor Konrad Adenauer and his defense minister Franz Josef Strauß were seriously considering acquiring an independent nuclear force (Schirrmacher 2007; Lorenz 2011). Jordan essentially became a one-man public scientific opposition to the Göttingen Eighteen, eventually entering politics by winning election to the West German parliament as a member of Adenauer's Christian Democratic Party and supporting arming West Germany with nuclear weapons—something that never happened.

Conclusion

By this time Pascual Jordan's liaison with National Socialism appeared forgotten. Although one can argue that almost every scientist working during the National Socialist period had to accommodate themselves to the regime, Jordan stands out because of his intensive collaboration with the Third Reich. Jordan was both a pioneer of modern physics and National Socialist propagandist. He triumphed over his "Aryan Physics" opponents because he was more useful to the National Socialist state at war. His biography reveals a symbiotic relationship between some of the most modern developments in science and the reactionary ideology of National Socialism. The 'better' Nazi, Pascual Jordan, is an illuminating example of a scientific career during the Third Reich.

References

Beyerchen, A. (1977). *Scientists under Hitler: Politics and the physics community in the Third Reich*. New Haven: Yale University Press.

Beyler, R. (1994). *From positivism to organism: Pascual Jordan's interpretations of modern physics in cultural context* (Ph.D. dissertation). Harvard University, Cambridge, MA.

Beyler, R. (1996). Targeting the organism. The scientific and cultural context of Pascual Jordan's quantum biology. *ISIS, 87*, 248–273.

Beyler, R. (2003). The demon of technology, mass society, and atomic physics in West Germany, 1945–1957. *History and Technology, 19*(3), 227–239.

Beyler, R. (2011a). Exhuming the three-man paper: Target-theoretical research in the 1930s and 1940s. In P. R. Sloan & B. Fogel (Eds.), *Creating a physical biology: The three-man paper and early molecular biology* (pp. 99–142). Chicago: University of Chicago Press.

Beyler, R. (2011b). Jordan alias Domeier: Science and cultural politics in late Weimar conservatism. In C. Carson, A. Kojevnikov, & H. Trischer (Eds.), *Weimar culture and quantum mechanics:*

Selected papers by Paul Forman and contemporary perspectives on the Forman thesis (pp. 487–503). London: Imperial College Press.

Beyler, R., Eckert, M., & Hoffmann, D. (2012). The Planck medal. In D. Hoffmann & M. Walker (Eds.), *The German physical society in the Third Reich* (pp. 169–186). Cambridge: Cambridge University Press.

Dahn, R. (2019). Big science, Nazified? Pascual Jordan, Adolf Meyer-Abich, and the abortive scientific journal *physis*. *ISIS, 109*, 68–90.

Dingler, H. (1937). Die "Physik des 20. Jahrhunderts". Eine prinzipielle Auseinandersetzung (Zu einem Buche von P. Jordan). *Zeitschrift für die gesamte Naturwissenschaft, 3,* 321–335.

Eckert, M., & Märker, K. (Eds.). (2004). *Arnold Sommerfeld. Wissenschaftlicher Briefwechsel.* Berlin: Diepholz.

Ehlers, J., & Schücking, E. (2002). Aber Jordan war der Erste. *Physik Journal, 11,* 71–72.

Heisenberg, W., & Heisenberg, E. (2011). *"Meine Liebe Li!." Der Briefwechsel 1937–1946.* St. Pölten: Residenz Verlag.

Hildebrandt, K. (1935). Positivismus und Natur. *Zeitschrift für die gesamte Naturwissenschaft, 1,* 1–22.

Hoffmann, D. (1982). Johannes Stark - eine Persönlichkeit im Spannungsfeld von wissenschaftlicher Forschung und faschistischer Ideologie. In K.-F. Wessel (Ed.), *Philosophie und Naturwissenschaft in Vergangenheit und Gegenwart. No. 22: Wissenschaft und Persönlichkeit* (pp. 90–102). Berlin: Sektion marxistisch-leninistische Philosophie der Humboldt-Universität zu Berlin.

Hoffmann, D. (1988). Zur Teilnahme deutscher Physiker an den Kopenhagener Physikerkonferenzen nach 1933 sowie am 2. Kongress für Einheit der Wissenschaft, Kopenhagen 1936. *NTM, 25,* 49–55.

Hoffmann, D. (2003). Pascual Jordan im Dritten Reich – Schlaglichter. *Preprint MPI für Wissenschaftsgeschichte Berlin* 248:35.

Hoffmann, D. (2005). Die Graue Eminenz der DDR-Physik. *Physik Journal, 4*(10), 56–58.

Hoffmann, D. (2010). Physikalische Forschung im Spannungsfeld von Wissenschaft und Politik. In H.-E. Tenorth (Ed.), *Geschichte der Universität Unter den Linden 1810–2010, Vol. 6: Selbstbehauptung einer Vision* (pp. 551–581). Berlin: De Gruyter.

Hoffmann, D. (2012). The Ramsauer Era and self-mobilization of the German physical society. In D. Hoffmann & M. Walker (Eds.), *The German physical society in the Third Reich* (pp. 126–168). Cambridge: Cambridge University Press.

Jordan, P. (1933). Die Wandlung der Universität. *Rostocker Universitäts-Zeitung, 4,* May 9, 1933.

Jordan, P. (1935). *Physikalisches Denken in unserer Zeit.* Hamburg: Hanseatische Verlagsanstalt.

Jordan, P. (1936a). Olympiade der Wissenschaft. *Der Student in Mecklenburg-Lübeck,* December 1936.

Jordan, P. (1936b). *Die Physik des 20. Jahrhunderts. Einführung in die Gedankenwelt der modernen Physik.* Braunschweig: Vieweg.

Jordan, P. (1941a). *Die Physik und das Geheimnis des organischen Lebens.* Braunschweig: Vieweg.

Jordan, P. (1941b). Naturwissenschaft im Umbruch. *Deutschlands Erneuerung, 25,* 452.

Klemperer, V. (1975). *LTI. Lingua Tertii Imperii: Die Sprache des Dritten Reiches.* Leipzig: Reklam.

Kramish, A. (1986). *The Griffin: Paul Rosbaud and the Nazi atomic bomb that never was.* Boston: Houghton Mifflin.

Litten, F. (2000). *Mechanik und Antisemitismus. Wilhelm Müller (1880–1968).* Munich: Institut für Geschichte der Naturwissenschaften.

Lorenz, R. (2011). *Protest der Physiker. Die "Göttinger Erklärung" von 1957.* Bielefeld: Transcript Verlag.

Rammer, G. (2012). "Cleanliness among Our Circle of Colleagues": The German physical society's policy towards its past. In D. Hoffmann & M. Walker (Eds.), *The German physical society in the Third Reich* (pp. 367–421). Cambridge: Cambridge University Press.

Ramsauer, R. (1935). Besprechung von Jordan, Pascual, Physikalisches Denken in der neuen Zeit. *Zeitschrift für die gesamte Naturwissenschaft, 1,* 342–343.

Rosenberg, A. (1930). *Der Mythus des zwanzigsten Jahrhunderts*. Munich: Hoheneichen.
Schirrmacher, A. (2007). Physik und Politik in der frühen Bundesrepublik Deutschland: Max Born, Werner Heisenberg und Pascual Jordan als politische Grenzgänger. *Berichte zur Wissenschaftsgeschichte, 30,* 13–31.
Schücking, E. (1999). Jordan, Pauli, Politics, Brecht, and a variable gravitational constant. In A. Harvey (Ed.), *On Einstein's path* (pp. 1–14). New York: Springer.
Walker, M. (1989). *German National Socialism and the quest for nuclear power, 1939–1949*. Cambridge: Cambridge University Press.
Walker, M. (1990). Legenden um die deutsche Atombombe. *Vierteljahreshefte für Zeitgeschichte, 38,* 45–74.
Walker, M. (1995). *Nazi science: Myth, truth, and the German atom bomb*. New York: Plenum Press.
Walker, M. (2003). "Nazi Science?" Natural science in National Socialism. In U. Hoßfeld, J. John, O. Lemuth, & R. Stutz (Eds.), *"Kämpferische Wissenschaft." Studien zur Universität Jena im Nationalsozialismus* (pp. 995–1012). Cologne: Böhlau.
Walker, M. (2006). Otto Hahn: Responsibility and repression. *Physics in Perspective, 8*(2), 116–163.
Walker, M. (2009). Nuclear weapons and reactor research at the Kaiser Wilhelm Institute for physics. In S. Heim, C. Sachse, & M. Walker (Eds.), *The Kaiser Wilhelm society during National Socialism* (pp. 339–369). Cambridge: Cambridge U. Press.
Wegener, P. P. (1996). *The Peenemünde wind tunnels*. New Haven: Yale University Press.
Wise, M. N. (1994). Pascual Jordan: Quantum mechanics, psychology, National Socialism. In M. Renneberg & M. Walker (Eds.), *Science, technology, and National Socialism* (pp. 224–254). Cambridge: Cambridge University Press.
Wolters, G. (1992). Opportunismus als Naturanlage: Der Fall Hugo Dingler. In P. Janich (Ed.), *Entwicklungen der methodischen Philosophie* (pp. 257–327). Frankfurt am Main: Suhrkamp.
(1938–1939). *Vorlesungsverzeichnis der Universität Rostock*. Rostock: University of Rostock.

Chapter 8
Relativity and Dialectical Materialism: Science, Philosophy and Ideology in Hans-Jürgen Treder's Early Academic Career

Raphael Schlattmann

Introduction

Especially during the Sixties and Seventies of the previous century, Hans-Jürgen Treder was one of the leading East German researchers working in GRG (General Relativity and Gravitation). He was amongst the most prominent physicists within the former GDR and shaped the landscape of East German gravitational research thematically and organizationally over decades. Belonging to the first postwar generation of students of Marxism-Leninism in the SOZ/GDR and being interested in philosophical questions of physics from an early age on, Marxist thought had a decisive influence on his views on science and society. Like his contributions to the history of science or natural-philosophy, his publications in physics show that his work and his understanding of physics were generally characterized by overarching principles and the inclusion of the historical and epistemological origins of the physical problems he treated. The question then arises as to how his research practice and his views on science came about and how they were related to social and cultural developments. Does his early biography offer any insights into these relations, and does it perhaps even carry exemplary features of the first postwar generation of young East German scientists or the connections between science and philosophy/ideology? In his works on the history of science Treder often faced similar questions. In 1983, he published an anthology entitled "Great physicists and their problems" (Treder 1983). The first chapter is dedicated to "Great Physicists" and covers biographical essays from Aristotle to Einstein, the second is devoted to "major problems" of physics. In the foreword he writes:

> For physicists the most accessible presentation of the history of their science is that of a systematic progression of problems, methods, and results in interaction with progressing

---

R. Schlattmann (✉)
Technical University of Berlin, Straße des 17. Juni 135, 10623 Berlin, Germany
e-mail: raphael.schlattmann@tu-berlin.de

© The Editor(s) (if applicable) and The Author(s), under exclusive licence to Springer Nature Switzerland AG 2020
C. Forstner and M. Walker (eds.), *Biographies in the History of Physics*,
https://doi.org/10.1007/978-3-030-48509-2_8

theory formation on the one hand, and increasing technical application on the other. This view [...] equates the epistemology of knowledge production [...] with historical development. [...] This is actually a 'retrospective prophecy'. The ultimate result of an intrinsically contradictory historical process, determined by the social and historical conditions on the one hand and by the very diverse personal abilities and possibilities of the personalities involved, is ideally reconstructed. The teleological conception of the history of physics not only contradicts its daily routine, but has often led and still leads to misleading views of the respective situation of physics in its time by contemporary scientists. [...] Out of the social conditions then arise the fundamental questions and the philosophical speculations. Expectations with which the physicists approach their tasks. It is obvious that fundamental changes in the mindsets of physicists cannot be deduced logically from physics alone. [...] the way in which the advances demanded by social necessity assert themselves in the knowledge of the objective laws of nature is, however, decisively influenced by the thought- and working style of the scholar who introduces this progress. Thus, the scientific biography of great physicists is not only an anecdotal part of the history of physics, but also says something about the content of physics and about the form of how this content presents itself. (Treder 1983, pp. 7–8)

Treder substantiates the ambiguous book title in this passage. In addition to the 'scientific' problems of historical actors, for a history of science that does not want to proceed teleologically but biographically, the 'personal' problems, i.e. the internal and external events and their conditions are also relevant. Relevant seems to mean; to be helpful for an understanding of how the considered knowledge came to be and how this was possible. Although Treder emphasizes the interconnection of content and context to be necessary for an adequate history of science, thus distancing himself from the often criticized 'classical' biography, his interest in the socially and culturally embedded subject seems limited and is dominated by his view of the scientist as the bearer and creator of 'ideas'. However, it is precisely this interest in the subject in all its personal and contextual complexity that has led to a resurrection of the biographical genre in the history of science and has been emphasized in studies of scientists since the 1990s (Klemun 2013). Referring to Helge Kragh, Margit Szöllösi-Janze described the possibility of an "integrative perspective on science" as the greatest advantage of a scientist's biography and, in addition to more general contextualization, paid special attention to very subjective issues such as work- and lifestyle, failure or physicality and illness (Szöllösi-Janze 2000). The following biographical case study of Hans-Jürgen Treder's early academic career aims for such an integrative narrative that illuminates the connections between his personal and cognitive developments, the conditions of Cold War science and the developments of science policy in the SOZ/GDR.

Relativity and Science Policy in the Post-War Period

In the early years after Einstein's formulation of general relativity, the public and scientific interest in the new theory was extraordinarily high. The transformed concepts of space and time changed Newton's classical ideas fundamentally and resulted in a great uncertainty about the physical and epistemological implications of the new theory. Similar to quantum theory in the mid-1920s, physical questions correlated with philosophical and ideological ones and were also discussed in a much wider context than only in the scientific community. After its initial successes it was not until the 1950s that, along with a more systematic study of gravitation, international institutionalization of general relativity research increased (Blum et al. 2015). The social and military relevance of the natural sciences and in particular physics was further emphasized due to the experiences made in World War II and manifested itself in personnel and financial increases of those disciplines. In addition, changes in external conditions correlated with internal community efforts aimed at rallying the small number of active researchers in the field of general relativity. As a result of these efforts and also due to other factors like new mathematical methods and experimental possibilities, through the sixties and seventies a small number of often separated scientists transformed into an established and very active field of research (Blum et al. 2016). However, this early process of international institutionalization as well as the interpretation and further development of general relativity were subject to permanent tensions. These originated in the unresolved epistemic status of the theory and the ideological dynamics of the cold war (Lalli 2017). Marxism-Leninism as it was understood within the Eastern Bloc states always portrayed itself as a scientific world view which advocated a spirit of progress, while having great difficulties integrating the statements of modern physical theories into its ideology. The reinterpretation of fundamental concepts such as matter, space and time or causality was immanent to science but also implied a possible reinterpretation of the social order since its philosophical foundations rested upon those concepts but were developed in the nineteenth century. In the early Soviet Union under Stalin the claim of a worldview developing in union with 'scientific progress' was always subordinated to ideological aims. The resulting scientific-philosophical tensions could bear immense consequences for supporters of allegedly 'idealistic theories'—including relativity and quantum theory—and ranged from closing entire institutions to the murder of individual scientists. In various forms, these tensions and subsequent negotiation processes also occurred in other socialist states such as the SOZ/GDR. However, socialism first had to be established in the eastern zone of Germany.

The upheavals of the post-war years in the wake of denazification and the reconstitution of political structures under Soviet administration are reflected in the transformation of higher education in the SOZ/GDR. The first years of reconstruction were largely characterized by autonomy of both institutional and cognitive aspects of science but increasingly limited from 1947 onward (Burrichter and Malycha 2001, pp. 21–26). The establishment of the Soviet-sponsored *Kulturbund* in mid-summer 1945 exemplified the initial possibility of intellectual pluralism in the spirit of humanistic reeducation, even though this atmosphere changed rapidly (Kapferer 1990, pp. 9–10). Dismantling the 'bourgeois' structures of higher education and research institutions was a key issue for policymakers, but proved to be more difficult and protracted than hoped. On the one hand, this was due to an initially inconsistent science-policy and, on the other, conditioned by the traditional science structures whose degree of organization and self-perception should be overcome. In addition, there were major impairments as a direct result of the war, like destroyed teaching and research facilities, as well as the consequences of National Socialism, such as the deportation and expulsion of Jewish scientists. Combined with the measures taken in the process of denazification and a steady emigration of scientific staff, this led to an immense shortage of personnel, which was also associated with a transfer of research practices and scientific equipment (Ash 2010, pp. 216–217). For these reasons, the political agenda was initially mostly limited to a personnel policy. Despite an existing consensus among the Allies to initiate a major process of re-education in Germany, the reasons and goals in East and West differed fundamentally (Kowalczuk 2003, p. 106). Due to the precarious staffing situation, a policy of funding resulted, which primarily tried to create a new intelligentsia of the working class, while privileging exceptional scientists of the bourgeois milieu at the same time (Malycha 2003, pp. 37–38). The traditional demands of a staff-selection based on scientific performance clashed with a socialist cadre policy which emphasized sociopolitical aspects as being decisive in the transformation process. The attachment of scientists to the SED system was therefore initially advanced with the help of scholarship programs, selection of assistants and ideological training but had to face great problems as well as open rejection (Jessen 1999, pp. 30, 52–53). Starting in the spring of 1947 by order of the SMAD, a program for the sponsorship of young scientists was established to support 200 candidates who had to have a higher education, to have demonstrated their aptitude for scientific and pedagogical activities, and also to be "active anti-fascists" (Befehl Nr. 55 der SMAD 05.03.1947).

Between Physics and Philosophy

Among these candidates was Hans-Jürgen Treder. Born in 1928 in Berlin Charlottenburg, he attended the Mommsen-Gymnasium until 1944 and was drafted in March 1945 to support the air defenses of the Volksturm (PRA Treder, pp. 88–89). At the time Treder lived with his mother and sister in Charlottenburg in the immediate vicinity of the Bonhoeffer family whose son Dietrich was executed shortly before the end of the war. Similarly, his uncle Max Treder, a member of the technical apparatus of the Rote Hilfe, died before the war ended from the consequences of a five-year prison sentence imposed on him by the Nazis (Sandvoß 2000, p. 158). In addition to his own immediate experiences during the war these events seem to have contributed significantly to his political views and his rejection of the National Socialist regime. From July 1945 to January 1946 Treder completed a newly introduced course to obtain his *Abitur* (high school degree) at the Charlottenburger Gymnasium. Through a conversation with Ernst Wildangel, the head of the Main Office of Education of Berlin and a KPD member, Treder became aware of the writings of R. Luxemburg and V. Lenin. During his examination in January 1946, his knowledge about the reign of Queen Victoria attracted the attention of the responsible British education officer, who gifted him Nietzsche's collected works despite Treder's lack of English language skills (Hoffmann 1989, p. 35). His successful diploma was also due to the fact that he already had an extraordinary knowledge of mathematics and natural sciences.

The physicist Robert Rompe, one of the politically most influential scientists of the SOZ/GDR, lifelong supporter and later co-author with Treder, reported on their first meeting in 1946:

> He brought with him an extensive correspondence, which he then led with Heisenberg and Weizsäcker about problems of relativistic physics. It was obvious that he was an unusually capable young person [...]. Striking was his great familiarity with literature beyond mathematics, physics. (Rompe 1989, p. 8)

Still a high-school student, Treder co-founded the Youth Committee Charlottenburg and became a Member of the SED and the FDGB. In 1946 he began to study physics at the Technical University, which was founded the same year (PRA Treder, p. 89) and also started to write his first articles. They were published in the supplement "Technology and Research" of the Berlin evening newspaper *Nacht-Express* and from 1947 onward in the *Einheit—Zeitschrift für Theorie und Praxis des Wissenschaftlichen Sozialismus*, the most prominent theoretical journal of the SED.

The *Nacht-Express* was the first evening paper readmitted in the SOZ after the war and was initially intended to be an impartial news release by the SMAD in response to Western papers (Strunk 1996, pp. 88–89). This strategic appearance of independence was one aspect of the short period of intellectual pluralism described above, which began to change as early as 1947. Thus, Treder's first contributions were characterized by presenting a wide variety of current physical issues in a concise, vivid and largely apolitical way. The contrast to his later works in the *Einheit* is striking. In his first contribution, "Das Schicksal des Universums" ("The Fate of the Universe") (Treder 1946a) the "heat death" of the universe—an issue he continued to address in later works—was discussed without any reference to the great philosophical and political implications surrounding this topic, but in connection with Robert A. Millikan's views on cosmic rays and a reconversion-hypothesis proposed by him in response to the heat death (Millikan and Cameron 1928). It is unclear, but seems unlikely, that Treder knew Millikan's religious motivation for introducing this hypothesis, although this was suggested in the original articles (Kragh 2004, pp. 90–91). The prospect of an end to the cosmos repeatedly forced Marxist as well as religious scientists to politically classify their scientific opinion, which Treder did not do in 1946. The same year he also explained in a neutral manner how to imagine the size of the *finite* and *expanding* universe—two very controversial topics for dialectical materialists at the time—by means of a pencil tip and the distance between the *Großer Stern* and the *State Opera* Unter den Linden in Berlin (Treder 1946b). Three years later the ideological situation had changed dramatically and Treder's rather sarcastic assessment read:

> The author of the theory of the temporal finiteness of the world is the Jesuit priest Abbé Lemaitre. In 1928 he introduced the physically incorrect theory of the expanding universe which implies that there must have been a time when the universe did not exist. This would then prove the necessity of a creation of the world by a God. (Treder 1949a, p. 1028)

This complete change of tone and opinion within a short period of time is exemplary of the volatility and the rapid transformation of the philosophical discourse in the SOZ, but also illuminates the political dimension of highly theoretical topics.

After three semesters of studying physics, Treder changed to the University of Berlin (from 1949 onward: Humboldt University of Berlin) to study philosophy in 1948. By his own account the reason was that the educational possibilities of the Technical University did not agree with his scientific goals (PA Treder, p. 7). At the University of Berlin his philosophical teachers were Walter Hollitscher and Klaus Zweiling. Hollitscher was an Austrian Marxist who had received his doctorate in 1934 from Moritz Schlick with a thesis on the principle of causality in quantum theory and who was since 1949 a visiting professor at the Humboldt University. Zweiling was chief-editor of the *Einheit* at the time and initiated an application of Treder for the above-mentioned scholarship for young scientists to the Deutsche Verwaltung für Volksbildung (German Administration for Education in the SOZ) in 1948. Like the then secretary of the Central Council of the FDJ, Peter Heilmann, Zweiling acted as guarantor in the scholarship application (PA Treder, p. 18). Zweiling had also studied physics and had received his doctorate in 1922 from Max Born in Göttingen. From 1924 onward he was the editor of various social-democratic and socialist newspapers and imprisoned by the Nazis from 1933 to 1936. In 1948 he received his venia legendi in Berlin with his thesis "Materialism and Natural Sciences" and was appointed full professor of dialectical materialism at the Humboldt University in 1955. With the appointment of Zweiling ("Philosophy and Natural Science") as a lecturer of the philosophical seminar led by Liselotte Richter in the summer semester of 1949 and its expansion by the addition of Kurt Hager ("Principles of Philosophical and Historical Materialism") and Walter Hollitscher ("Natural Philosophy") in the winter semester of 1949/50, the presence of Marxist-Leninist philosophy grew (Humboldt-Universität zu Berlin 1949a, b). Until then it had been only represented by the young Wolfgang Harich ("Introduction to Dialectical and Historical Materialism"). Treder thus encountered a cognitive environment that corresponded to his education in mathematics and physics when he changed to philosophy. It was also programmatic for the future development of the discipline at the Humboldt University, since the later founded section of Marxist-Leninist philosophy was primarily devoted to philosophical problems of the sciences. However, the constellation at the University of Berlin was also a distinctive one. Richter, who in 1948 was appointed the first female philosophy professor in Germany, was at times the faculty's only lecturer with a university degree in philosophy, since all other senior professors either were called to the Western Zones and emigrated a few years after the end of the war or died (Rauh 2017, p. 203). This vacuum in East German philosophy was mostly filled by antifascists and exile communists who were for the most part not educated as philosophers and had sometimes only graduated from two-year courses in Marxism-Leninism (Maffeis 2007, p. 53). Kapferer points out that many of the philosophical debates in the SOZ until 1948/49 were therefore primarily conducted by non-philosophers and that a "preponderance of 'physicists' [existed] in the formation of Marxist-Leninist philosophy", including Klaus Zweiling, Gerhard Harig, Georg Klaus, Hermann Ley, Robert Havemann and also Hans-Jürgen Treder (Kapferer 1990, p. 17).

Discussing Modern Physics

Treder addressed and discussed philosophical-epistemological questions of physics raised by Zweiling in various articles such as "Der dialektische Aufbau der Materie" ("The Dialectical Structure of Matter") (Treder 1947), "Dialektik und Kausalität" ("Dialectics and Causality") (Treder 1948a) or "Die Kopenhagener Schule" ("The Copenhagen School") (Treder 1948b). The development of Zweiling's and Treder's contributions to the *Einheit* can be viewed as symptomatic for the direction the development of the philosophical interpretation of modern physics was taking. GDR historiography has described the contributions in the *Einheit* during the years 1946–1949 as a representation of the state of theoretical discussions on the relationship between philosophy and the natural sciences, which seems plausible and concurs with Sachse's findings (Sachse 2006, p. 31), even though conditions and content were hardly critically assessed (Wrona and Richter 1979, p. 96). The basic understanding of science of the KPD/SED in the years 1946–1948 consisted in a more or less conscious division of knowledge into ideologically relevant and irrelevant areas. Since it was officially assumed that policies were generally constructed with regard to dialectical and historical materialism, political action itself was communicated as scientifically justifiable. Thus, the social sciences, which were thought to be closely interlinked with political practice, initially appeared to be more ideologically relevant than the natural sciences (Schulz 2010, pp. 25–26). Therefore philosophical reflection of questions of natural-philosophy were politically less constrained than questions of other societal issues tied to the content of SED policy (Mocek 2001, p. 182). This demonstrated, for example, the possibility of discussing in a less biased way topics of modern physics that were condemned publicly in the Soviet Union at the same time. Unlike in the USSR the results of the special and general theory of relativity as well as quantum theory were not publicly rejected in the GDR. Demands to dismiss the theory of relativity and replace it with a materialistic 'theory of rapid movements', as was the case in the early fifties by the official party philosophy of the USSR, were not heard in the GDR (Sachse 2006; Mocek 2001; Böhme 1984). However, the ever-increasing Sovietization of the SED was also reflected in the articles of the *Einheit*, even though neither specific Soviet, nor former Marxist discussions of modern physics had been taken up. Although doubts were voiced, both constructive and destructive criticism mostly referred to the so-called interpretation or 'idealistic appropriation'. Exemplary for this is an article Treder published in 1949, which, while characterized by the view of "idealism" as the philosophical enemy, described Einstein as the "greatest living physicist" but also stressed that Einstein had been influenced by the "truly idealistic philosophers" David Hume and Ernst Mach (Treder 1949b, p. 267).

Treder's first contributions to the *Einheit* in 1947 also mark his initial attempts to interpret modern physics from the perspective of dialectical materialism, but still have little of their later pungency. Even the rejection of mechanistic materialism and "idealism" is descriptive and historicizing, rather than dominated by expressions of class struggle. Treder discussed modern physics in the form of quantum mechanical complementarity as being an example of dialectical leaps and Heisenberg's uncertainty relation as the "physical formulation of the possibility of the passage from quantity to quality" (Treder 1947, p. 988). The roles of Heisenberg and Jordan in Treder's articles are particularly important for a clarification of the direction the changing discourse was talking. Having declared both of them to be among the great scientists of the twentieth century in 1947, and acknowledging Jordan's efforts to explain the expansion of the universe and highlighting Heisenberg's mathematical genius, this portrayal changed rapidly from 1948 onwards. In May 1948 Treder responded to an article by Alex Rieck who had tried to link the statistical nature of quantum theory to human free will (Rieck 1948), one of the most central arguments of the ostensibly idealistic conception of material indeterminism. Treder noted that this extension neither permitted any meaningful statements, nor was an interpolation of this kind admissible, and remarked that Rieck thereby referred to Pascual Jordan: "the most radical representative of the idealistic physical conception of the so-called Copenhagen School" (Treder 1948a, p. 452). Treder's analysis of quantum theoretical statements was further characterized by a clear rejection of their Copenhagen interpretation which would contain a "kind of fideistic mysticism" (Treder 1948b, p. 1192) and whose development he ascribed to the ideological constraints of bourgeois philosophy. It is important to stress that this rejection was not a permanent one, as other opinions, views and portrayals naturally also changed or transformed over time. Karl Popper, with whom he later corresponded, supposedly called him a "dogmatic Copenhagener", since Treder did not share Poppers critique of the Copenhagen Interpretation, which he in return called dogmatic (Schröder 1998, p. 351). However, at the time Treder, similar to other authors, argued in a direction which would in later debates result in a distinction between the concepts of *law* and *causality*.

Treder argued that according to their idealistic assumptions, it would be overlooked that a measurement would result in a real intervention at the atomic level which would produce a qualitative change in the atomic structure. This change in quality through the experimental intervention would be decisive for the fundamental loss of predictability of the processes, which were in themselves deterministic. An isolated consideration, however, would not be permissible in a quality-altering intervention; the measuring instrument and the object would have to be considered as a unit (Treder 1948a, p. 455). Treder criticized the subjectification of the concept of causality: in the Copenhagen interpretation causality would be understood as the ability to predict a process in its properties. At the nuclear level, however, the measurement process would destroy the predictability. But this would not mean a general failure of causality, if differently understood than via predictability (Treder 1948b, p. 1193). He thus differentiated between determination, predictability and causality, with a specific reference to Stalin's interpretation of the passage of quantitative into qualitative changes.

Initially, impulses for a materialistic view on certain topics often came from Zweiling. Influential was mainly his early contribution "Perspektive der Wissenschaft" ("Perspective of Science") (Zweiling 1946) which was an important reference for Treder but only during the years 1947–1948. This also applied to Zweiling's assessments of general relativity. With references to Zweiling, Treder stated that Einstein's theories contained the "mathematical form of Engel's thesis that space and time are the forms of existence of matter" (Treder 1949b, pp. 266–267) as a necessary consequence of both the special and general principles of relativity. Since Treder repeatedly sought the argument of an implicit materialism in his discussions of general relativistic questions, his contributions in this area are much less characterized by materialistic arguments against an allegedly idealistic interpretation. Like some authors he rather argued that Einstein's general theory of relativity implicitly would be based on the principles of dialectical materialism, but contrary to others he did not claim that Einstein himself consciously represented any particular materialistic philosophy.

According to Treder the four-dimensionality and consequently the principle of special relativity would automatically combine the concept of movement with the concept of mass to that of an "Energiemassefaden" (world line of a material object), so that this generalized concept of motion would be inherent in the concept of matter itself. The principle of general relativity then would associate this concept with the geometrical properties of the 'world'. All properties of matter would depend on their position in space and time but these in turn would depend on the distribution of matter (Treder 1949b, p. 266). According to this argument the Newtonian absolute space, which acts but cannot be acted upon, was abolished. An effect could occur only from a cause which would lie in the past light cone of the relevant event and since the shape of the light cones would be given by the distribution of matter, the principle of causality as an independent principle was abolished as well and became an intrinsic property of matter itself.

The topics of causality and the principle of general relativity were also an essential part of later attempts at a revised interpretation and illustrate how seriously Treder strived for a fusion of worldview and physical interpretation during those years. The above-mentioned 'preponderance of physicists' and the resulting presence of physical topics in philosophical discussions as well as the increasing Stalinization was furthermore demonstrated by the first conference of Soviet and German philosophers in May 1949. On the 40th anniversary of Lenin's "Materialism and Empirio-criticism", the House of Culture of the Soviet Union in Berlin, over five days, hosted lectures about the "application of dialectical materialism of Marx, Engels, Lenin and Stalin in the various fields of natural sciences, the humanities and social sciences" (Neues Deutschland 1949). Here again physical issues were given great attention. The lecture Zweiling held on "dialectical materialism in the light of modern physics" was featured in the May issue of the *Einheit* in a detailed article by Treder entitled "Gipfel der Wissenschaft" ("The Peak of Science"). The title again referred to Lenin's main philosophical work, which Treder called a "mighty weapon" in the struggle between materialism and idealism and subsequently equated this struggle with that between "science and fideism" (Treder 1949a, p. 458). Treder's last article in the *Einheit* appeared at the end of 1949 and presented Carl Friedrich von Weizäcker as a "warmonger" (Treder 1949c, p. 1032), Zweiling's last article appeared in early 1950. He was replaced in March 1950 in his function as the chief-editor and subjected to a party review and so the topic of modern physics gradually disappeared from the *Einheit* during the following years (Sachse 2006, p. 35).

Unintended Changes

Against this background, the entire process of the scientific sponsorship of Treder is instructive in terms of both structure and content. In his application for the scholarship, suggested by Zweiling, Treder added a description of his future scientific plans and a corresponding chapter of a manuscript titled "Die Bedeutung der Ergebnisse der Physik für die allgemeine Naturdialektik" ("The Significance of the Results of Physics for a general Dialectics of Nature") (PA Treder, pp. 19–32). In October 1949 he received the scholarship and was admitted to the program with Walter Hollitscher as his supervisor. Coming from Vienna Hollitscher held a comprehensive cycle of lectures on natural philosophy at the Humboldt University in 1949 in which he attempted to present and expand Engels' idea of a dialectics of nature in a coherent fashion (Laitko 2001, p. 420). Having been admitted to the program under Hollitscher and the end of his activity for the *Einheit* denotes the beginning of a transformation and scientification of Treder's way to philosophize, which presumably can be partly attributed to Hollitscher.

For the admission to the *Aspirantur* (post-graduate studies and doctorate) Treder tried to implement parts of the previously worded plans and wrote a manuscript in which he tried "[...] to represent the epistemological and ontological foundations and basic principles of modern physics on a dialectical-materialistic basis" (PA Treder, p. 42) in 1951. At the time, Treder was deeply involved in the social and scientific structures that were being further expanded by the SED and described himself in a questionnaire as a "functionary of the Socialist Unity Party of Germany and the Free German Youth". As such he had founded a "local peace committee" in West Berlin and had been awarded the peace medal of the GDR (PA Treder, p. 58). With the founding of the GDR in 1949 and in the context of the second reform of higher education in 1951, further and more consistent changes in the sense of ideological and political rapprochement with the Soviet Union and thus Stalinist structures were gradually implemented into the field of education. This was reflected in the introduction of compulsory basic studies of the social-sciences and the Parteilehrjahre, a program for members of the SED, which was introduced in 1950. These courses served as an education of the masses, whereby the Marxist-Leninist training was the clear focus and corresponded to the educational principles of the CPSU (Rauh 2017, p. 74). Within this program, Treder became an instructor of the West Berlin Circle and therefore knew the book "Kurzer Lehrgang der Geschichte der KPdSU (B)" ("Short Course of History of the CPSU (B)"), which represented the theoretical foundation of those courses (PA Treder, pp. 58, 60). Until 1956 this book was a propagandistic compilation used for political education and also contained Stalin's writings "On Dialectical and Historical Materialism" (Zentralkomitee d. KPdSU 1951). In the above-mentioned two manuscripts and in other articles, Treder repeatedly referred to this work and in particular to the so-called "Grundzug a)", which was one of the sub items of dialectical methodology. The manuscripts would remain in their unfinished status and were never published, but are intricately connected to some of Treder's early philosophical assumptions, especially identifying the general principle of relativity with the dialectical concept of cause (PA Treder, p. 47). The manuscript of 1951 marks a turning point between Treder's publications in the *Einheit* and the *Nacht-Express*, which ended in 1949, and his first publications in scientific journals as a doctoral student of physics in 1955. For the intervening six years and apart from two book reviews, it is Treder's only written work and, until 1963, the last one with explicit references to "Marxist-Leninist" sources.

Regarding Treder's cognitive development this episode seems extremely relevant as his application to the *Aspirantur* was rejected by the reviewing commission, which also included Hollitscher. This was also based on the assessment of the submitted manuscript which was rejected by Robert Rompe among others as being "incomprehensible and confused" (PA Treder, p. 60). His competences were generally seen in the field of physics and mathematics. Following this assessment, the State Secretariat for Higher Education wrote to Treder, stating that

> the previous form of your studies and your education did not take the right paths. Based on your professional knowledge and predisposition, we believe that you should not deal with philosophy or dialectics of nature, but with physics. (PA Treder, p. 61)

After his rejection Treder became seriously ill and could only pursue his studies after a prolonged stay in a hospital, which lasted about one and a half years until 1952. Robert Rompe suggested a change to Achilles Papapetrou who at the time worked at the Institute for Pure Mathematics of the German Academy of Sciences in Berlin (DAW) on general relativistic issues. Papapetrou, through the mediation of the former director of the Einstein Institute in Potsdam-Babelsberg Erwin Finlay-Freundlich, came in 1952 from England to the DAW to take up a position in theoretical physics with a focus on general relativity and also to take over the supervision of aspirants. He had started to work on problems of relativity in the mid-thirties while earning his doctorate at the Technical University of Stuttgart and was one of the few researchers worldwide who did so during and shortly after World War II. His generally anti-fascist and also Marxist attitude was also expressed in his attempts to interpret modern physics on the basis of dialectical materialism. Along with other factors, this led to his dismissal and stigmatization in post-war Greece (Hoffmann 2017). During his employment in Berlin Papapetrou dedicated himself exclusively to mathematical and physical issues and avoided philosophical-political discussions, although the theory of relativity was strongly attacked particularly in the early fifties by the Marxist-Leninist orthodoxy. This apolitical attitude, combined with a high level of mathematical sophistication, decisively influenced Treder's work in the years leading up to Papapetrou's departure to Paris in 1961.

The main topics of the newly established working group were the so-called *Bewegungsproblem* as well as time-dependent and in particular non-singular solutions of the general field equations. Among other things, these solutions were relevant because Einstein's program of a unified field theory necessitated the idea that elementary particles had to be representable as singularity-free solutions of the field equations. Even before his time with Papapetrou, Treder had dealt with approaches to unified field theories and continued to do so in his doctoral thesis. Contrary to Einstein's approach of an asymmetric generalization of the metric tensor, Treder shifted the coupling of the fields to a general affine connection whereby the connection reduced in a special case to Weyl's geometry, who had formulated his field theory over 30 years earlier (Treder 1956). However, due to his illness and change of disciplines there were some administrative complications that ultimately resulted in the fact that Treder, who was approved by many as being "extraordinary" or "undoubtedly highly talented" (PRA Treder, pp. 99, 113), was admitted to the doctorate without a diploma. His thesis was rated as "good" by all reviewers, although in particular his former teacher Friedrich Möglich had expressed reservations. As Papapetrou emphasized, the reason for this was not the quality of the work, but "because it shares the fate of all the works of famous people over the past 30 years, 'that no new physical effects are predicted'" (PRA Treder, p. 116). With the completion of his dissertation, Treder was employed by the Institute for Pure Mathematics at the DAW and became a scientific supervisor in 1961. In the years following his graduation, Treder developed a very high level of productivity, which he was to maintain until the end of his life. In the five years leading up to his habilitation in 1961, he published about twenty articles and gained international recognition, often publishing together with Papapetrou. Until shortly after the construction of the Berlin Wall Treder still lived in West Berlin but resettled to Berlin-Karlshorst in the course of his habilitation, for which he worked on an extensive study on gravitational shockwaves (Treder 1962). After completing his habilitation at the Humboldt University during the fall of 1961, Papapetrou left after 10 years in the GDR for Paris. This opened up a space for the young Treder, promising a successful academic career and giving him the opportunity to establish himself as the leading relativist of the GDR. His scientific excellence, and the fact that many other scientists and young talents had left the country before the Berlin Wall was completed, supported his fast rise. In addition to politically influential scientists such as Robert Rompe and Gustav Hertz, who had sponsored him early on, he also had the support of the official science administration, whose cadre political profile he fitted very well. In 1962 he was appointed lecturer at the Humboldt University and a professor in the following year. His appointment was based on evaluations by leading relativists like L. Infeld (Warsaw), M. A. Tonnelat (Paris) and A. Petrov (Moscow), who strongly supported the appointment and characterized him in Infeld's case as the "best relativist in the GDR" (PA Treder, pp. 99–103). In 1963 he furthermore joined the directorate of the Institute for Pure Mathematics which in 1966 culminated in his admission into the DAW and marked his rise to the center of the scientific stage of the GDR.

Conclusion

Hans-Jürgen Treder's early years in science carry many features of a unique career and individual thought, but can be equally instructive for the creation of a narrative in another case study of the first post-war generation of GDR science. The rapid formation of Marxist-Leninist philosophy after the war was at first in a volatile, yet constantly planned institutional, cognitive and personnel development and for a short period Treder actively participated in its intellectual formation. His support by individual scientists such as Rompe, Zweiling and Hollitscher, but also his presence in publications at a very young age, cannot be understood without taking his personal capabilities into account. But they are furthermore related to the precarious staffing situation during the first years of the SOZ and his early political involvement. From a cadre political perspective he was initially an example of the ideal socialist junior scientist. Consciously biased, anti-fascist, committed and capable; merging the politically intended world view with modern science. This template falls somehow short on a biographical level, especially since alongside the described conformism Treder's early writings also contained intra-materialist controversies and conceptual elasticity. The increasing tendencies towards Sovietization and respectively Stalinization of the sciences in the GDR can nonetheless be structurally traced throughout his years of education. Margit Szöllösi-Janze has described the knowledge of a "generation-specific norm" of scientist biographies as decisive in determining the room for maneuver or the individual scientific capacity, innovation and creativity of the respective actors. Moreover, the societal and cultural characteristics of the period under consideration would only be recognizable against the background of a "biographical normal form" (Szöllösi-Janze 2000, p. 25). Since normalization of biographical aspects means a de-individualization, such a comparison can be regarded as an invitation for different forms of confirmation bias. However, in terms of the milieu considered, comparison of structural aspects of normalized and actual biographies can still sharpen both perspectives. In this respect Treder's case study substantiates and specifies the characterizations of the early GDR philosophy also identified by other authors. Firstly, this is the question of periodization. In spite of the diverging research questions the first post war years were generally described as a brief phase of intellectual pluralism transitioning into a beginning Stalinization (Böhme 1984; Kapferer 1990; Mocek 2001; Sachse 2006; Maffeis 2007). This phase outlines the reconstruction of scientific structures and the formation of Marxist-Leninist philosophy during the years 1945–1951 and was characterized by an increased rejection of the so-called "bourgeois" or "decadent" science. The philosophical enemy of these years naturally was "idealism" used as an often undifferentiated, collective phrase for all "subjectivist" philosophies. Secondly, this involved the way the central concept of matter and consequent concepts were understood. They were almost exclusively determined by the writings of the omnipresent authorities Engels, Lenin and Stalin and allowed little room for interpretation, even though it had been debated internationally for some years (Böhme 1984, p. 163). Both aspects, the permanent recourse to the 'classics' and an increasing politicization of philosophical discourses, are also characteristic for the development of Treder's writings during his years of higher education, thereby reflecting a cognitive key aspect of a "biographical normal form". But already Treder's involuntary change back to physics means deviation from this typification. This coincided with the second reform of higher education in 1951/52, which extended the centralization of the GDR's higher education system and correlated, as shown above, with further changes in the philosophical landscape in which Treder then no longer played any part. The shift marked a biographical and intellectual disruption that becomes apparent in Treder's abstinence from writing about philosophical topics up to the year 1963 and was preceded by serious health issues. Similar to Marxist-Leninist philosophy, East-German gravity research was at the time of Treder's debut in a process of institutional and personnel formation but distinguished itself—contrary to Marxist-Leninist philosophy—through a long tradition of scientific practice. This was expressed in particular in its methodological approaches and its international orientation but also in its essentially unpolitical attitude, which was also due to Papapetrou's influence during his stay at the DAW. Publication on topics of natural-philosophy only resumed after Papapetrou had left Berlin and Treder was fully integrated into the science system of the GDR. Treder's early scientific biography is therefore cognitively and structurally intricately entangled with the general development of Marxist-Leninist philosophy, as well as the early years and commencing institutionalization of general relativistic research in the SOZ/GDR.

References

Ash, M. G. (2010). Konstruierte Kontinuitäten und divergierende Neuanfänge nach 1945. In M. Grüttner, R. Hachtmann, K. H. Jarausch, J. John, & M. Middell (Eds.), *Gebrochene Wissenschaftskulturen: Universität und Politik im 20. Jahrhundert* (1st ed., pp. 215–246). Göttingen: Vandenhoeck & Ruprecht.
Befehl Nr. 55 der SMAD über "die Verbesserung der materiellen Lage der Mitglieder der Deutschen Akademie der Wissenschaften in Berlin und über die Ausbildung des wissenschaftlichen Nachwuches" vom 5. März 1947, Bundesarchiv, DR 2/871. Dokument 28. In A. Malycha (Ed.), *Geplante Wissenschaft. Eine Quellenedition zur DDR-Wissenschaftsgeschichte 1945–1961*, 163. Beiträge zur DDR-Wissenschaftsgeschichte Reihe A, Dokumente (Vol. 1). Leipzig: Akademische Verlagsanstalt.
Blum, A., Lalli, R., & Renn, J. (2015). The reinvention of general relativity: A historiographical framework for assessing one hundred years of curved space-time. *ISIS, 106*(3), 598–620. https://doi.org/10.1086/683425.
Blum, A., Lalli, R., & Renn, J. (2016). The renaissance of general relativity: How and why it happened. *Annalen der Physik, 528*(5), 344–349. https://doi.org/10.1002/andp.201600105.
Böhme, G. (1984). Physik im weltanschaulichen Spannungsfeld: Eine Analyse der Diskussion "Über philosophische Fragen der modernen Physik" in der DDR 1952–1957. In C. Burrichter (Ed.), *Ein kurzer Frühling der Philosophie: DDR-Philosophie in der "Aufbauphase"* (pp. 157–187). Paderborn: Schöningh.
Burrichter, C., & Malycha, A. (2001). "Produktivkraft Wissenschaft": Das Verhältnis zwischen Wissenschaft und Politik in der SBZ/DDR. *hochschule ost leipziger beiträge zu hochschule & wissenschaft, 10*(2), 7–26.
Hoffmann, U. (1989). Hans Jürgen Treder. *Quadriga: Berliner Journal, 1*, 34–38. (Berlin-Information, Berlin).
Hoffmann, D. (2017). In den Fußstapfen von Einstein: Der Physiker Achilles Papapetrou in Ost-Berlin. In M. Hillemann & M. Pechlivanos (Eds.), *Deutsch-griechische Beziehungen im ostdeutschen Staatssozialismus (1949–1989): Politische Migration, Realpolitik und interkulturelle Begegnung* (pp. 179–204). Edition Romiosini: Epubli.
Humboldt-Universität zu Berlin. (1949a). Sommersemester 1949. Personal- und Vorlesungsverzeichnis. Berlin. Accessed April 17, 2018. http://edoc.hu-berlin.de/18452/1577.

Humboldt-Universität zu Berlin. (1949b). Wintersemsester 1949/50. Personal- und Vorlesungsverzeichnis. Berlin. Accessed April 17, 2018. http://edoc.hu-berlin.de/18452/1578.

Jessen, R. (1999). *Akademische Elite und kommunistische Diktatur. Die ostdeutsche Hochschullehrerschaft in der Ulbricht-Ära* (1st ed.). Kritische Studien zur Geschichtswissenschaft (Vol. 135). Göttingen: Vandenhoeck & Ruprecht.

Kapferer, N. (1990). *Das Feindbild der marxistisch-leninistischen Philosophie in der DDR 1945–1988*. Darmstadt: Wissenschaftliche Buchgesellschaft.

Klemun, M. (2013). 'Living Fossil'—'Fossilized Life'? Reflections on biography in the history of science. *Earth Sciences History, 32*(1), 121–131. https://doi.org/10.17704/eshi.32.1.0446820220487244.

Kowalczuk, I.-S. (2003). *Geist im Dienste der Macht. Hochschulpolitik in der SBZ/DDR 1945 bis 1961* (1st ed.). Forschungen zur DDR-Gesellschaft. Berlin: Ch. Links.

Kragh, H. (2004). *Matter and spirit in the universe. Scientific and religious preludes to modern cosmology*. History of Modern Physical Sciences (Vol. 3). Singapore, Hackensack, NJ, London: Distributed by World Scientific Pub; Imperial College Press.

Laitko, H. (2001). Walter Hollitscher und seine Naturdialektik-Vorlesung in Berlin 1949/50. In V. Gerhardt & H.-C. Rauh (Eds.), *Anfänge der DDR-Philosophie: Ansprüche, Ohnmacht, Scheitern* (1st ed., pp. 420–455). Forschungen zur DDR-Gesellschaft. Berlin: Ch. Links.

Lalli, R. (2017). *Building the general relativity and gravitation community during the cold war. Springer Briefs in History of Science and Technology*. Cham: Springer.

Maffeis, S. (2007). *Zwischen Wissenschaft und Politik. Transformationen der DDR-Philosophie 1945–1993*. Campus-Forschung (Vol. 922). Frankfurt am Main: Campus-Verlag.

Malycha, A. (Ed.). (2003). *Geplante Wissenschaft. Eine Quellenedition zur DDR-Wissenschaftsgeschichte 1945–1961*. Beiträge zur DDR-Wissenschaftsgeschichte Reihe A, Dokumente (Vol. 1). Leipzig: Akademische Verlagsanstalt.

Millikan, R. A., & Cameron, G. H. (1928). The origin of the cosmic rays. *Physical Review, 32*(4), 533–557.

Mocek, R. (2001). Marxistische Naturphilosophie in der Diskussion. In V. Gerhardt & H.-C. Rauh (Eds.), *Anfänge der DDR-Philosophie: Ansprüche, Ohnmacht, Scheitern* (1st ed., pp. 180–193). Forschungen zur DDR-Gesellschaft. Berlin: Ch. Links.

Neues Deutschland. (1949). Materialismus oder Idealismus: Diskussion sowjetischer und deutscher Wissenschaftler im Haus der Kultur der Sowjetunion, May 24, 1949. http://zefys.staatsbibliothek-berlin.de/ddr-presse/ergebnisanzeige/?purl=SNP2532889X-19490524-0-4-0-0. Accessed April 17, 2018.

PA Treder: Universitätsarchiv der Humboldt-Universität zu Berlin. HUB, UA, Personalakte Hans-Jürgen Treder.

PRA Treder: Universitätsarchiv der Humboldt-Universität zu Berlin. HUB, UA, Promotionsakte Hans-Jürgen Treder, Promotionen 5.12.1956.

Rauh, H.-C. (2017). *Philosophie aus einer abgeschlossenen Welt. Zur Geschichte der DDR-Philosophie und ihrer Institutionen* (1st ed.). Forschungen zur DDR-Gesellschaft. Berlin: Ch. Links.

Rieck, A. (1948). Kausalität - Notwendigkeit - Freiheit. *Einheit, 3*(5), 445–451.

Rompe, R. (1989). Laudatio. (Hans-Jürgen Treder zum 60. Geburtstag.). In H.-H. Emons (Ed.), *Hans-Jürgen Treder zum 60. Geburtstag: Vorträge auf dem Kolloquium des Forschungsbereiches Geo- und Kosmoswissenschaften, der Klasse Physik und der Klasse Geo- und Kosmoswissenschaften der AdW der DDR am 8. Sept. 1988* (pp. 8–14). Sitzungsberichte der Akademie der Wissenschaften der DDR N, Mathematik, Naturwissenschaften, Technik (Vol. 10). Berlin: Akademie-Verlag.

Sachse, C. (2006). *Die politische Sprengkraft der Physik. Robert Havemann im Dreieck zwischen Naturwissenschaft, Philosophie und Sozialismus (1956–1962)*. Diktatur und Widerstand (Vol. 11). Berlin: LIT.

Sandvoß, H.-R. (2000). *Widerstand in Prenzlauer Berg und Weißensee*. Schriftenreihe über den Widerstand in Berlin von 1933 bis 1945 (Vol. 12). Berlin: Gedenkstätte Deutscher Widerstand.

Schröder, W. (Ed.). (1998). *From Newton to Einstein. A Festschrift in honour of the 70th birthday of Hans-Jürgen Treder*. Mitteilungen des Arbeitskreises Geschichte der Geophysik (Vol. 17). Bremen-Roennebeck: Science Edition.

Schulz, T. (2010). *"Sozialistische Wissenschaft". Die Berliner Humboldt-Universität (1960–1975)*. Zeithistorische Studien (Vol. 47). Köln: Böhlau.

Strunk, P. (1996). *Zensur und Zensoren. Medienkontrolle und Propagandapolitik unter sowjetischer Besatzungsherrschaft in Deutschland* (1st ed.). Edition Bildung und Wissenschaft (Vol. 2). Berlin: De Gruyter.

Szöllösi-Janze, M. (2000). Lebens-Geschichte - Wissenschafts-Geschichte. Vom Nutzen der Biographie für Geschichtswissenschaft und Wissenschaftsgeschichte. *Berichte zur Wissenschaftsgeschichte, 23*(1), 17–35. https://doi.org/10.1002/bewi.20000230104.

Treder, H.-J. (1946a). Das Schicksal des Universums. *Nacht-Express: die illustrierte Berliner Abendzeitung,* April 9, 1946.

Treder, H.-J. (1946b). Rätsel der Einstein-Welt. *Nacht-Express: die illustrierte Berliner Abendzeitung,* August 13, 1946.

Treder, H.-J. (1947). Der dialektische Aufbau der Materie. *Einheit, 2*(10), 987–989.

Treder, H.-J. (1948a). Dialektik und Kausalität. *Einheit, 3*(5), 452–456.

Treder, H.-J. (1948b). Die Kopenhagener Schule. *Einheit, 3*(12), 1187–1193.

Treder, H.-J. (1949a). Gipfel der Wissenschaft: Zum 40. Jahrestag des Erscheinens von Lenins "Materialismus und Empiriokritizismus". *Einheit, 4*(5), 451–458.

Treder, H.-J. (1949b). Materialismus und Relativitätstheorie: Zum 70. Geburtstag Albert Einsteins. *Einheit, 4*(3), 265–268.

Treder, H.-J. (1949c). Mißbrauch der Wissenschaft: C. F. v. Weizsäcker im Dienste amerikanischer Kriegshetze. *Einheit, 4*(11), 1027–1032.

Treder, H.-J. (1956). *Eine einheitliche affin-geometrische Darstellung des allgemeinen Feldes mit Hilfe symmetrischer Affinitäten*. Dissertation, Humboldt-Universität zu Berlin.

Treder, H.-J. (1962). *Gravitative Stoßwellen. Nichtanalytische Wellenlösungen der Einsteinschen Gravitationsgleichungen*. Berlin, Humboldt-Univ., Habil., 1961. Schriftenreihe der Institute für Mathematik bei der Deutschen Akademie der Wissenschaften zu Berlin, Reihe A: Reine Mathematik (Vol. 11). Berlin: Akademie-Verlag.

Treder, H.-J. (1983). *Große Physiker und ihre Probleme. Studien zur Geschichte der Physik*. Berlin: Akademie-Verlag.

Wrona, V., & Richter, F. (Eds.). (1979). *Zur Geschichte der marxistisch-leninistischen Philosophie in der DDR. Von 1945 bis Anfang der sechziger Jahre*. Zur Geschichte der marxistisch-leninistischen Philosophie in Deutschland (Vol. 3). Berlin: Dietz.

Zentralkomitee d. KPdSU (B). (1951). *Geschichte der Kommunistischen Partei der Sowjetunion. (Bolschewiki). Kurzer Lehrgang* (8th ed.). Bücherei des Marxismus-Leninismus (Vol. 12). Berlin: Dietz.

Zweiling, K. (1946). Perspektive der Wissenschaft. *Einheit, 1*(5), 272–287.

Chapter 9
Biography and Autobiography in the Making of a Genius: Richard P. Feynman

Christian Forstner

Introduction

In 1965 the Nobel Foundation honored Sin-Itiro Tomonaga, Julian Schwinger, and Richard Feynman for their fundamental work in quantum electrodynamics and the consequences for the physics of elementary particles. However, only Richard Feynman was perceived as a genius by the public. In his autobiographies he managed to connect his behavior, which contradicted several social and scientific norms, with the American myth of the 'practical man'. This connection led to the image of a common American with extraordinary scientific abilities and helped enhance the public image of Feynman as genius. Would a biographer accept this image, which has resulted from Feynman's autobiographies? This question is the starting point for a deeper historical analysis that tries to put Feynman and his actions back into historical context.

Richard Feynman was "half genius and half buffoon", his colleague Freeman Dyson wrote in a letter to his parents in 1947 shortly after having met Feynman for the first time (Feynman 1999 Foreword by Freeman Dyson). It was precisely this combination of outstanding and talented scientist with clown that has allowed Feynman to appear as a genius amongst the American public. There is a discrepancy between Feynman's image as a genius, which was created to a significant degree through Feynman's autobiographical writings, and the historical perspective on his earlier career as a young aspiring physicist. Because a critical discussion of this discrepancy is necessary, it is essential to historicize and contextualize Feynman's contribution to his image as genius through his autobiographical writings, interviews, and anecdotes. A biographical analysis will analyze how closely Feynman's character as a scientist, his theories, and their genesis were interwoven with the thought style of

C. Forstner (✉)
Friedrich-Schiller-Universität Jena, AG Wissenschaftsgeschichte-Ernst-Haeckel-Haus,
Kahlaische Str. 1, Jena 07745, Germany
e-mail: Christian.Forstner@uni-jena.de

his social community and the American physics community, as well as how Feynman himself contributed to his image as a genius in his autobiographical writings. The Feynman papers at the California Institute of Technology as well as the series of oral history interviews by the American Institute of Physics serve as source material.

Feynman and the Community of American Physicists

In June 1947 an exclusive circle of just under 30 leading physicists from the United States of America met at Shelter Island, a small island outside of New York, including John von Neumann, J. Robert Oppenheimer, Isidor I. Rabi und John A. Wheeler. In a casual atmosphere, they discussed the newest discoveries in atomic physics and quantum electrodynamics. Both of these sub domains of physics were implicitly determined as the two main fields of research for postwar physics. The Shelter Island Conference and the following conferences in Pocono and Oldstone were the most significant conferences for the development of physics after the Second World War (Schweber 1986a) (Fig. 9.1).

Fig. 9.1 Photograph of the participants of the Shelter Island Conference (Fermilab Photograph, courtesy AIP Emilio Segrè Visual Archives, Marshak Collection)

Richard Feynman, who in contrast to later meetings did not play a central role at this conference, stands hidden at the edge of the participant photo. According to the terminology of the sociologist Ludwik Fleck, in this image we see a thought collective, including Feynman.

In his 1935 essay *The Genesis and Development of a Scientific Fact*, Fleck developed his epistemic and sociological conclusions (Fleck 1983). Fleck understood science as a social enterprise of a thought collective, a community, that communicates, interacts, and shares the same thought style, which is based on the historical and cultural context of the group and offers the theoretical conditions upon which the group builds their knowledge.

The participant photo of the conference illustrates such a thought collective, more precisely it shows the esoteric circle of the thought collective of US physicists. This esoteric circle establishes an elite within the thought collective that decisively shapes the development of the field. The participants at the Shelter Island Conference did this during the conference. They discussed the two main fields of research for postwar physics of the USA: quantum electrodynamics and particle physics. These discussions shaped the two central research topics, the research questions, and the methods, or using Fleck's words: the thought style of the collective.

Feynman, who received the Nobel Prize in physics in 1965, apparently developed his own concept, not as part of this collective, but instead as a genius disassociated from all social ties. This image of Feynman was significantly shaped by his autobiographical narratives and anecdotes that appeared after Feynman was awarded the Nobel Prize. Feynman appears not to have belonged to a thought collective or community, rather to have been a genius free of any context. Therefore, this historical study will critically analyze Feynman's image as a genius and put the genius Feynman back into his historical community and his social context.

The Style of US Physics: Practical Men und Cultivated Persons

In the nineteenth century the spirit of the 'practical man', who used natural science to better acquire natural resources, predominated at public colleges. At private colleges, the liberal arts determined the canon of the student's education. Natural sciences were frequently delegated or separated to technical schools, such as at Harvard the Lawrence School of Science or at Yale the Sheffield School. At the beginning of the 1870s, a turning point came in public awareness. Someone who could participate in a discourse about natural science was now regarded as a 'cultivated person'. As a result, many new private universities such as Cornell University and Johns Hopkins University were founded, which followed a clear research imperative. The private investors who had acquired their assets in business and industry were regarded as 'practical men' and through the funding of pure natural sciences, enhanced their public status. By means of funding abstract natural sciences, they gained the status of a 'cultivated person' (Kevles 1995, 3–24).

Next to the development of a clear research imperative, a tradition of precision measurement had emerged in the USA in the nineteenth century that went above and beyond a simple gathering of facts (Bourguet et al. 2002). One example is the failed attempt to verify the ether by Albert A. Michelson and Edward W. Morley. In contrast, theoretical physics existed only rudimentarily and was represented by individuals without exhibiting a larger structural context. In the course of the reception of the Bohr-Sommerfeld quantum theory, the necessity for the development of theoretical physics was recognized. Leading European ambassadors of the theory were invited to lecture tours through the US among them Stephan Boltzman, Max Planck, Arnold Sommerfeld, and Hendrik A. Lorentz, and postgraduates and post doctoral students were delegated to research centers in Europe and then returned to establish new schools of theoretical physics in the USA. The European research style was not transferred 1:1 (Ash 2006). In fact, theory was integrated into a new thought collective and adjusted to the thought style of US physicists. When theoretical physics in the USA emerged out of the shadowy existence of experimental physics at the end of the 1920s, a close connection was maintained between both branches. As a rule, every physicist had to pass through an experimental basic education and research courses were carried out collectively. It was the duty of the theorist, not to engage in any philosophical speculation, rather exclusively to carry out calculations which would be verified experimentally. Because of this pragmatic approach, US physicists reflected on the philosophical aspects of their work much less frequently than their colleagues in Germany did (Schweber 1986b; Kevles 1995, 4–90; Forman 1989, 99–100). An example of this can be seen in a 1938 quotation from the theoretical physicist John C. Slater, in which he elevates the development of experimental prognoses to the main task of a theorist:

> A theoretical physicist in these days asks just one thing of his theories: if he uses them to calculate the outcome of an experiment, the theoretical prediction must agree within limits, with the results of the experiment. He does not ordinarily argue about philosophical implications of this theory. Almost his only recent contribution to philosophy is the operational idea, which is essentially only a different way of phrasing the statement I have just made, that the one and only thing to do with a theory is to predict the outcome of an experiment [...] Questions about a theory which do not affect its ability to predict experimental results correctly seem to me quibbles about words, rather than anything substantial, and I am quite content to leave such questions to those who derive some satisfaction from them. (Slater 1938, 277)

In the postwar era, this pragmatic research style, which was identified by characters such as Slater who joined with the younger theoretical physicists, worked very closely together with experimental physicists, and knew their needs exactly, changed. In particular, in particle physics and military research the work of experimenters and theorists merged together so closely that the old style that Slater's generation had labeled pragmatic was developed further into a 'pragmatic, utilitarian, instrumental style'.

Feynman and the Thought Collective of US Physics

Scientific education aims not only to impart knowledge and work techniques to students, but also to mediate a certain method of thought and research, a thought style, to potential young academics. This initiation process is accompanied by pressure on the individual to adapt in the form of assignments, tests, etc., until, after successful knowledge acquisition and adaption, the individual is accepted into the scientific community through a variety of initiation rituals (Fleck 1983).

The young Feynman appeared to have no problems with the integration mechanisms into the collective. On the contrary: he passed the adaption measures so quickly, that he still had time and space left over for his independent studies. Feynman's adaption of the thought style can already be seen during his school days. Outside of regular class, he dedicated considerable time to his independent studies and quickly took a pragmatic stance, which showed that Feynman did not care much about the approach, as long as the outcome was correct. (Feynman 1966a, 16–53; Gleick 1993, 55–60; Schweber 1994, 374; Mehra 1994, 22–43) Or according to Feynman's words:

> I wouldn't give a damn – I know I didn't care – to find out what way you had to do it, because it seemed to me, if I did it, I did it. (Feynman 1966a, 27)

Within the scope of his independent studies, Feynman did not remove himself from the thought style of the US American physical community (Forstner 2007, 59–76; Schweber 1986b, 1989), rather assimilated their way of thinking outside of the usual mechanisms. This also becomes clear during his undergraduate studies at Massachusetts Institute of Technology (MIT), where he enrolled in mathematics, then following the American myth of the practical man switched to engineering sciences, before he found his ultimate major in physics. This myth also showed up in Feynman's physical works, when he rhetorically placed experiment above theory. Feynman was introduced to theoretical physics through classic American textbooks. The textbook by John C. Slater and Nathaniel H. Frank (Slater and Frank 1933), which the courses at MIT were based on, had the professed goal to educate the students as productive physicists. In the American understanding of a theoretical physicist, productive meant that measurable prediction fits the experimental data without engaging in philosophical speculation. The numerous assignments at MIT did not present Feynman with any problems. He executed them, adopted the thought style-appropriate knowledge, and was integrated into the collective in this way. Together with a fellow student, he took up independent studies. They set aside the abstract textbook by English physicist Paul A. M. Dirac and instead chose the textbook of Americans Linus Pauling and E. Bright Wilson (Pauling and Wilson 1935), which was oriented towards the practical use of quantum mechanics.

Independently, they derived the Klein-Gordon equation, but after applying it to the hydrogen atom, had to conclude that the practical test of the equation had failed. Feynman later called this experience a crucial moment, in which he recognized that the beauty of an equation was not what was decisive about physics, rather the test of the predicted results against the experiment. Once again, we can detect Feynman's reference to the thought style of American physics. In his paper "Forces in Molecules"

(Feynman 1939) Feynman directly calculated the working forces instead of making the standard observation about energy minima. He thereby concluded that, according to the quantum mechanical calculation of charge distribution, only the classical electrostatic forces appeared in the equations. In a direct way, Feynman had found a method to link time saving and higher accuracy of the results (Forstner 2007, 155–59). As a graduate student at Princeton, Feynman, continued to study the fundamental problems of quantum electrodynamics. At that time, he worked closely together with John A. Wheeler, his supervisor and later friend (Halpern 2017). Feynman was enthusiastic about finally having arrived at the frontline of contemporary research, meaning he was thrilled by the questions specified by the thought collective and thought style. This is a further indication of Feynman's integration into the thought style. Feynman's pragmatic position, that the way to the goal did not matter, so long as the experimental values were consistent, appeared at numerous opportunities at Princeton as well. Together with Wheeler, Feynman researched the scattering matrix theory and pair production. Independently, Feynman once again started research ideas on the infinite self-energy of the electron, which he had drawn from one of Dirac's books in his undergraduate studies at MIT. First of all, he tried to get a handle on the problem through the elimination of the field concept in classical theory. For this purpose, Feynman and Wheeler regarded the entire space-time path chosen by a particle. The emphasis is on 'chosen' because, according to the principle of least action that both physicists adopted, a particle teleologically 'chooses' the path for which the least amount of energy is needed. The classical theory appeared to function well, but Wheeler and Feynman extended the theory to quantum electrodynamics (Forstner 2007, 159–68).

Already incorporated into wartime research, Feynman summarized the results of these works in his dissertation and arrived at the first application of the principle of least action in quantum mechanics. Feynman described the path to this point more than 20 years later as if it was a dream. During the day, he had worked with the German emigrant Herbert Jehle on an article of Dirac, and during the night, Feynman then lay in bed and suddenly saw the solution appear in front of him:

> I was lying in bed thinking about this thing, and thought, "What would happen if I wanted to get the wave function at one time, and at finite interval later suppose that the interval was divided into a large number of small steps?... I could represent the coordinates that I was integrating over a succession of positions through which the particle was supposed to go, and then this quantity, this sum, would be like an integral the integral of L, which is in fact the action... I saw the action expression, suddenly, so to speak... In the air, in the head. Yeah. You see the action coming on. And I said, "My God, that's the action! Wow!" I was very excited. So I had filed a new formulation of quantum mechanics in terms of action, directly. I got up and wrote everything out, and checked back and forth, and made sure it was all right, and so on. (Feynman 1966b, 166)

Feynman's story of dreaming about the paths is also in his Nobel Prize speech. This story has both the function of emphasizing his intuition, and at the same time is a break with tradition (Feynman 1966b, 155, 1972, 166–67; Mehra 1994, 137). This narration is strongly reminiscent of Friedrich A. Kekulés' famous daydream about the discovery of the benzene ring. More than 20 years after the discovery, he made a similar report (Rocke 1985):

> I turned the chair toward the fireplace and dozed off. Once again, the atoms were mocking me in front of my eyes. This time, smaller groups lingered modestly in the background. My mind's eye, sharpened through repeated faces of similar types, was now able to distinguish larger structures from manifold configuration. Long rows, multiple times more densely joined together, everything in movement, curling and turning in a snake-like manner. And look, what was that? One of the snakes grasped its own tail and tauntingly whirled the structure before my eyes. As if being awakened by a lightning bolt I woke up; once again, I spent the rest of the night working out the consequences of the hypothesis. (Schultz 1890, 1306; quoted from Kekulé's talk)

In contrast to his dream of the path integral formulation, which broke with the research tradition of the community, Feynman purposefully tied in the prior works of the community into his dissertation and emphasized the usefulness of his research for the current quantum electrodynamics and particle physics research programs. Similarly, adhering to the accepted research style, he pointed out that the experimental testability of the theories was significant criteria for their accuracy:

> The final test of any physical theory lies, of course, in experiment. No comparison has been made in this paper. The author hopes to apply these methods to quantum electrodynamics. It is only out of some such direct application that an experimental comparison can be made. (Feynman 1942)

During the Second World War, Feynman worked in Los Alamos as the leader of a theoretical research group in the theoretical division under Hans Bethe on the central project of the construction of the atomic bomb. The war research in Los Alamos was characterized by the specific work style of the scientists. Instead of working on single fundamental theoretical problems, they worked in an application-oriented manner, taking small steps towards a solution to the problem. Moreover, the theorists worked closely together with experimental physicists and engineers. In place of exact solutions, general estimations of the results within clearly defined tolerances sufficed, and well-defined work instructions made these results usable for further groups of people. This work style also shaped Feynman's personal style in the immediate postwar period, which reinforced his openness and receptiveness for the predetermined guidelines of a social community. Furthermore, Los Alamos functioned as a career network. After 1945, Feynman was nearly overwhelmed by job offers. This demonstrated his strong foothold in the community (Galison 1998; Gleick 1993, 227–300; Schweber 1994, 397–405).

For Feynman, the theorist and German emigrant Hans Bethe was a central person, someone promoted him in the postwar period. While still in Los Alamos, Bethe had successfully enlisted Feynman as a professor at Cornell University. Bethe sponsored Feynman, and established contacts between Feynman and the leading-edge of the physical community. After 1945, Feynman got involved in the efforts to ensure civilian control of atom research through public lectures in the scope of the Federation of Atomic Scientists (Feynman 1966c, 68–69).[1] This is contrary to the image of the

---

[1] FBI-File of Richard Feynman, received under the Freedom of Information Act, Memorandum [ca. 1946, not dated].

genius who is detached from all social activities, which is how Feynman portrays himself in his later autobiographical writings.

We can discover a different image of genius in Feynman's story about his quantum electrodynamics research: As a young physicist burdened with high expectations and competitive stress, Feynman suffered a burn-out at Cornell and decided to only do what was fun for him. In his subsequent autobiographical narrative, it was after this decision that he arrived at his version of quantum electrodynamics through the just-for-fun analysis of the precession movement of a plate:

> It was effortless. It was easy to play with these things. It was like uncorking a bottle: Everything flowed out effortlessly. I almost tried to resist it! There was no importance to what I was doing, but ultimately there was. The diagrams and the whole business that I got the Nobel Prize for came from that piddling around with the wobbling plate. (Feynman 1997, 174)

The path from the rotating plate to quantum electrodynamics remains a secret to the non-professional. It is no longer traceable and similar to the daydream about the path integral formulation, appears utterly 'ingenious'. As a matter of fact, Feynman's work about his version of quantum electrodynamics consisted of a constant process of translation between Feynman's language and that of the collective. Bethe accompanied Feynman during this translation process and supported him. Feynman presented the modern version of the path integral formulation in 1947 at the Shelter Island Conference without receiving a positive response from his colleagues. His formulations were too far away from the standard operator theory. A friend and colleague suggested that he prepare a publication of his work, which ultimately appeared in *Reviews of Modern Physics* in 1948 (Feynman 1948). The publication contained in an axiomatic formulation a version of the path integral, in which every possible virtual path is allocated amplitude and phase, so that every conceivable virtual path delivers a contribution to the cumulative feasible amplitude. This publication was also molded by discussion between Feynman and members of the collective (Gleick 1993, 334–39; Schweber 1994, 409–13).

After the Shelter Island Conference, Feynman continued to work on his formulation of quantum electrodynamics. Bethe supported him in the calculations of the Lamb-Shifts, a displacement of energy levels in hydrogen atoms, which can only be explained through effects of quantum electrodynamics. As his mentor, Bethe motivated Feynman to write publications and helped him with the translation of the results into the language of the community, as Feynman later remembers:

> So I made the corrections. I tried to translate it back in the language that other people use. And then I went in, the next day. He [Bethe] showed me how to calculate the self-energy of an electron, and I showed him what the correction ought to be. I tried to translate my principles into the other language that he was explaining this thing in. (Feynman 1966c, 29)

Feynman compiled his results while under significant competitive strain. His colleague Julian Schwinger always seemed to be a step ahead of him and in addition, was better positioned in the community due to his numerous publications. At the follow up conference to Shelter Island in Pocono, Feynman introduced his results on quantum electrodynamics. Feynman's unusual approach was met with surprise and Freeman Dyson's transfer of Feynman's approach into a style consistent with existing knowledge allowed for the first time a widespread reception of Feynman's theory. In 1965 Feynman, along with Schwinger and Tomonaga, received the Nobel Prize for his work on quantum electrodynamics.

After the bestowal of the Nobel Prize, a genius cult developed around Feynman. The development of this cult was strengthened through autobiographical anecdotes, which later appeared as autobiographical publications, which will be analyzed in the following section. In contrast to Feynman his competitor Julian Schwinger seemed to be just a highly gifted physicist in the public view (Forstner 2007, 186–96).

Feynman as a Genius

On the 21st of October, 1965, Feynman received a telegram from Stockholm that congratulated him, Julian Schwinger, and Sin-Itiro Tomonaga for receiving the Nobel Prize. Feynman's life changed with the Nobel Prize: the young, promising physicist from Los Alamos, who had competed with Schwinger in quantum electrodynamics, ultimately became a public character and reached an unforeseen popularity like no American physicists had before him. A cult developed around Feynman's character, which persists up to the present day. The drumming physicist Feynman adorns posters and book jackets. Recordings of his drum pieces can be purchased; his written works became bestsellers, both autobiographical and popular science, as well as science for specialists. Feynman-Diagrams decorate the church windows of the Saint Nikolai Church in Kalkar in Germany; Feynman pops up in comics. The cult around the 'genius Feynman' is surpassed solely by the cults around Albert Einstein, whose image, in the meantime, is available to buy in the form of Albert Einstein action figures and life-size standing cardboard figures.

In the introduction already quoted above, Freeman Dyson compared the relations of his character to Feynman with the English authors Johnson and Shakespeare. Johnson mastered his craft as an author, while Shakespeare distinguished himself through the 'genius'. In this comparison, Dyson identifies with Johnson and Feynman with the genius of Shakespeare (Feynman 1999, viii Foreword by Dyson). This comparison followed Feynman's death by more than ten years and is the only one known to me, in which a scientist explicitly depicts the scientist Feynman as a genius. Scientists depict one another as highly gifted or talented, but do usually not describe each other with the word genius. This term arises from the interdependency between scientists and the associated public in which they move, and can be summarized in three different points:

The magician is an element of the genius image, because the way the scientific results have been achieved is no longer comprehensible for most people, so that this magic formula appears to have been created free of context. The above cited dream stories of Feynman and Kekulé are fundamental to this element. The genius image in Feynman's stories is further characterized by the fact that he apparently detached himself from the norms of his scientific as well as social environment and through this was separated from the world. The 'revolutionary man' is the last element of Feynman's image as a genius. His research results provided a useful new perspective on an existing subject area for the scientific community and the public.

All three points are necessary to comprehend the 'genius Feynman'. For example, if the third point was missing, a lunatic remains, but not a scientific genius. The first two points of the definition of a genius can be strengthened by the self-portrayal of scientists to the public, while in the third point the popular portrayal of natural science gains in meaning. In contrast to Schwinger and Tomonaga, Feynman only acquired the reputation of a genius amongst the American public. In his biographical self-portrayal, he was successful in connecting contradictory social and scientific norms with the American myth of the 'practical man'. For the public, this presented him as a 'typical' American with uncommon scientific capabilities and significantly contributed to his crowning as a genius in the public's awareness. In his Feynman biography the journalist James Gleick discusses the question "what is a genius?" with a strong focus on the genius as magician and his originality, but disregards the role of the public in the process of making a genius, something that will be incorporated in the story of this paper (Gleick 1993, 450–77).

Feynman as a Magician

The scientific genius as a magician 'sees' solutions, while the typical scientist has to derive them from established laws. Feynman systematically built up this image of a genius in his most well-known autobiographical anecdote volume titled *Surely you're Joking Mr. Feynman* (Feynman 1997). This begins with a story about his High School algebra club. The time for a solution to the assignments posed was too short for a conventional approach. Therefore it was necessary to arrive at a solution with the help of tricks, or as even Feynman reports: "Sometimes I could see it in a flash." (Feynman 1997, 23). During his graduate studies at Princeton, he discovered that "monster minds" such as John A. Wheeler, John von Neumann and Albert Einstein could see the solution, while he still had to tediously calculate them (Feynman 1997, 78). Later he separated himself from the scientists at the Institute of Advanced Studies at Princeton. Because these scientists exclusively sat in offices to gather their thoughts, they did not receive inspiration and were no longer in a position to develop new ideas (Feynman 1997, 165). The value of these scientists, up until that point portrayed as 'monster minds', now diminished. In the following chapter, Feynman refers to his own creativity that, among other things, led to his formulation of quantum mechanics through the precession movement of a plate. During a conference in Japan, Feynman ultimately saw the solution of physical problems for himself (Feynman 1997, 244–45). Therefore the argumentation proceeded in three phases: First, the establishment of the 'monster minds', then their degradation based on their lack of creativity, and finally the assertion of his own creativity and physical intuition. With this sequence, he strengthened the image of Richard Feynman as an exceptional personality and ingenious scientist.

The Genius Removed from the World

The image of the genius removed from the world was depicted through the apparent detachment of the actor from his social context. This occurs in Feynman's stories through the thematic of breaching the norms on two levels: scientific and social.

The autobiographical anecdote volume, *Surely you're Joking Mr. Feynman*, includes many such examples of breaching norms. Indeed the title of the book comes from this: Feynman had requested tea with milk and lemon at the dean's teatime, to which he received the response "surely you're joking, Mr. Feynman". A genius does not request tea with milk and lemon, but Feynman's presentation of this small mishap illustrates how he made fun of the conventions of Princeton's academic elite (Feynman 1997, 60). This becomes even clearer, when he describes the language of Princeton's academics, which he characterizes as an imitation of the English colleges of Cambridge and Oxford. Here Feynman distances himself from the conventions of a social group to which he himself belonged, and consciously separates himself from Princeton's elite (Feynman 1997, 59–97).

Feynman repudiated the norms of the social framework of science. Thus, he did not publish his results frequently, did not participate as a referee in the system of self-assessment of scientific journals and did not compose any review articles. Similarly, he rarely took part in faculty meetings (Feynman 1966c, 131, 169–70). Feynman also consistently rejected honorary doctorates and acceptance into scientific societies. Prizes and honors in science are to be understood as recognition for the work of the researcher, who ultimately leaves society his results and receives recognition in return. Initially, Feynman did not react to his appointment to the American Philosophical Society. Only after he received an assertive request to get in touch, did he reject his appointment two and a half months later.[2] Feynman's reason for the rejection of the membership was not because of a rejection of the respective society, rather a general rejection of the system of scientific recognition. For example, he justified his withdrawal from the National Academy of Science with the following letter to the President in 1961:

> The care with which we select "those worthy of the honor" of joining the Academy feels to me like a form of self-praise. How can we say only the best must be allowed to join those who are already in, without loudly proclaiming to our inner selves that we who are in must be very good indeed? Of course I believe I am very good indeed, but that is a private matter and I cannot publicly admit that I do so, to such an extent that I have the nerve to decide that this man, or that, is not worthy of joining my elite club.[3]

In his volume of anecdotes, he similarly picks out the central theme of breaching norms that stand up against the norms of civil society. These include his regular visits to topless bars, the design of a painting for a brothel, and the testimony in favor of the topless bar's owner in court. Feynman not only rejected civil norms that to him

---

[2]Letter from George W. Corner to Richard Feynman, April 19, 1968, Richard Phillips Feynman Papers (RPFP), California Institute of Technology, Institute Archives, Mail Code 015A–74 Pasadena, CA 91125, USA, Folder 1.1. and Feynman's refusal July 1, 1968.

[3]Letter from Richard Feynman to Detlev W. Bronk, August 10, 1961, RFPF, Folder 1.13.

seemed to be part of a double standard, but also made this rejection public (Feynman 1997, 270–75). Similarly, Feynman's experiences with women and gambling in Las Vegas appeared as incompatible with a civil society (Feynman 1997, 221–31). The same is true of Feynman's description of hallucinations that he experienced. These took place in a special studio, where one would be locked into water filled tanks in order to experience hallucinations (Feynman 1997, 335–37). These experiences and Feynman's experimentation with testing his own limits, allowed him to appear as a lunatic to the average American.

In spite of his continual breaching of norms that affected not only American civil society, but also the social framework of science, Feynman did not experience any kind of sanctions. Since Los Alamos, Feynman was strongly anchored in the scientific thought collective. This clarifies, for instance, the numerous job offers that he received after 1945, as well as his participation at exclusive postwar conferences, such as the Shelter Island Conference and the Pocono Conference. The bestowal of the Nobel Prize to Feynman also depicts the high regard in which the community held Feynman. Furthermore, Feynman's approach significantly contributed to quantum electrodynamics. His work style always remained within the frame of the thought style of the community and did not question their systematic basic principles. In spite of his extraordinary behavior, Feynman was often able to connect to the public.

In his popular science writings, a rejection of the behavior of the educated civil elite goes along with the transfiguration of practical manual tasks and a rejection of philosophical works. For example, Feynman writes: "In the early fifties I suffered a disease from the middle age: I used to give philosophical talks about science." (Feynman 1997, 279) It is clear why Feynman was not outlawed for breaching norms, but in order to understand why he was praised as a physics genius, it is necessary to examine the genius as a revolutionary.

Genius as Revolutionary

Feynman's formulation of quantum electrodynamics, similar to Schwinger and Tomonaga's operator formalism, was not revolutionary in terms of Kuhn's paradigm shift. A break from previous physics did not follow, rather the new theory established ties to quantum mechanics (Kragh 1999, 332–36). The revolution, understood as a central innovation in Feynman's approach, was part of a different domain, which his former competitor Julian Schwinger characterizes in the following: "Like the silicon chip of more recent years, the Feynman diagram was bringing computation to the masses." (Schwinger 1983, 343). Feynman's method enabled physicists to arrive at results in a quick way, and in comparison to Schwinger and Tomonaga's, without complex calculations (Kaiser 2005). A colleague of Feynman's already determined this after receiving an advanced copy:

I want to thank you for forwarding prepublication copies of your papers on electrodynamics. [...]. I also gave a series of three or four seminars on them. They certainly are wonderful in making calculations easier.[4]

Through the translation of highly complex mathematical theory into simple language, tied in with a graphic viewing apparatus, the popularization of this theory was made possible. Therein consists the second 'revolution' of Feynman's works. Not only did he make quantum electrodynamics more widespread among the physics community, but he also made modern physics popular, like very few other physicists of the twentieth century, through his numerous lectures that were later publicized as essays and in books.

The Character of Physical Law (Feynman 1965) was created from a series of lectures that Feynman had held in 1960. In this volume, Feynman leads the reader to complex and fundamental questions from Newton's law of gravitation to the concept of symmetry in modern physics, which he visualizes without sliding into the trivial. In the process, Feynman also ventures on to complex themes. The volume *QED-The Strange Theory of Light and Matter* similarly emerged from a popular lecture series and, to this day, is a unique popular scientific introduction to Feynman's formulation of quantum electrodynamics (Feynman 1985). His capability to make such complex data accessible to others, made Feynman famous as a teacher and textbook author. The famous *Feynman Lectures* emerged from a multiple semester lecture series and fortified Feynman's reputation as a brilliant teacher (Feynman et al. 1989a, b, c).

Conclusion

A similar cult of genius has never expanded around any other American physicist like the one around Richard Feynman. His autobiographical writings played an essential role in the genesis of this cult of genius. These publications put numerous anecdotes that circulated around his character into written form, and thereby significantly contributed to the efforts of brilliant physicists and Nobel Prize laureates to lift him up into the gallery of genius. Quite contrary to Feynman's own staging of his career, the historical biographical case study in the first part of this essay demonstrates that his physics by no means was created out of context, but instead was a deliberate effort, influenced by the prevailing thought style, to make his results available to the scientific community. The cult of genius around Feynman, which began to grow even more after his Nobel Prize, increasingly concealed Feynman's foothold in the thought collective. Feynman significantly contributed to the growth of the cult of genius with his stories and autobiographical writings. His biographical self-portrayal tied together his subsequent defiance of social and scientific norms with the American myth of the 'practical man'. He therefore appeared in public as a typical American with extraordinary scientific abilities that were incomprehensible to the average Joe and were eventually revealed to be genius.

---

[4]Letter from Robert F. Christy to Richard Feynman, September 12, 1949, RPFP, Folder 1.20.

References

Ash, M. (2006). Wissens- und Wissenschaftstransfer. Einführende Bemerkungen *Berichte zur Wissenschaftsgeschichte, 29*, 181–189.
Bourguet, M.-N., Licoppe, C., & Sibum, H. O. (Eds.). (2002). *Instruments, travel and science. Itineraries of precision from the seventeenth to the twentieth century*. London.
Feynman, R. P. (1939). Forces in molecules. *Physical Review, 56*, 340–343.
Feynman, R. P. (1942). *The principle of least action in quantum mechanics*. PhD Thesis, Princeton University, Princeton.
Feynman, R. P. (1948). Space-time approach to non-relativistic quantum mechanics. *Reviews of Modern Physics, 20*, 367–387.
Feynman, R. P. (1965). *The character of physical law*. London.
Feynman, R. P. (1966a). Part I on March 4, 1966 interview by Charles Weiner. Niels Bohr Library and Archives, One Physics Ellipse College Park, MD.
Feynman, R. P. (1966b). Part II on March 5, 1966 interview by Charles Weiner. Niels Bohr Library and Archives, One Physics Ellipse College Park, MD.
Feynman, R. P. (1966c). Part III on June 27, 1966 interview by Charles Weiner. Niels Bohr Library and Archives, One Physics Ellipse College Park, MD.
Feynman, R. P. (1972). The development of the space-time view of quantum electrodynamics. Nobel lecture, December 11, 1965. In *Nobel lectures. Physics, 1963–1970* (pp. 155–178). Amsterdam.
Feynman, R. P. (1985). *QED. The strange theory of light and matter*. Princeton.
Feynman, R. P. (1997). *"Surely you're joking Mr. Feynman!" Adventures of a curious character*. In R. Leighton & E. Hutchings (Eds.). New York.
Feynman, R. P. (1999). *The pleasure of findings things out*. Cambridge.
Feynman, R. P., Leighton, R. B., & Sands, M. L. (1989a). *The Feynman lectures on physics. Mainly electromagnetism and matter* (Vol. 2). Redwood City, CA.
Feynman, R. P., Leighton, R. B., & Sands, M. L. (1989b). *The Feynman lectures on physics. Mainly mechanics, radiation, and heat* (Vol. 1). Redwood City, CA.
Feynman, R. P., Leighton, R. B., & Sands, M. L. (1989c). *The Feynman lectures on physics. Quantum mechanics* (Vol. 3). Redwood City, CA.
Fleck, L. (1983). *Erfahrung und Tatsache. Gesammelte Aufsätze*. Frankfurt am Main.
Forman, P. (1989). Social niche and self-image of the American physicist. In M. de Maria, M. Grilli, & F. Sebastiani (Eds.), *The restructuring of physical sciences in Europe and the United States 1945–1960* (pp. 96–104). Singapore, New Jersey.
Forstner, C. (2007). *Quantenmechanik Im Kalten Krieg*. Diepholz: GNT-Verlag. http://www.worldcat.org/oclc/488544816.
Galison, P. (1998). Feynman's war. Modelling weapons, modelling nature. *Studies in the History and Philosophy of Modern Physics, 29*, 391–434.
Gleick, J. (1993). *Richard Feynman. Leben und Werk des Genialen Physikers*. München.
Halpern, P. (2017). *The Quantum Labyrinth: How Richard Feynman and John Wheeler revolutionized time and reality*. New York: Basic Books.
Kaiser, D. (2005). *Drawing theories apart. The dispersion of Feynman diagrams in postwar physics*. Chicago.
Kevles, D. (1995). *The physicists. The history of a scientific community in Modern America* (3rd ed.). Cambridge.
Kragh, H. (1999). *Quantum generations. A history of physics in the twentieth century*. Princeton.
Mehra, J. (1994). *The beat of a different drum. The life and science of Richard Feynman*. New York.
Pauling, L., & Wilson, E. B. (1935). *Introduction to quantum mechanics. With applications to chemistry*. New York.
Rocke, A. J. (1985). Hypothesis and experiment in Kekulé's Benzene theory. *Annals of Science, 42*, 355–381.
Schultz, G. (1890). Bericht über die Feier der Deutschen Chemischen Gesellschaft Zu Ehren August Kekulés. *Berichte der Deutschen Chemischen Gesellschaft, 23*, 1265–1312.

Schweber, S. S. (1986a). Shelter Island, Pocono, and Oldstone. The emergence of American quantum electrodynamics after World War II. *Osiris, 2*, 265–302.

Schweber, S. S. (1986b). The empiricist Temper Regnant: Theoretical physics in the United States, 1920–1950. *Historical Studies in the Physical and Biological Sciences, 17*, 55–98.

Schweber, S. S. (1989). Some reflections on the history of particle physics in the 1950s. In L. M. Brown, M. Dresden, & L. Hoddeson (Eds.), *Pions to quarks. Particle physics in the 1950s* (pp. 668–693). Cambridge.

Schweber, S. S. (1994). *QED and the men who made it. Dyson, Feynman, Schwinger and Tomonaga.* Princeton.

Schwinger, J. (1983). Renormalization theory of quantum electrodynamics. In L. Brown & L. Hoddeson (Eds.), *The birth of particle physics* (pp. 329–353). Cambridge.

Slater, J. C. (1938). Electrodynamics of ponderable bodies. *Journal of the Franklin Institute, 225*, 277–287.

Slater, J. C., & Frank, N. H. (1933). *Introduction to theoretical physics.* New York.

Part II
Collectives

Chapter 10
A Biography of the German Atomic Bomb

Mark Walker

The "German Atomic Bomb" was not a person, in fact it never existed. This story is really a collective biography of different people in different contexts perceiving the potential of such a weapon in different ways. This includes how they were affected by this potential weapon, how they perceived its threat or benefit, and how they responded, with this sometimes changing over time. Therefore the German Atomic Bomb has been a constantly shifting metaphor for the collective fear and hope created by nuclear fission. It had a beginning, it developed and changed over time, and it should have had an end.

Nuclear Fission

This story begins with a classic history of science scenario: an international race during the 1930s to understand the mysterious phenomena displayed by uranium when it was bombarded by neutrons. One group of competitors included the German chemist Otto Hahn and the Austrian physicist Lise Meitner, who were subsequently joined by the younger chemist Fritz Strassmann. The collaboration in Germany was suddenly interrupted in 1938 by the German absorption of Austria and subsequent expansion of Nazi anti-Semitic law, policy, and persecution to cover people like Meitner (Sime 1996, pp. 161–183). When Carl Bosch, president of the Kaiser Wilhelm Society, asked Minister of the Interior Wilhelm Frick for an exception, the response was chilling:

> ... political considerations oppose providing Professor Meitner with a travel passport. It is undesirable when well-known Jews travel from Germany to other countries in order to act

---

M. Walker (✉)
Department of History, Union College, Schenectady, NY 12308, USA
e-mail: walkerm@union.edu

© The Editor(s) (if applicable) and The Author(s), under exclusive licence to Springer Nature Switzerland AG 2020
C. Forstner and M. Walker (eds.), *Biographies in the History of Physics*,
https://doi.org/10.1007/978-3-030-48509-2_10

as representatives of German science or even to use their name and experience ... to work against Germany.

Frick's representative noted that the Kaiser Wilhelm Society could certainly find a way for Meitner to continue her work in Germany, even if she had to leave the Society. Most unsettling was the added note that: "In particular, the head of the SS and chief of the German police [Heinrich Himmler] has expressed this view ..."[1] Meitner had already endured other discrimination, but now feared for her safety and fled the country (Sime 1996, pp. 184–209).

This did not end the collaboration, which continued at a distance through letters between Hahn and Meitner. When Hahn and Strassmann bombarded uranium with neutrons and found that radioactive barium, an element much smaller than uranium, had been produced as a result, Hahn shared the news with Meitner:

> It could still be a very strange coincidence. But more and more we come to the shocking conclusion: our radium isotopes are not behaving like radium, rather like barium... Strassmann and I have agreed that we only want to tell you this for now. Perhaps you can suggest some fantastic explanation. We know that it actually cannot burst into barium...[2]

Meitner responded encouragingly:

> For the time being the assumption of such a far-reaching explosion is very difficult for me, but we have experienced so many surprises in nuclear physics, that one cannot simply say: it is impossible.[3]

Meitner in turn discussed it with her physicist nephew Otto Frisch. Together they worked out the physics of nuclear fission, including naming it, and predicted that this reaction would in turn release additional neutrons and a great deal of energy (Sime 1996, pp. 231–258). At this time a joint publication between the German chemists and the two émigré physicists was out of the question. Hahn and Strassmann revealed the chemical discovery to the world in December of 1938, while Meitner and Frisch explained the physics in a subsequent article (Hahn and Strassmann 1979, pp. 65–76, Meitner and Frisch 1979, pp. 97–100).

These publications, as well as word of mouth, unleashed an even more intense international competition to understand and exploit nuclear fission. It is striking how much information that would later prove important for nuclear weapons was published before the veil of secrecy fell over the research. Niels Bohr and John Wheeler demonstrated that probably only the rare isotope uranium 235 could easily be fissioned by neutrons, and that when the more common isotope uranium 238 absorbs a neutron, it would probably transmute into a transuranic element that also would fission like uranium 235. Moreover, any transuranic element would be even easier to fission than 235 (Bohr and Wheeler 1979, pp. 156, 171).

---

[1] Frick to Bosch, (18 April 1938), NL 080/101-01, Walther Gerlach Papers, Archives of the German Museum (Munich).

[2] Hahn to Meitner (19 December 1938), DJ-29 023, David Irving Microfilm Collection, Archives of the German Museum (Munich).

[3] Meitner to Hahn, (21 December 1938), NL 080/127-01, Walther Gerlach Papers, Archives of the German Museum (Munich).

Several scientists demonstrated that nuclear fission in fact released both energy and neutrons. Siegfried Flügge, a physicist at Hahn's institute in Berlin, subsequently introduced the readers of the journal *Die Naturwissenschaften* to the concept of a nuclear fission chain reaction: if a fissioning uranium nucleus released more than one neutron, and these neutrons caused further fissions, then this might cause an exponentially-increasing chain reaction. Flügge noted that if, for example, every uranium atom in a meter cube of uranium oxide were split, then the energy liberated would raise a cubic kilometer of water twenty-seven kilometers high! (Flügge 1979, p. 121) This fantastic image reflected the apparently unbounded potential of nuclear energy. Nuclear fission had unlocked a new world for nuclear physics.

The Lighting War

At the same time scientists were investigating nuclear fission, Germany was taking one step after another towards war. Almost as soon as nuclear fission had been discovered, the possibility of a German atomic bomb cast a foreboding shadow over the exciting and potentially transformative potential of nuclear energy. Scientists in several countries contacted their governments about the possible military applications of a nuclear fission chain reaction. Several German scientists independently brought the potential of nuclear weapons to the attention of different military and political authorities.

After the invasion of Poland the "nuclear physics working group" began work in Germany under the command of Army Ordnance. Very quickly the theoretical physicist Werner Heisenberg pointed out that there were two basic applications of nuclear fission chain reactions, controlled reactions that could be used to produce heat and electricity, and uncontrolled reactions, which would be nuclear explosives. It was also clear that there were two paths to these applications: what the Germans called "uranium machines" (nuclear reactors), and isotope separation. When Carl Friedrich von Weizsäcker realized in 1940 that an operating uranium machine would produce transuranic elements, it became clear that both paths could lead to nuclear explosives and atomic bombs. A year later, Weizsäcker submitted a patent application explaining how such transuranic elements could be used in a 'bomb', while experiments in Heisenberg's institute demonstrated the feasibility in principle of a nuclear reactor.

In September of 1941 Heisenberg and Weizsäcker took advantage of an opportunity to participate in a propaganda event, an astrophysics workshop in German-occupied Copenhagen, to visit their mentor Niels Bohr. It is clear that they were concerned about Bohr's safety, and for this reason urged him and his colleagues to cooperate closely with the German occupation officials. Bohr did not take this advice (Walker 2019). This visit is best known for a conversation in which Heisenberg revealed to Bohr that he was researching the possible military applications of nuclear fission. A few years after the end of the war Heisenberg claimed that:

When I spoke with Niels Bohr in Copenhagen in the autumn of 1941, I asked him the question whether a physicist had the moral right to work on atomic problems during the war. Bohr asked in turn, whether I believed that a military application of atomic energy was possible, and I answered: yes, that I knew. I then repeated my question, and to my amazement Bohr replied that military service for physicists in all countries was unavoidable and therefore also justified. Bohr obviously considered it impossible, that here the physicists from all peoples would so to speak join together against their governments...[4]

When a similar account was published in 1958 by Robert Jungk's book *Brighter than a thousand Suns* (Jungk 1956, pp. 108–111), Bohr responded by drafting, but never sending letters to Heisenberg refuting his version of events:

It had to make a very strong impression on me that at the very outset you stated that you felt certain that the war, if it lasted sufficiently long, would be decided with atomic weapons. I had at that time no knowledge at all of the preparations that were under way in England and America. You added, when I perhaps looked doubtful, that I had to understand that in recent years you had occupied yourself almost exclusively with this question and did not doubt that it could be done. It is therefore quite incomprehensible to me that you should think that you hinted to me that the German physicists would do all they could to prevent such an application of atomic science. (Dörries 2005, p. 175)

When Bohr subsequently fled to the United States, he brought this message with him: the Germans know atomic bombs can be made, and they are working hard to make them.

Allied Fear of a Nazi Bomb

Émigrés from Hitler's Germany also brought the military potential of nuclear fission to the attention of governments. In the United States, the Hungarian physicists Edward Teller, Eugene Wigner, and Leo Szilard drafted a letter in 1939 to President Roosevelt warning about the German interest in uranium and persuaded Albert Einstein to send it under his own name.

In the course of the last four months it has been made probable ... that it may become possible to set up nuclear chain reactions in a large mass of uranium, by which vast amounts of power and large quantities of new radium-like elements would be generated. Now it appears that this could be achieved in the immediate future... it is conceivable... that extremely powerful bombs of a new type may thus be constructed...

I understand that Germany has actually stopped the sale of uranium... the son of the German Under-Secretary of State, von Weizsäcker, is attached to the Kaiser Wilhelm Institute in Berlin, where some of the American work on uranium is now being repeated. (Rowe and Schumann 2007, pp. 359–361)

This was followed by several further letters from Einstein to Roosevelt.

---

[4] Werner Heisenberg to Bartel van der Waerden (28 April 1948) Goudsmit Papers, https://repository.aip.org/islandora/object/nbla%3AAR2000-0092, accessed April 24, 2019.

10 A Biography of the German Atomic Bomb

Parallel to Teller, Wigner, and Szilard's efforts, in 1940 two other émigré physicists, Frisch and Rudolf Peierls, wrote a striking memorandum for the British government:

> The possible construction of 'super-bombs' based on a nuclear chain reaction in uranium... Effective methods for the separation of isotopes have been developed recently, of which the method of thermal diffusion is simple enough to permit separation on a fairly large scale.
>
> This permits, in principle, the use of nearly pure 235U [uranium isotope 235] in such a bomb... one might think of about 1 kg as suitable size for a bomb... The energy liberated by a 5 kg bomb would be equivalent to that of several thousand tons of dynamite...
>
> In addition to the destructive effect of the explosion itself, the whole material of the bomb would be transformed into a highly radioactive stage. The energy radiated by these active substances will amount to about 20% of the energy liberated in the explosion, and the radiations would be fatal to living beings even a long time after the explosion... Effective protection is hardly possible. Houses would offer protection only at the margins of the danger zone.

The clear message was that nuclear weapons were feasible and could be produced within a few years, so that there was a real danger that Nazi Germany would make them first. As a result of these émigré initiatives, nuclear weapon projects began both in Britain and the United States, and subsequently were consolidated in America.

Immediately after the war, the American physicist Henry Smyth described the atmosphere in December of 1941:

> Perhaps more important than the [scientific] change was the psychological change. Possibly Wigner, Szilard, and Fermi were no more thoroughly convinced that atomic bombs were possible than they had been in 1940, but many other people had become familiar with the idea and its possible consequences. Apparently, the British and the Germans, both grimly at war, thought the problem worth undertaking. Furthermore, the whole national psychology had changed. Although the attack on Pearl Harbor was yet to come, the impending threat of war was much more keenly felt than before, and expenditures of effort and money that would have seemed enormous in 1940 were considered obviously necessary precautions in December 1941. (Smyth 1945, pp. 73–74)

As Germany conquered most of Europe and was trying to bomb Britain into submission, the specter of Nazi nuclear weapons drove British, American, and especially émigré scientists, many of whom would otherwise have been ambivalent about working on such weapons, to join the effort to create atomic bombs as quickly as possible. The work in Britain was transferred to Canada and the United States, where massive investments in resources and manpower created the secret "Manhattan Project". Whole new cities were created to produce fissionable materials (nuclear explosives), and a secret laboratory was built in Los Alamos, New Mexico (Rhodes 1986).

All of this was justified and spurred on by the specter of Nazi nuclear weapons. Indeed even when it appeared that the Allies had the upper hand in the war, the other German 'wonder weapons' like cruise missiles, jet aircraft, and ballistic rockets continued to stoke the fear that, if Germany also got the atomic bomb, it could snatch victory from the jaws of defeat. Only when Germany had surrendered in the spring of 1945 did this fear finally fade away. However, by then the atomic bomb had taken on a life of its own.

Selling Uranium

By the end of 1941 Heisenberg saw an open road to atomic bombs. This was a momentous time. During the first two years of Germany's successful lightning war the research could proceed at an unhurried pace, so that 'wonder weapons' like atomic bombs did not appear necessary. The end of the German advance in the Soviet Union, combined with the German declaration of war against the United States, suddenly called into question many research projects that did not promise immediate military applications.

It is important to understand that the military potential of nuclear fission was not limited to bombs. Nuclear reactors and power plants would also be welcome. Powerful new sources of electricity were needed as Germany struggled to boost war production. Possible applications in submarines, ships, and other vehicles also appeared promising. Most far off, but also most fantastic, were nuclear explosives many times more powerful than existing chemical explosives.

When Army Ordnance took stock of the German research in February, 1942, uranium isotope separation through a 'separation tube' had failed. Model uranium machines with metal uranium and heavy water as moderator appeared promising, but relatively little metal uranium and heavy water were available. During the first two years of the Lightning war, atomic bombs were interesting possibilities for the future; when the fortunes of war turned, German military officials wanted to know whether they would be available in time to help win the war.

The change in fortune of the war in the winter of 1941/1942 had immediate consequences for the German "working group". Two senior scientists who had already made very important contributions, the physical chemist Paul Harteck and von Weizsäcker, were called up in January of 1942. Heisenberg personally intervened with Army officials to allow the two scientists to continue working on uranium instead of going to the front, but the threat to the entire project was clear, especially when Army Ordnance subsequently decided that the research would not be decisive for the war and transferred it to a civilian authority (Walker 1989, pp. 46–60).

This was the context for two secret lecture series for military and political officials in 1942. Otto Hahn, Harteck, and especially Heisenberg impressed their audiences by dangling the possibility of nuclear weapons before them. Speaking in February before the Reich Research Council, Heisenberg said:

> Therefore the pure isotope uranium 235 undoubtedly represents an explosive of utterly unimaginable effect. However, this explosive is difficult to obtain... The American research appears to be working in this direction with special emphasis... As soon as such a machine [nuclear reactor] is in operation, according to an idea from Weizsäcker, the question of how to obtain the explosive takes a new turn. Namely the transformation of the uranium in the machine produces a new substance (element with atomic number 94 [plutonium]), which very probably, just like pure uranium 235, is an explosive of the same unimaginable effect.[5]

---

[5] Werner Heisenberg, "Die theoretischen Grundlagen für die Energiegewinnung aus der Uranspaltung" (February 26, 1942) AMPG. I. Abt., Rep. 34, Nr. 119. Max Planck Society Archives.

10 A Biography of the German Atomic Bomb

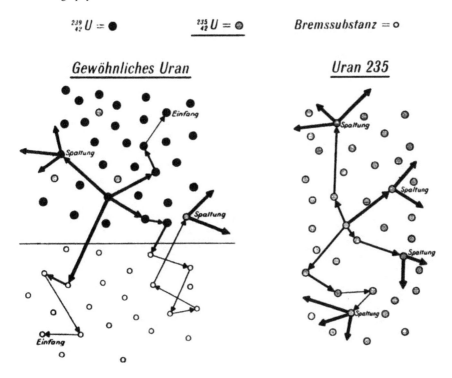

Fig. 10.1 Chain reactions in uranium

A schematic diagram Heisenberg used in his talk shows (on the left) a controlled nuclear fission chain reaction in an arrangement of ordinary uranium and moderator, that is, a nuclear reactor, as well as (on the right) an uncontrolled chain reaction in a mass of pure uranium 235, which constitutes a nuclear explosive (Fig. 10.1).

Four months later Kaiser Wilhelm Society president Albert Vögler arranged a series of lectures for Armaments Minister Albert Speer, Air Force Field Marshal Eberhard Milch, and their staffs. Once again, Heisenberg 'sold' his research:

> … according to the positive results achieved up until now, once the nuclear reactor has been built, it appears that we can also follow the path blazed by von Weizsäcker to explosive substances [plutonium] which are millions of times more powerful than any available now.
>
> But even if this does not happen in the foreseeable future, the construction of a nuclear reactor opens up a field of technical applications almost too large to survey. On one hand I am thinking of the construction of sea vessels, eventually perhaps aircraft able to travel great distances, also of the infinitely diverse applications of the radioactive substances produced in the reactor for many technical and scientific problems…
>
> Even if one considers the difficulty of such development work, one must grasp that in the next few years new land of the greatest significance for technology can be discovered. Since we know that in America this problem has been tackled by many of the best laboratories, here in Germany we can hardly afford not to investigate these questions. Even if one recognizes that such developments usually take a long time, if the war with America lasts several years

more, one has to reckon with the possibility that one day the technical utilization of atomic energy will suddenly be able to play a decisive role in the war.[6]

In a letter to his wife in June of 1942, Heisenberg described how well the talks had gone, and how enthusiastic Speer, Milch, and others had been.

> Those were eventful days in Berlin. The important people in the country were truly gathered in the Harnack House–practically the entire armaments council and many other powerful individuals. It went better than all expectations. The talks by Bothe and myself obviously made a strong impression; in any case afterward personally and professionally we were treated fabulously. This is all very strange to me; suddenly I do not have to worry at all about the Lenard-Stark clique [a group of Nazi physicists] any more and can push through almost everything that is important to me. (Heisenberg and Heisenberg 2011, p. 201)

Although the scientists involved in the working group continued to worry about being called up for frontline service, and the project manpower was not expanded significantly, the prospect of a German atomic bomb provided job security for the scientists by allowing them to contribute to the war effort in this way. Unfortunately, just as support for the research solidified, the war began to crumble.

Total War

In the summer of 1942, Heisenberg was optimistic that he and his colleagues could create and sustain a chain reaction in a nuclear reactor by the end of the year. Indeed this is when a group working in the United States under Enrico Fermi achieved this. However, as the war increasingly came home to Germany, this goal gradually slipped away.

The casting of metal uranium in pieces and forms suitable for an uranium machine became progressively more difficult, and finally impossible. Heavy water, the choice of moderator in the German uranium machines, was too expensive to produce in Germany, and sabotage and Allied bombing stopped production in occupied Norway, freezing the amount of heavy water. Moreover, as the Germans experimented with the heavy water, it became less and less pure, which degraded its moderating effect. By this time this was recognized, it was no longer possible to get a device for purifying the heavy water online. By the end of the war, the Germans had a prototype double centrifuge that could slightly enrich the amount of isotope 235 in small samples of uranium, and a nuclear reactor consisting of a lattice of uranium cubes immersed in heavy water, a very impractical design, which came close to, but did not reach criticality (Walker 1989, pp. 129–152).

It is important to understand that the modest German achievement when compared with the American Manhattan Project was not due to a lack of support, or perhaps one must clarify what is meant here by support. As mentioned above, in February of 1942 Army Ordnance came to the conclusion that nuclear energy or nuclear weapons

---

[6]Werner Heisenberg, "Die Arbeiten am Uranproblemen" (June 4, 1942), AMPG. I. Abt., Rep. 34, Nr. 93. Max Planck Society Archives.

would not be ready in time to influence the outcome of the war, and transferred it to a civilian authority. This decision could have been reversed if the researchers subsequently had demonstrated sufficient progress, but they did not. As a result, no attempt was made to expand the work up to the industrial scale.

On the other hand, the researchers continued to receive exceptional support from the Reich Research Council, the Kaiser Wilhelm Society, and most importantly, Speer's Ministry of Armaments. This support was clearly the result of a hope that the research might nevertheless yield tremendously powerful new weapons and allow Germany to stave off defeat. Perhaps most important, these scientists worked very hard to achieve as much progress as soon as possible.

Scientists like Harteck and Heisenberg also continued to tread an increasingly thin line. As the possibilities of building a self-sustaining nuclear reactor and to separate out uranium 235 became more and more remote, scientists and administrators concerned about losing their exemptions from frontline service nevertheless gave false hope to increasingly desperate Nazi officials by suggesting that miracles might still happen. Indeed the Nazi leadership was impressed, as can be seen in a diary entry from Propaganda Minister Josef Goebbels from March of 1942:

> I have received a report on the newest results of German science. The researches in the area of atomic disintegration are so far advanced that in certain circumstances their results will be able to be used in this war. Here the smallest input can result in such an immense destructive effect, that one can look towards the course of the war, if it is still going on, and a subsequent war, with a certain horror. Modern technology gives man destruction that is unimaginable. Here German science is at its peak, and it is also necessary that we are the first in this area; for the side that introduces a revolutionary innovation into this war has an ever greater chance to win it. (Walker 1990, p. 77)

In December of 1944, when the war had become desperate, Minister Speer wrote personally to Walther Gerlach, the head of the nuclear physics working group:

> However, I ascribe an extraordinary significance to research in the area of nuclear physics and follow your work with great expectations... I am certain that you are pushing the experiments forwards intensely. You can always count on my support for overcoming difficulties which would obstruct the work. Despite the exceptional demands on all forces for armaments, the relatively modest resources for your work can always be provided...[7]

In the end, Goebbels, Speer, and others were disappointed. The German atomic bomb was a chimera, not Germany's salvation.

Scapegoats

As the Allied armies pushed east through France into Germany, the Alsos Mission, an American scientific intelligence gathering unit, searched for evidence of a German atomic bomb, but found none. The head scientist from this mission, Samuel

---

[7] Albert Speer to Walther Gerlach (December 19, 1944) R3/1579 Federal German Archives Koblenz.

Goudsmit, who had been close to Heisenberg as a colleague, learned while in Europe that his parents had died in Auschwitz.

Immediately after the war, Goudsmit displayed an unsympathetic attitude towards his German colleagues:

> ... let the Germans themselves solve their problems of revival of pure and applied science and of scientific industry. Let them reopen their universities and research institutions when they can do so without outside material help. Let them publish their results again if they can find the paper and presses ...[8]

Goudsmit used the example of the German attempts to harness nuclear fission in several publications to criticize German scientists' collaboration with the Third Reich. In Goudsmit's hands, the German atomic bomb became a scapegoat for German science's collaboration with Nazism.

> With but slight exaggeration, I believe that German progress was primarily inspired by one man, Heisenberg surrounded by a group of scientists who hardly ever doubted his judgment ... their notion of a uranium bomb was so far off the right track that it could never have succeeded ...
>
> I feel certain that Heisenberg would have been one of the leading figures on the atomic energy project, if he had been working on our side. In Germany, Heisenberg worked in a vacuum, the only atmosphere present being the one created by his own ability. Even in the case of a great scientist like Heisenberg, this ability was not enough. (Goudsmit 1946, pp. 98–99)

The centerpiece of Goudsmit's argument was a fundamentally flawed description of the German understanding of how an atomic bomb would work.

> In fact, the whole German idea of the bomb was quite different from ours and more primitive in conception. They thought that it might eventually be possible to construct a pile [nuclear reactor] in which the chain reaction went so fast that it would produce an explosion. Their bomb, that is, was merely an explosive pile and would have proved a fizz compared to the real bomb. (Goudsmit 1983, p. 177)

Goudsmit had found misleading information in liberated Strasbourg at the beginning of the mission that suggested the Germans conceived of an atomic bomb as an entire reactor going out of control. Subsequently the Alsos Mission gathered additional reports and arrested scientists, but Goudsmit did not investigate their conception of nuclear weapons further. He had already found evidence confirming what he wanted to find: Nazism, and in particular the collaboration between the Nazis and German scientists, had ruined German science. Heisenberg fought back through an interview in the *New York Times*:

> Such was our alleged ignorance that we did not know the difference between a bomb and a furnace [nuclear reactor]. The truth is that the general principles which must of necessity underlie the design and construction of an atomic bomb were well known to German physicists. So was the possibility of transmuting uranium into plutonium in a furnace. (Heisenberg 1948)

---

[8] Samuel Goudsmit to Vannevar Bush (August 23, 1945) Goudsmit Papers, https://repository.aip.org/islandora/object/nbla%3AAR2000-0092, accessed April 24, 2019.

Goudsmit was forced to recant and he did so grudgingly:

> In my book 'Alsos' I have stated that the German scientists did not know the difference between a pile [nuclear reactor] and a bomb. This is perhaps an oversimplification, as far as a few key men were concerned, but I used it to avoid technical explanations. The German reports demonstrate that my oversimplification nevertheless conveys how far behind were the German colleagues. (Goudsmit 1949)

However, Goudsmit's original claim remains influential today (Popp 2016; Walker 2017).

Heisenberg was concerned with more than his scientific reputation. In a 1948 letter to Goudsmit, he defended his conduct during the Third Reich as well:

> Perhaps I should write about the political stance, which you consider right to call compromise with Nazism. During the entire time I never doubted for an instant that a considerable portion of the German government at the time consisted of fools and rascals. But I also knew that if we Germans did not succeed in undermining this system from within and finally eliminating it, then a great catastrophe would erupt, which would cost the lives of millions of innocent people in Germany and other countries…
>
> I have never had the least understanding for the people who withdrew from all responsibility and then in a safe table conversation asserted: 'You see, Germany and Europe will perish, I have always said that.'
>
> As far as the world is concerned, I would have found it better if National Socialism had been replaced from within with something better, instead of being eliminated from without through the force of arms…
>
> One cannot easily turn the thoughts of people to good things through the force of arms, and because of the indescribable misery in Germany the conditions are now very unfavorable for influencing thought in the direction that all serious people consider necessary. What we need in Germany is not a hateful reckoning with the past, rather a quiet reconstruction and the slow beginning of a life worthy of human beings.[9]

Goudsmit in turn clearly had a change of heart. In an article published later that year, he sounded very different from his harsh immediate postwar stance:

> We need a positive program to guide events in German scientific circles in a more favorable direction. This means that we must morally support those German colleagues in whose integrity we have confidence. There are many of them. We do not have to agree with all their opinions and should make allowances for the disturbing circumstances under which they have lived and are still living. We must again communicate with them as in the days before Hitler. (Goudsmit 1948, p. 106)

Rehabilitation

Beginning with their detention after the war at Farm Hall, a country house in England, von Weizsäcker and Heisenberg developed the argument that, although they had never been in a position to build atomic bombs, they would not have done so in any case

---

[9] Werner Heisenberg to Samuel Goudsmit (January 5, 1948) Goudsmit Papers, https://repository.aip.org/islandora/object/nbla%3AAR2000-0092, accessed April 24, 2019.

(Walker 1995, pp. 207–241). These arguments were popularized in Robert Jungk's book *Brighter than a Thousand Suns*, which was published in Germany in 1956. In particular, Jungk publicized the meeting Bohr and Heisenberg had had in 1941:

> [Heisenberg] ... gradually and cautiously steered the conversation towards the question of the atom bomb. But unfortunately he never reached the stage of declaring frankly that he and his group would do everything in their power to impede the construction of such a weapon if the other side would consent to do likewise.
>
> The excessive prudence with which both men approached the subject caused them in the end to miss it altogether. When Heisenberg asked whether Bohr considered it probable that such a bomb could be constructed, the latter... replied with conviction 'No!'
>
> Thereupon Heisenberg nerved himself to assure the other with all the emphasis at his command that he knew it to be perfectly possible to produce such a weapon and that it could actually be manufactured, if a very great effort were made within a very short time.
>
> Bohr himself was much troubled by Heisenberg's account of the position, so much so that he paid less than the right amount of attention to Heisenberg's remarks later about the dubious moral aspect of such a weapon. (Jungk 1956, p. 110)

The second edition, which appeared in 1958, was also translated into several languages, including Danish and English (Jungk 1958).

In fact Jungk's main purpose was to criticize the American and émigré scientists who had built the atomic bombs used in Japan. He used Weizsäcker's and Heisenberg's apparent resistance to both building atomic bombs and Hitler to make their colleagues who created the first nuclear weapons look bad by comparison:

> It seems paradoxical that the German nuclear physicists, living under a saber-rattling dictatorship, obeyed the voice of conscience and attempted to prevent the construction of atom bombs, while their professional colleagues in the democracies, who had no coercion to fear, with very few exceptions concentrated their whole energies on production of the new weapon. (Jungk 1958, p. 105)

Although Heisenberg was not completely happy with Jungk's book, he did appreciate its positive portrayal of him and his colleagues. Heisenberg sent Jungk a private letter correcting some statements in the book, but did not distance himself from it in public. With regard to Copenhagen, Heisenberg wrote Jungk essentially the same account he had sent van der Waerden in 1948. Jungk included excerpts from Heisenberg's letters in the second edition of his book, using them to support his interpretation of Copenhagen.

The effects of Jungk's book can be seen in the subsequent response to the 1957 "Göttingen Manifesto", when eighteen leading West German scientists, including Hahn, Heisenberg, and von Weizsäcker, published an open letter stating that they would not support the development of West Germany's own nuclear weapons.

> We do not feel competent to make concrete suggestions for the politics of the great powers. For a small country like the Federal Republic [of Germany], we believe that today it best protects itself and promotes world peace if it expressly and voluntarily renounces the possession of atomic weapons of every type. In any case none of the undersigned would be prepared to participate in the manufacture, testing, or use of atomic weapons in any way. At the same time we emphasize that it is extremely important to promote the peaceful application of atomic energy with all means, and we want to continue contributing to this challenge. (Göttingen Declaration 1982)

For many Germans, the manifesto was merely a continuation of what these scientists had done during the war; they had denied nuclear weapons to Hitler, now they also were denying them to Federal Republic politicians like Franz Josef Strauß and Konrad Adenauer.

Comparison with the USA

The carefully cultivated claim, that if necessary a small group of physicists around Heisenberg could and would have controlled the research in order to forestall nuclear weapons, is very hard to accept. Once the Second World War began, but especially once the tide turned, German scientists had a powerful motivation: to help avoid catastrophic defeat, to protect their country, their friends and family, and their own lives. If it had appeared possible to build atomic bombs, many, if not most of the German scientists who were researching nuclear reactors and isotope separation probably would have tried just as hard to succeed as did their American and émigré colleagues.

In fact, the German scientists were never in the position where they could have helped build nuclear weapons, but what if they could have? The Manhattan Project suggests some answers to this question. As mentioned above, originally the threat of Nazi nuclear weapons was overpowering and justified a herculean effort by scientists, some of whom otherwise would have not participated. Victory in Europe finally came in May of 1945, months before the first successful American test of an atomic bomb. Surprisingly few of the scientists who had been working to beat Hitler to the bomb now even questioned whether or not they should keep working. Only the Polish physicist Joseph Rotblatt left the Manhattan Project.

Several scientists, including some like Szilard who had been so instrumental in getting the project started, now tried to avoid using it against the Japanese, but their petitions and arguments were listened to politely, then set aside. On June 16, 1945, the advisory Science Panel of Arthur Holly Compton, Ernest Lawrence, J. Robert Oppenheimer, and Enrico Fermi, wrote:

> You have asked us to comment on the initial use of the new weapon. This use, in our opinion, should be such as to promote a satisfactory adjustment of our international relations. At the same time, we recognize our obligation to our nation to use the weapons to help save American lives in the Japanese war...
>
> The opinions of our scientific colleagues on the initial use of these weapons are not unanimous: they range from the proposal of a purely technical demonstration to that of the military application best designed to induce surrender. Those who advocate a purely technical demonstration would wish to outlaw the use of atomic weapons, and have feared that if we use the weapons now our position in future negotiations will be prejudiced. Others emphasize the opportunity of saving American lives by immediate military use, and believe that such use will improve the international prospects, in that they are more concerned with the prevention of war than with the elimination of this specific weapon. We find ourselves closer to these latter views; we can propose no technical demonstration likely to bring an end to the war; we see no acceptable alternative to direct military use...

With regard to these general aspects of the use of atomic energy, it is clear that we, as scientific men, have no proprietary rights. It is true that we are among the few citizens who have had occasion to give thoughtful consideration to these problems during the past few years. We have, however, no claim to special competence in solving the political, social, and military problems which are presented by the advent of atomic power. (Cantelon et al. 1991, pp. 47–48)

Echoing Jungk, although without moralizing, one can ask: if American and émigré scientists could not stop the American government from dropping atomic bombs on Japan, is there any reason to believe that, if nuclear weapons had been available to Hitler's government, Heisenberg, Weizsäcker, or anyone else could have stopped the Nazis from using them?

The Zombie Bomb

The German atomic bomb remains controversial. It is a zombie bomb, it is undead, and cannot be put to rest. Michael Frayn's play Copenhagen, which consists of conversations between the ghosts of Niels Bohr, Margarethe Bohr, and Werner Heisenberg, is a perhaps unintentional metaphor for this (Frayn 1998). Even though the German atomic bomb never existed, it has profoundly affected many lives, meaning very different things to different people, with this meaning also changing over time. The German atomic bomb is still influential today, not really because of what happened, but instead because of what might have been. It blurs the line between history and myth, between hero and villain, perhaps even between good and evil.

References

Bohr, N., & Wheeler, J. (1979). The mechanism of nuclear fission. In H. Wohlfarth (Ed.), *40 Jahre Kernspaltung. Eine Einführung in die Originalliteratur* (pp. 142–190). Darmstadt: Wissenschaftliche Buchgesellschaft.
Cantelon, P., Hewlett, R., & Williams, R. (Eds.). (1991). *The American atom: A documentary history of nuclear policies from the discovery of fission to the present* (2nd ed.). Philadelphia: University of Pennsylvania Press.
Dörries, M. (Ed.). (2005). *Michael Frayn's Copenhagen in debate: Historical essays and documents on the 1941 meeting between Niels Bohr and Werner Heisenberg*. Berkeley: Office of the History of Science and Technology, University of California, Berkeley.
Flügge, S. (1979). Kann der Energieinhalt der Atomkerne technisch nutzbar gemacht werden? In H. Wohlfarth (Ed.), *40 Jahre Kernspaltung. Eine Einführung in die Originalliteratur* (pp. 119–140). Darmstadt: Wissenschaftliche Buchgesellschaft.
Frayn, M. (1998). *Copenhagen*. New York: Anchor Books.
Frisch-Peierls Memorandum. (1940). Atomicarchive.com. http://www.atomicarchive.com/Docs/Begin/FrischPeierls.shtml. Accessed September 17, 2018.
Göttingen Declaration. (1982). *Mililtärpolitik Dokumentation 25*, 41–42.
Goudsmit, S. (1946). Secrecy or science? *Science Illustrated, 1,* 97–99.
Goudsmit, S. (1948). Our task in Germany. *Bulletin of the Atomic Scientists, 4,* 106.

Goudsmit, S. (1949). German war research. *New York Times*, January 9, 1949.
Goudsmit, S. (1983). *Alsos* (2nd ed.). Los Angeles: Tomash.
Hahn, O., & Strassmann, F. (1979). Über den Nachweis und das Verhalten der bei der Bestrahlung des Urans mittels Neutronen entstehenden Erdalkalimetalle. In H. Wohlfarth (Ed.), *40 Jahre Kernspaltung. Eine Einführung in die Originalliteratur* (pp. 65–76). Darmstadt: Wissenschaftliche Buchgesellschaft.
Heisenberg, W. (1948). Nazis spurned idea of an atomic bomb. *New York Times,* December 28, 1948.
Heisenberg, W., & Heisenberg, E. (2011). *"Meine Liebe Li!" Der Briefwechsel 1937–1946.* St. Pölten: Residenz Verlag.
Jungk, R. (1956). *Heller als tausend Sonnen.* Stuttgart: Scherz & Goverts.
Jungk, R. (1958). *Brighter than a thousand suns.* New York: Harcourt, Brace, & Co.
Meitner, L., & Frisch, O. (1979). Disintegration of uranium by neutrons: A new type of nuclear reaction. In H. Wohlfarth (Ed.), *40 Jahre Kernspaltung. Eine Einführung in die Originalliteratur* (pp. 97–100). Darmstadt: Wissenschaftliche Buchgesellschaft.
Popp, M. (2016). Misinterpreted documents and ignored physical facts: The history of "Hitler's Atomic Bomb" needs to be corrected. *Berichte zur Wissenschaftsgeschichte, 39,* 265–282.
Rhodes, R. (1986). *The making of the atomic bomb.* New York: Simon and Schuster.
Rowe, D., & Schulmann, R. (Eds.). (2007). *Einstein on politics: His private thoughts and public stances on nationalism, zionism, war, peace, and the bomb.* Princeton: Princeton University Press.
Sime, R. L. (1996). *Lise Meitner: A life in physics.* Berkeley: University of California Press.
Smyth, H. (1945). *Atomic energy for military purposes.* Princeton: Princeton University Press.
Walker, M. (1989). *German National Socialism and the quest for nuclear power, 1939–1949.* Cambridge: Cambridge University Press.
Walker, M. (1990). *Die Uranmaschine. Mythos und Wirklichkeit der deutschen Atombombe.* Berlin: Siedler Verlag.
Walker, M. (1995). *Nazi science: Myth, truth, and the German atom bomb.* New York: Perseus Publishing.
Walker, M. (2017). Physics, history, and the German atomic bomb. *Berichte zur Wissenschaftsgeschichte, 40,* 271–288.
Walker, M. (2019). Copenhagen Revisited. In M. Björkman, P. Lundell, & S. Widmalm (Eds.) *Intellectual Collaboration with the Third Reich: Treason or Reason?* (pp. 230–247). London: Routledge.

Chapter 11
The Multiple Lives of the General Relativity Community, 1955–1974

Roberto Lalli

Introduction

The history of general relativity presents intriguing challenges for historians and philosophers of science. Its origins and receptions have figured prominently in debates about conceptual transformations, the dynamics of theory change, relationships between theory and experiment and, of course, physical and philosophical notions of space and time (see e.g., Holton 1969; Cassidy 1986; Glick 1987; Hentschel 1990; Ryckman 2005; Renn 2007; Gutfreund and Renn 2017). This scholarly relevance and the radical reconceptualization the theory entailed have traditionally obscured a crucial aspect of the evolution of Einstein's theory of gravitation. While today the theory is overwhelmingly presented as one of the building blocks of physics in popular and scientific writings (e.g., Aspelmeyer et al. 2017), the theory held only a marginal position in physics research for many years after the famous 'confirmation' pursued by eminent English astronomers in 1919 (Earman and Glymour 1980; Sponsel 2002; Stanley 2003; Kennefick 2012). As shown by science historian Eisenstaedt (1986, 1989), starting from the mid-1920s, the theory underwent a period of stagnation, corresponding to a marginalization of research on general relativity with respect to the main research strands in theoretical physics—a period Eisenstaedt called the "low-water mark" of general relativity. While the essential features and causes of this historical dynamics are still a matter of debate, there is strong consensus among scholars that Einstein's theory of gravitation became part of the mainstream of physics only in the post-World War II period, sometime between the mid-1950s and the early 1970s (Kaiser 1998; Kragh 1999; Kennefick 2007). The actors themselves underlined the exceptional character of the process they were involved in, describing this momentous change in the status of the theory

---

R. Lalli (✉)
Max Planck Institute for the History of Science, Berlin, Germany
e-mail: rlalli@mpiwg-berlin.mpg.de

© The Editor(s) (if applicable) and The Author(s), under exclusive licence to Springer Nature Switzerland AG 2020
C. Forstner and M. Walker (eds.), *Biographies in the History of Physics*,
https://doi.org/10.1007/978-3-030-48509-2_11

as a "renaissance" or "golden age" of general relativity (Will 1989, 1993; Thorne 1994).

If one relies on this narrative, major questions emerge concerning the exact relationship between the social and conceptual aspects of the passage between two distinct phases: the marginalization of the theory between the mid-1920s and the mid-1950s and its re-emergence as a major physical theory from the mid-1950s onward. Previous historical accounts have convincingly undermined the belief that the 'renaissance' of general relativity was a simple consequence of new serendipitous discoveries in the astrophysical domain in the 1960s. It has been correctly argued that the interest toward the theory increased before and independently from these discoveries (Kaiser 2000; Wright 2014; Kaiser and Rickles 2018). Yet the exact features of this complex phenomenon remain uncertain. In collaboration with Alexander Blum and Jürgen Renn, the author contributed to this debate by elaborating a historiographical framework to interpret the long-term evolution of the theory in terms of changing socio-epistemic networks. This framework takes a general historiographical approach with a focus on specific factors and case studies (Blum et al. 2015, 2016, 2017, 2018).

One of those aspects is the changing social composition of the relativity community. While some of the few experts in general relativity in the low-water-mark period started identifying a specialized figure of the 'relativist' in their correspondence network, the network remained quite small, and strongly limited to specific institutional settings.[1] Scientists were dispersed both geographically and intellectually, separated by national borders and by the fact that they pursued research agendas perceived as unconnected. In later recollections, many actors complained about the status of the field during the low-water-mark period, which was characterized by isolated endeavors. In doing so, they explicitly stressed the importance of the establishment of an international community between the 1950s and 1960s (Eisenstaedt 1986; Lichnerowicz 1992; Newman 2005).

I take these views seriously and propose to carefully analyze the role of community-building activities pursued at the international level in the return of Einstein's theory of gravitation to the mainstream of physics. These activities rapidly led to the establishment of an informal elite organization called the *International Committee on General Relativity and Gravitation* (ICGRG) in 1959. As discussed at length elsewhere (Lalli 2017), the focus on the community-building activities allows one to argue that different research agendas—previously slightly related only by the fact that they relied on similar mathematical tools and formalisms—came to be understood as part of a unique field named "general relativity and gravitation" as the formation of the ICGRG institutionalized this new research domain.

In this essay, I explore the possibility of understanding the ICGRG as the living embodiment of a scientific domain by approaching the history of the committee from a biographical perspective (Greene 2007). The conception of the ICGRG, its birth, its adolescence, its coming into maturity and its metamorphosis into a new entity in 1974,

---

[1] See correspondence between Howard P. Robertson and J. L. Synge, H. P. Robertson Papers, 100240MS, Caltech Archives, California Institute of Technology, Pasadena, CA, box 5, folder 25.

the *International Society of General Relativity and Gravitation* (ISGRG), are not simply understood as metaphors for anthropomorphizing an institutional entity as the protagonist of a compelling historical novel (Shortland and Yeo 1996). Rather, these 'stages' are employed as analytical tools to explore how the organization's function as the embodiment of a scientific field interfaced with other functions the ICGRG came to fulfill, mainly in its role as a political actor in the Cold War context. This approach allows one to take into account the different 'lives' of this institution into one story and pinpoint the elements of this narrative that are essential in determining the formation of a research field and its developments in connection with the changing socio-political contexts. At the same time, this attempt shows the limitations of biographical approaches and terminologies for institutional settings that, as in this case, are informal and international agglomerates of scientists working in different contexts, having different agendas, both scientific and political, and coming from different cultural and ideological backgrounds.

From Conception to Birth: International Conferences as the Gestation Phase of the Relativity Community

Biographies of individuals usually start with the parenthood, or at times even with older family heritage. For obvious reasons, they do not address the actions and intentions of the protagonist before birth, while still an embryo. For institutions, it appears instead that just this phase is a particularly relevant one. It is then that the major identity traits of the new institution are being formed as in the case of the ICGRG. In the early, unstructured phase prior to its foundation, major epistemic moves allowed, first, to reconsider the theory of general relativity as a relevant and understudied scientific topic and, then, to identify it as a specific research domain worth of further scrutiny. This move was a social action, shared by a sizeable number of practitioners who started constructing an international community on these bases. The major events in this dynamic, which was intellectual and social at the same time, were the initial international conferences dedicated exclusively to general relativity.

In the historiographical debates on the renaissance of general relativity, one of the most controversial topics is which conference might be considered the watershed moment signaling renewed interest in general relativity theory: the Bern conference in 1955 or the Chapel Hill conference in 1957 (Lichnerowicz 1992; Mercier 1992; Schweber 2006; Rickles 2011; Kaiser and Rickles 2018). If one focuses on the institutional ramifications, one might even argue that the conference held in Royaumont, near Paris, in 1959 was the central event, as it is there that the ICGRG was established. Whatever view one adopts, international conferences played a relevant role in the process through which general relativity came to be recognized as an important part of physics. As I shall show, the three above-mentioned conferences were intimately related. One has to consider them as a connected set of events in order to fully understand the establishment of the ICGRG as well as the parallel and related creation of a commonly shared international research field called "general relativity and gravitation" (GRG). The three of them together constituted the gestation phase of the ICGRG, of the related community and of the field this community was working on.

This perspective is based on the fact that there was a radical difference between these post-World War II institutionalized gatherings and the social encounters of relativity experts occurring during the low-water-mark period. Before the mid-1950s, scientists working on topics related to general relativity pursued varied research agendas mostly related to programs on the unification of electromagnetism and gravitation, search of cosmological models, and early attempts at unifying quantum mechanics and general relativity (North 1965; Kragh 1996; Goenner 2004, 2014; Blum and Rickles 2018). Only a few of those working in these research agendas considered themselves as part of a community of relativity experts. Mostly, this was only the case for those connected to the research environment of the Institute for Advanced Study in Princeton, where Einstein pursued his unsuccessful and marginal attempts at producing a unified theory of gravitation and electromagnetism from 1933 up to his death in 1955 (van Dongen 2010).

The question is whether or not there were any attempts at creating a shared platform to discuss these different agendas during the low-water-mark period. Indeed attempts happened only rarely, when the organizers wanted to develop a specific research agenda and address specific research questions in cosmology or unified field theories with a restricted group of specialists (Glick 1987; Bonolis 2017). Furthermore these rare events did not lead to any subsequent efforts to create a community involving the participants. Quite the contrary, experts working in different national contexts openly challenged research agendas pursued elsewhere (see e.g., Lalli 2016).

What happened in the 1950s, starting from the Bern conference, was completely different. In 1953, the isolated chair of theoretical physics at the University of Bern, André Mercier, decided to organize a large meeting in Bern in 1955 in order to celebrate the fiftieth anniversary of Einstein's formulation of the special relativity theory. In spite of the negative reception by his physics colleagues in Bern, Mercier was able to enroll the eminent Nobel laureate Wolfgang Pauli as the chairman of the conference (Mercier 1979, 1992). Mercier was a theoretical physicist with broad philosophical interests as well as a strong commitment to education and international relations. In his philosophical writings, he had declared that science was the primary human activity to achieve peace among nations (Mercier 1950). By the time of the conference, he had also taken part, as the President of the Swiss Physical Society, in the successful negotiations leading to the creation of CERN, which would be located in his beloved country (Hermann et al. 1987).[2] At the same time, his various activities had not left him much time to pursue original research in theoretical physics. The

---

[2]"Conférence pour l'Organisation des Etudes concernant la Création d'un Laboratoire européé de Physique nucléaire, Paris, 17–21 Décembre 1951" UNESCO/NS/NUC/4, http://unesdoc.unesco. org/images/0015/001540/154028fb.pdf. Accessed 13 February 2017; and "Minutes of the First Session of the CERN European Council for Nuclear Research," Paris, 5–8 May 1952, http://cds. cern.ch/record/19494/files/CM-P00075404-e.pdf. Accessed 13 February 2017.

other main organizer of the conference, Wolfgang Pauli, was certainly interested in the foundational problems of the theory of general relativity, but had not made it a central part of his research programs either. It was in this specific local context, where foundational questions in theoretical physics met broader philosophical and social concerns about the role of science in a world conditioned by political tensions of the Cold War, that the activities to build an international community of relativists started.

The importance of this specific local context in the events that led to the first international conference entirely devoted to relativity has several implications in our understanding of the renaissance process. While physics was experiencing an exponential growth in terms of funding and personnel in the post-World War II period (Kaiser 2002, 2012), this dramatic transformation did not lead to the re-emergence of gravitation research, at least not alone. An increasing number of students in theoretical physics were certainly available to address problems in gravitation. However, in order to actively pursue these still marginal topics, it was necessary to establish new forms of institutional support. For theoretical research on gravitation, help in this direction came from idiosyncratic places. In the United States, support came from eccentric businessmen who cultivated the hope that research on general relativity would sooner or later lead to the production of anti-gravitational devices (Kaiser and Rickles 2018). In Europe, institutional support was obtained thanks to the activities of Swiss physicists who were either interested in foundational issues or, in the case of Mercier, involved in the attempts to use science as a way to improve political relations.

The local context also constrained the rationale behind the choice of the speakers. Pauli imposed an elitist conception of the conference, with main plenary lectures given by renowned experts in relativity-related topics, plus a restricted number of other speakers.[3] The invited lecturers were chosen on the ground of personal acquaintance of the organizers. Many of them worked in Europe, or were representative of a pre-World War II European scientific culture. Apart from the authoritative invited speakers, the other presenters were chosen by national academies of scientific societies that were invited to send their delegates to the conference (Pauli 1999, pp. 129–133).[4] This procedure indicates that the organizers were following institutional models that were regulating the re-organization of international scientific cooperation in the early Cold War arena: those models were provided by established nongovernmental organizations such as the *International Council of Scientific Unions* (ICSU) and the international unions. Representation within these institutional bodies was based on the notion of national membership. Their members were chosen as delegates by recognized national institutions such as Research Councils or Academies (see e.g., Fennell 1994; Greenaway 1996).

---

[3]A. Mercier, Sur la Théorie de la Gravitation et de la Relativité Générale GRG, p. 15. Handakten Prof. André Mercier (1934–1998), Staatsarchiv Bern, Bern, folder BB 8.2 1556, Dossier on GRG.

[4]See also Chandrasekhara Venkata Raman to Werner Heisenberg, 9 November 1954, Nachlaß Werner Heisenberg, Rep. 93, Abteilung III, Max Planck Archiv, Berlin, folder 1704.

The time and place of the event, as well as the organizational structure based on the invitation of national delegates, were ideal to fully take advantage of the change in the international relations following the death of Joseph Stalin in 1953 and the subsequent change in USSR foreign policy (Ivanov 2002). In a period of slight détente between the superpowers, Soviet scientists—supervised by the Soviet central administrative and political structures—were slowly undertaking collaborative initiatives with their colleagues on the other side of the Iron Curtain (Hollings 2016). Neutral Switzerland was the ideal place to host political as well as scientific East-West encounters (Strasser 2009), and the Soviet Academy of Sciences agreed to send two scientific delegates to the conference, one of whom was the esteemed theoretical physicist Vladimir Fock.

In this particular political context, the Bern conference embodied a multitude of meanings, which went beyond the purely scientific gathering, as was explicitly recognized by some of the conference participants. It was a political event, in which scientists proclaimed the need to pursue cooperation in order to maintain contacts between scientists working in different countries and, implicitly, to improve East-West relations (Mercier and Kervaire 1956, pp. 25–37). But, the event had important scientific content too. The initial design of the conference mixed an elitist conception informed by philosophical concerns and a politically motivated attempt to improve East-West scientific relations in a research domain with no apparent military and industrial applications. The actual scientific outcome went far beyond the expectations of the organizers. Conference participants recognized, first, that important advances had been accomplished in the understanding of general relativity and, second, that there were important unsolved questions concerning the physical predictions of general relativity that needed to be answered before one could make significant progress in the different attempts to go beyond the theory (McCrea 1955; Pauli 1956; Bergmann 1956).

The Bern conference might be considered the fertilization event of international community-building activities related to relativity. The success of the initiative made scientists aware of the variety of functions of these kinds of events. From the scientific side, it became possible to recognize crucial unsolved problems and reconfigure research agendas accordingly. From the political perspective, general relativity was recognized as a non-sensitive research domain and, therefore, ideal for experiments in East-West international scientific cooperation.

Although the idea spread that something was coming to life, the questions were what and how to continue, especially because the Bern conference, in spite of being a successful initiative, had relevant limits. The elitist conception of the conference made it impossible for a new generation of American relativity experts to attend the meeting. On the other hand, the American community of physicists was the most active in the field, both in terms of numbers and in the efforts to draw connections with the physics discipline at large. Furthermore, American scientists perceived their academic culture as being much more democratic than the hierarchical atmosphere pervading most European institutional settings. From their perspective, the Bern conference had many shortcomings, both in terms of the presented scientific approaches and the possibility of open and fruitful discussions (Mercier 1979).[5] The creation of a research center completely dedicated to gravitation theory at the University of North Carolina at Chapel Hill provided the occasion to organize a conference on this topic (Rickles 2011; Kaiser and Rickles 2018). The conference was perceived both as a continuation of the Bern conference and as a challenge to it. Whereas the Bern conference was mostly European, the Chapel Hill conference had a strong American character. Whereas the Bern conference was also characterized by political motivations, the Chapel Hill conference was designed as purely scientific by the organizers. Whereas the Bern conference had an elitist character, the major feature of the Chapel Hill conference was the possibility for graduate students or early postdocs (mostly American) to discuss their work in front of an international audience (DeWitt and Rickles 2011; DeWitt-Morette 2011).

As far as the internationality of the gathering is concerned, however, the Chapel Hill conference had shortcomings too. A mixture of political and financial motivations prevented those working in Eastern European countries to attend the meeting. Those few coming from Europe could join the conference only thanks to the flight transportation support provided by the US Air Force, but this help had strong political constraints (Goldberg 1992). Even some Western European scientists involved in the activities of national communist parties had difficulties in attending the conference.[6]

In spite of these limitations, the conference was perceived as instrumental to the rapid progress of the field. The need to continue with these kinds of collective experiences was so clear that before the conference ended, the French participants—mathematician André Lichnerowicz and theoretical physicist Marie-Antoinette Tonnelat—proposed to host a new international conference two years later in France. This conference was held at Royaumont, near Paris, in 1959 (Lichnerowicz and Tonnelat 1962). In this conference the organizers tried to use the experience from the previous two conferences and combine the best of both. And it worked. The conference became a platform in which a sizeable number American, Western European and Eastern European experts in general relativity could meet. At the same time, the organizers maintained the democratic character of the Chapel Hill conference, with no hierarchical division among plenary and shorter talks. Many of the around 120 attendees could enjoy the atmosphere of excitement of being a (small) community interested in similar problems, which could hardly find a similar audience in the respective countries. As Mercier put it, at Royaumont, "in practice all the relativists' schools of the world were represented."[7]

In this favorable social and intellectual atmosphere, the applied mathematician Hermann Bondi made the proposal to formally institutionalize these attempts of community building (Mercier 1970). He suggested the creation of an official international

---

[5] See also A. Mercier, *Sur la Théorie de la Gravitation et de la Relativité Générale GRG*. Handakten Prof. André Mercier (1934–1998), Staatsarchiv Bern, Bern, folder BB 8.2 1556, Dossier on GRG.

[6] See the correspondence concerning the Chapel Hill conference in Cécile DeWitt-Morette Papers, Briscoe Center for American History, University of Texas at Austin, Box 4RM235.

[7] "Pratiquement toutes les écoles de relativistes du monde furent représentées." A. Mercier, *Leçons sur la Théorie de la Gravitation et de la Relativité Générale GRG*, p. 15, Handakten Prof. André Mercier (1934–1998), Staatsarchiv Bern, Bern, folder BB 8.2. 1556, Dossier on GRG.

group of scholars who would organize the activities for promoting research related to general relativity and gravitation, especially through the organization of international conferences every three years. Bondi's idea was enthusiastically accepted by authoritative experts working in different countries, thus giving birth to the ICGRG (Mercier and Schaer 1962).

In forming this selected (or better self-selected) elite group the need of maximum representativeness became the guiding principle. The membership of the committee was understood as representative of the different scientific approaches and agendas as well as different national traditions. In spite of this idealistic desire to create a body truly representative of most scientific instances and national communities, the ICGRG could not properly fulfill this function. West and East Germany were excluded, while they both had very active research centers in the field. Given the uncertain political status of Germany, it was likely preferred to exclude both countries in order to avoid any kind of political recriminations. India was also absent, possibly because of the financial difficulties of Indian scientists in attending international conferences. From the scientific perspective, experimental work was not represented, even though Robert Dicke was already about to establish himself as the leading figure in testing between competing gravitational theories (Peebles 2017).

When it was established in 1959, the ICGRG was a semi-formal institution with sixteen members, without any written agreement, which tried to model itself according to more structured international scientific organizations. The officers of the ICGRG were the President(s) and the Secretary. While the President(s) remained in charge for only three years—from one conference to the next—the Secretary could be indefinitely re-elected. André Mercier, the main organizer of the Bern conference, was chosen as the permanent secretary of the ICGRG. In this capacity, he had the possibility to greatly influence the activities of the committee until 1974, when the ICGRG transformed itself into a formal institution of quite a different nature, the *International Society on General Relativity and Gravitation* (ISGRG). In this period, the ICGRG was the main institutional actor in the development of general relativity in the international arena.

Struggles in Shaping Its Scientific Identity: Strategies and Tensions in the ICGRG Activities

We can interpret the activities of ICGRG in the period following its establishment as attempts of this new institutional body to self-determine its own scientific identity, which necessarily entailed the efforts to define the epistemic status of the theory of general relativity. The discussions among members revolved around a series of issues about the best strategies to follow in order to coordinate the field. The main strategy was to solidify the tradition of international conferences. This apparently uncontroversial task involved a series of subtle decisions. One of these was whether participation to such international gatherings should be by-invitation only or open to those willing to attend. In spite of the resistance of American scholars who were proposing a more open participation, the by-invitation-only rule endorsed by physicists Léon Rosenfeld and Christian Møller, both based in Copenhagen, was accepted.[8] As the ICGRG was the main body in charge of drafting the list of invited participants, the ICGRG members gave themselves enormous power in defining the most relevant lines of research to be discussed in international conferences.

A different topic of debate was the sponsorship of international conferences from more structured and bureaucratized international scientific organizations, such as the international unions. Since the Chapel Hill conference, the *International Union for Pure and Applied Physics* (IUPAP) had sponsored the international conferences on relativistic theories of gravitation (DeWitt and Rickles 2011). It remained the main institutional sponsor for the conferences organized by the ICGRG in 1962 in Poland and in 1965 in London. During the London conference, IUPAP officials required a more formal connection with the ICGRG in order for IUPAP to guarantee the continuation of its sponsorship in the future. The proposal that the ICGRG should be officially affiliated with IUPAP, however, had much deeper epistemic implications than a simple organizational reconfiguration. While some American physicists were eager to promote this affiliation, mathematicians from continental Europe voiced their doubts. For them, an institutional affiliation with IUPAP would have implied that general relativity would be associated only with physics, while, they argued, it was intimately related to many different disciplinary domains. Lichnerowicz, among others, argued that IUPAP was just one of the various international unions to which their committee could adhere.[9]

Professional identities and interests were at stake in an organizational reconfiguration of this sort. The need of some leading physicists, like American theoretical physicist John A. Wheeler, who wished to see general relativity accepted by his peers as a branch of physics (Wheeler and Ford 2000; Misner 2010; Bartusiak 2015) clashed with the needs of other scholars who preferred maintaining the interdisciplinary character of the field. In view of the strong disagreements among members, the ICGRG remained formally independent from international unions, but IUPAP continued to be its major institutional sponsor until the creation of the new society in 1974.

Another strategy soon pursued by the ICGRG to improve the communication between scientists working in "General Relativity, Gravitation, and related subjects" was to issue a periodical bulletin, called *Bulletin on General Relativity and Gravitation*. As described by its editor, Mercier, the endeavor was a way to challenge the dispersion of the field both at the epistemic and the social level by providing a census of individuals and publications (Mercier and Schaer 1962; Mercier 1970). Subtly, this publication venue also implied that the different research agendas related

---

[8]Rosenfeld and Møller to members of the Committee on General Relativity and Gravitation, 2 January 1961, Vladimir Fock Papers, 1919–1974, Archive of the Russian Academy of Sciences. St. Petersburg Branch, St. Petersburg, Fond 1034, Inventaire 2, folder 180.

[9]Minutes of the Meeting of the ICGRG, 30 June and 7 July 1965, Papers of the International Society on General Relativity and Gravitation (hereafter PISGRG), Library of the Max Planck Institute for the History of Science, Berlin, folder 1.1.

to general relativity were artificially unified within an overreaching research domain, defined as GRG.

The *Bulletin* was published from 1962 to 1970 when it was absorbed into a new scientific periodical called *General Relativity and Gravitation*. Published under the auspices of the ICGRG, *General Relativity and Gravitation* became the first scientific periodical entirely devoted to this recently created scientific domain. The origins of the journal are to be found in the early discussions within the ICGRG. Polish physicist and former assistant of Albert Einstein, Leopold Infeld, had proposed to establish a journal devoted to relativistic theories of gravitation in the early 1960s, but the proposal encountered the objections of those ICGRG members who feared that such a specialized journal would have further isolated this kind of research from other branches of physics. As mentioned above, Wheeler and others were instead willing to increase the communication with a broader community of physicists, rather than to create a specific platform for 'relativists.' After Infeld's premature death in 1968, Mercier felt the duty to realize this idea in spite of the reluctance of several members of the committee (Mercier 1979).

In 1969, Mercier sent a questionnaire to the ICGRG members as well as to other scholars included on the list he prepared for the *Bulletin* enquiring whether they agreed on the need for a new international journal on general relativity and gravitation.[10] The majority of the responses welcomed the idea, but only a portion of the contacted scientists replied. Different scientists gave opposing interpretations of the outcome of this initiative. What Mercier saw as an endeavor supported by "a majority greater than what any parliament would require,"[11] was considered by American relativity expert and former collaborator of Albert Einstein, Peter Bergmann, as an undue appropriation of the ICGRG mission by Mercier. Supported by many American exponents of the GRG community, Bergmann tried to convince Mercier to abandon his plans or, at least, not to publish the new periodical under the auspices of the ICGRG.[12] He failed. Mercier pursued his vision supported only by a part of the relativity community. In March 1970, the first issue of *General Relativity and Gravitation* appeared and the journal shortly became the major publishing venue for research results of this scientific domain. Besides the organization of international conferences, the activities concerning the publications were the main operations actuated by the ICGRG to promote and organize a field that was gradually taking shape. But, while the conferences were a result of community efforts, publications were mainly the outcome of Mercier's own individual initiatives.

The various organizational and scientific conflicts displayed in this early phase of the ICGRG were a ramification of contrasts between different national or regional cultures. The major contrast was between Americans and continental Europeans who held different views on the fruitful directions of research, the relationships with

---

[10] Mercier to Scientists throughout the world who work in the field of GRG, 27 October 1969, Peter Bergmann Papers (unprocessed collection), Syracuse University Archives, Syracuse, NY (hereafter PBP).

[11] Mercier to ICGRG members, 22 September 1969, PBP.

[12] Bergmann to Bondi, 8 October 1969, PBP.

scientific disciplines, and the character of social gatherings. The early differences between the Bern and the Chapel Hill conferences in the 1950s still shaped the ICGRG activities in the 1960s. This original conflict came to the fore in very explicit ways during the mid-1960s when it was decided to number the conferences. The problem arose as to which one was the first, whether the Chapel Hill or the Bern conference. To solve the conflict Mercier accepted to concede the name GR1 to the Chapel Hill conference, provided that the Bern conference would maintain its historical significance with the official name GR0 (Mercier 1979).

During the 1960s, the ICGRG thus struggled to define its own identity in the scientific landscape. These attempts strongly depended on the initial gestation phase that resulted in two different cultural approaches to community building embodied by the opposite styles of the Bern and Chapel Hill conferences. When the ICGRG tried to self-determine its status by constructing its own imagined past (Wilson 2017) based on the early international conferences, the gestation phase was again a source of conflict as different parts recognized different generative traditions at odds with one another. The same occurred in the attempts to define which broader discipline general relativity belonged to (whether physics, mathematics, etc.). While the conflicts were not really solved, the ICGRG in this period contributed nonetheless to stabilize and solidify a field with networking activities and conference organizations. Whether it was the fruit of individual initiatives, like Mercier's, or the fruit of the organizational structure of the ICGRG, the research domain was finally recognized as a specific branch of the scientific enquiry and the ICGRG was identified as its main institutional embodiment.

The ICGRG as a Political Actor: The GR5 Conference in the Soviet Union

Tensions related to different academic and scientific cultures characterized the first eight years of the ICGRG. At the same time, the committee was able to work quite well within the political divide of the Cold War. Successful international conferences had been organized on both sides of the iron curtain and scientists working in Soviet-Bloc countries were involved in the activities of the committee together with their Western colleagues. This is not to say that the ICGRG was completely indifferent to politics. Quite the contrary, many facets of the committee clearly depended on the political situation, such as the decision to have an identical number of Soviet and American scientists as well as the exclusion of scientists working in either part of Germany. The decision concerning the need to fairly represent most national contexts was a covert way to accept the 'national membership' rule governing the work of international unions and the like. Since this rule suited the centralized administrative and political structure of the Soviet Union well, the structure itself of the ICGRG favored the implementation of East-West scientific cooperation. In summary, until 1967 the political context was a factor favoring the activity of the ICGRG, as some scientists, like Mercier, were involved in this activity mainly because they were aware of the diplomatic implications of international scientific cooperation (Mercier 1959, 1968, 1983).[13]

The situation was so rosy from the political standpoint that in 1965 the ICGRG planned to host the 1968 conference in the Soviet Union under the chairmanship of Vladimir Fock (Martinez 2019).[14] This was at about the same time when general relativity was gaining momentum within the physics and astrophysics communities after the discovery of quasars in 1963 and cosmic microwave background radiation in 1964 (Longair 2006). The major goal of those members of the ICGRG who wanted to strengthen the links between general relativity and the rest of physics was being achieved through the establishment of the fields of relativistic astrophysics and observational cosmology (Robinson et al. 1965; Kragh 1996; Schucking 2008; Bonolis 2017). In 1967, the discovery of pulsars opened new exciting research prospects for gravitational physics (Peebles 2017). Astrophysical discoveries were rapidly transforming the field, and this change provided both an opportunity and a challenge for the ICGRG, which was constructed when gravitation research was in a completely different state. The ICGRG members tried to exploit the new opportunities by increasing the number of members and including some exponents of the newly born field of relativistic astrophysics.[15]

This was the exciting scientific scene in which Fock and four Soviet colleagues were organizing the GR5 conference in the capital of Georgia, Tbilisi, when dramatic events had a disruptive impact on the organization of the conference and on the international relativity community at large. In June 1967, the Six-Day War completely changed the political landscape in the Middle East. One of the consequences was that the Soviet Union disrupted diplomatic relations with Israel. After the interruption of diplomatic relations it became a common Soviet policy to prevent the participation of Israeli scholars in conferences held in the Soviet Union. The same occurred for the GR5 conference. The ICGRG had included three Israeli scholars to be invited at the Tbilisi conference, Asher Peres, Moshe Carmeli and ICGRG member Nathan Rosen.[16] Less than four months before the conference none of them had received the expected invitation.[17]

Peter Bergmann acted as the intermediary between the Israeli physicists and the then President of the ICGRG, Hermann Bondi, voicing the suspicion that this delay was due to political reasons.[18] After Bondi had uncovered that this was the case, he threatened Fock to withdraw ICGRG sponsorship of the conference under the

---

[13] See also "Mercier Claims Science May Unite World Politics," *The Times Picayune, New Orleans*, 14 December 1968, Handakten Prof. André Mercier (1934–1998), folder BB 8.2. 1579.

[14] Minutes of the Meeting of the ICGRG, 30 June and 7 July 1965, PISGRG, folder 1.1.

[15] See Footnote 14.

[16] Bondi to all the members of the ICGRG, 12 July 1968, PBP.

[17] Peres to Bergmann, 14 May 1968, PBP.

[18] Bergmann to Bondi, 17 May 1968, PBP. See also Bergmann to Bondi, 3 May 1968, The Papers of Sir Hermann Bondi, GBR/0014/BOND, Churchill Archives Centre, Churchill College, Cambridge, UK (hereafter BOND), folder 4/4 A.

principle that "political difficulties must not stand in the way of scientific meetings." A strenuous negotiation followed, which ended with the agreement that the ICGRG would not withdraw its sponsorship of the Tbilisi meeting provided that at least one Israeli scientist, Peres, was invited.[19] Some ICGRG members vociferously protested against the terms of this agreement, but it was reluctantly accepted that one invitation would have been enough to safeguard the principle explicitly stated by Bondi.[20]

Amidst these dramatic discussions, on August 20, 1968, Warsaw Pact armed troops invaded the Czechoslovak Socialist Republic with the aim of putting an end to the liberal reforms enacted by Alexander Dubček. This event had also immediate drastic consequences for the relativity community, especially in Europe. Mercier, who had until then been one of the most active scientists in building intellectual and social relations between scientists working on different sides of the Iron Curtain, decided to send a telegram to the ICGRG members inviting them to boycott the meeting.[21]

A few days before the conference, Bondi received the confirmation that Peres had not received any invitation yet. As a consequence, Bondi officially declared that the Tbilisi conference could not be considered as a "truly" international conference sponsored by the ICGRG. The conference had rather the status of "a Soviet-organized conference to which numerous foreign scientists have been invited." In Bondi's official statement, he also declared that in order to avoid misunderstandings, he, as ICGRG President, would not partake in the conference. Bondi, however, invited all other colleagues to attend the conference stressing that no other "political or non-scientific issues", meaning the invasion of Czechoslovakia, could be considered acceptable reasons for boycotting the conference.[22]

Even though Peres finally received the invitation, it was impossible for him to get the visa in time and no Israeli scholar attended the meeting.[23] However, this formal invitation was welcomed by Bondi as a way to solve the "crisis" and to proclaim that the Tbilisi conference could be officially held under the sponsorship of the ICGRG.[24] It was too late. Only a small portion of Western European and American relativity experts eventually attended the Tbilisi conference. Many colleagues had

---

[19] Bondi to the Members of the ICGRG, 12 July 1968, PBP.

[20] Utiyama to Bondi, 18 July 1968; and Kilmister to Bondi, 24 July 1968, BOND, folder 4/4A, Bergmann to Bondi, 24 July 1968, PBP.

[21] Telegram from Mercier to Bergmann, 26 August 1968, PBP. Mercier also sent a letter to all the scholars invited to the Tbilisi conference in which he reported the text of the telegram. Mercier to Scientist invited to partake in the Tiflis-Conference on Gravitation and the Theory of Relativity, 27 August 1968, BOND, folder 4/4A.

[22] Telegram from Bondi to Members of the ICGRG, 29 August 1968, PBP, also in BOND, folder 4/4A.

[23] Telegram from Peres to Bondi, 5 September 1968; telegram from Kereselidze to Bondi, 4 September 1968; Miss Speathe to Bondi, 4 September 1968, BOND, folder 4/4A; Rosen to Fock, 15 October 1968, PBP.

[24] Telegram from Bondi to Bergmann, 6 September 1968, PBP. Various telegrams to ICGRG members, members of the Organizational Committee of the Tbilisi conference and to the Secretary of the Soviet Academy of Sciences, 6 September 1968, BOND, folder 4/4A.

not followed Bondi's suggestion and withdrew their participation either in protest against the Czech crisis or in dismay at the treatment of their Israeli colleagues.[25] The dramatic entering of political issues within the international activities of the growing international relativity community seriously endangered the continuation of these activities and the existence of the ICGRG itself.

The situation implied that the ICGRG members could not avoid explicitly discussing political matters in the future. This became already evident in the discussions during the ICGRG meeting at the Tbilisi conference. Out of twenty-four members, only eight were present, and only three of them came from Western countries: Bryce DeWitt, Møller and Wheeler.[26] During the meeting and in the subsequent correspondence exchange, every single issue became politically laden. Consequently, all the rules that had been informally adopted up to that moment had to be carefully renegotiated. The rule that the chairman of the conference would become the next ICGRG president was now criticized by some who did not approve that Fock would become the President of the committee after what they perceived as an unfair, politically motivated treatment of highly valued scientists.[27] Moreover, it was clear that the political context had to be taken into consideration in the selection process of future venues for the conference. Rosen had officially proposed to host the next conference in Israel. The Soviet ICGRG members strongly opposed this option, and those few present at the Tbilisi meeting agreed that the conference should be held in a "sufficiently neutral place." For this reason, Møller was asked to organize it in Copenhagen.[28]

The renewal of the ICGRG through the replacement of old members with new ones also became a matter of explicit political conflict. Discussions on this point had already occurred beforehand, but they had never led to any official resolution. After the Tbilisi conference, it became clear that it was necessary to immediately make a decision, but this implied yet another conflict between opposing academic and political cultures. Soviet scholars deemed it essential that the ICGRG should remain representative of national communities and that the choice of national delegates should be left to local centralized structures such as the Soviet Academy of Sciences or the Soviet Gravitation Committee. They also requested that the balance between geo-political areas be maintained in the replacement of members, which implied that retiring Soviet members should be replaced by an equal number of Soviet scholars. Other national communities had very different concerns about the replacement of committee members.[29] American physicists educated in the post-World War II

---

[25] Peter Havas to Bondi, 29 August 1968; Goldberg to Organizing Committee, 3 September 1968; telegram from Bergmann to Fock, 3 September 1968, PBP; Penrose to Bondi, 28 August 1968, Jules Géhéniau to Bondi, 30 August 1968; and telegram from Alfred Schild to Bondi, 30 August 1968, BOND, folder 4/4A.

[26] Minutes of the ICGRG during the GR5 conference held in Tbilisi on 12 September 1965, PBP.

[27] Robinson to Mercier, 5 April 1969, Engelbert Schucking Papers, Library of the Max Planck Institute for the History of Science, Berlin, Box 5, folder Ivor Robinson.

[28] Møller to Bondi, 15 October 1968, Christian Møller Papers, Correspondence 1971–81, Niels Bohr Archive, Copenhagen (hereafter CMP), Box A-D, folder 3.

[29] Bergmann to all members of the ICGRG, undated, probably November 1968 ca., PBP.

period were very vocal in asking a complete restructuration of the committee, as they considered it a self-appointed group that had no right to hold the decisional power the ICGRG members had appropriated (Held et al. 1978; Mercier 1979).[30] They conceded that the committee had been playing a crucial role in the promotion of international cooperation on general relativity, but they deemed it necessary to replace the committee members with younger and more active scientists.[31]

These new politically laden conflicts overlapped with the various tensions between academic cultures and different views on the direction for future research that had characterized the previous period. As a result, the ICGRG became a platform for experimenting in boundary-work (Gieryn 1999) between scientific and political matters in the complex arrays of activities pursued by the ICGRG. On these matters, the participants' views differed enormously, as the different reactions to the events preceding the Tbilisi conference had clearly shown. The former President of the ICGRG, Bondi, tried to explicitly define what were unacceptable political interferences on scientific activities in a letter to the ICGRG members: only the refusal to grant participation to scholars for political reasons was to be considered unacceptable by the ICGRG. The boycott of the meeting for political reasons was instead unacceptable within such an international body.[32] With this statement, Bondi, was de facto following a rule that was being formally adopted by more structured institutions, like ICSU and various international unions (Fennell 1994; Greenaway 1996). Other participants, however, did not agree with this stance, and the Tbilisi affair remained a matter of harsh debate in the following years. This debate also challenged the structure of the ICGRG itself. Especially scholars who felt excluded from important decisions started to question it and to demand its restructuring.

A New Birth: Metamorphosis into the International Society on General Relativity and Gravitation

All these tensions exploded at the following GR conference, held in Copenhagen in 1971. American ICGRG members were outspoken on behalf of younger colleagues and requested an immediate restructuring of the ICGRG.[33] Radical proposals for a complete replacement of the committee were made at the ICGRG meetings in Copenhagen. To these proposals, Soviet ICGRG members reacted very negatively

---

[30]In a personal communication, Joshua Goldberg similarly stated that the ICGRG was increasingly seen by the younger scholars as a "self-appointed group" of experts without any right to administrate the GRG international community at the institutional level.

[31]DeWitt to Mercier, 28 April 1971, GR6 Conference in Copenhagen, NORDITA Archive, Niels Bohr Archive, Copenhagen, 1971 (hereafter GR6P), Box A-L.

[32]Bondi to the members of the Committee on GRG, "The events of summer 1968," undated handwritten note, probably 4 September 1968. It is likely that it was actually sent to the ICGRG members. See Bondi to Mrs. Browne, 4 September 1968. See also Bondi to Mercier, undated handwritten note, BOND, folder 4/4A.

[33]DeWitt to Mercier, 28 April 1971, GR6P, Box A-L.

and tried to maintain the privileged status of the committee, with support coming from other European colleagues. After heated discussions between those who wanted to innovate the ICGRG and those who took a conservative stance, a majority vote accepted a proposal made by Peter Bergmann to create an altogether different scientific organization: a democratic society on the model of the American Physical Society. Within this larger society—in principle open to all experts who wished to subscribe—the ICGRG would become the executive committee. In this capacity, the ICGRG would have continued the tradition initiated in 1959 and would have maintained an authoritative position for its members. The problems remained, however, as to how to regulate the replacement of the committee members. The final decision was to substitute one third of the ICGRG every three years through a democratic vote of all society members. The need for geopolitical representativeness of the ICGRG members was to be conserved by dividing the society members in geographical groups and by imposing that retiring ICGRG members could only be substituted by other scholars belonging to the same geographical area. During these discussions, a second delicate decision had to be made: where to hold the next conference. Rosen again officially proposed Israel, and this time the attending members voted in favor of the proposal against the will of Soviet scholars who strongly opposed it.[34] In order to immediately involve all those attending the Copenhagen conference, it was decided to organize an ad hoc general assembly, in which all the participants could discuss and vote on the decisions made by the ICGRG as well as to immediately elect the eight new ICGRG members. In preparation of this meeting, Bondi and Mercier drafted a preliminary statute of the new society called *International Society on General Relativity and Gravitation* (ISGRG).[35]

All the decisions made before the ad hoc assembly were problematic for the scientists working in Soviet-bloc countries. Soviet authorities had asked the Soviet delegates at the Copenhagen conference to firmly reject Israel as the hosting country of the next conference (Khalatnikov 2012). The foundation of a new institutional structure, where an international community at large could elect all members of the executive committee, also constituted a threat to the power of centralized organizations in Soviet-bloc countries to determine who should sit in international decision-making bodies. For Soviet physicists, the fact that the new institution would maintain the principle of geopolitical representativeness was not sufficient. In this tense situation, at the ad hoc assembly, British mathematician Ivor Robinson vociferously attacked the organizers of the Tbilisi conference for their treatment of Israeli colleagues.[36] Feeling under attack in this international setting and under political pressure in their

---

[34] "Minutes of the Meetings on the Committee on GRG, held on occasion of the GR6-Conference in Copenhagen," p. 5, Papers of George Dautcourt, personal collection, Berlin (hereafter DAUT).

[35] Mercier to members of the sub-committee on the foundation of the society, 18 May 1972; and Mercier to A. A. Sokolov and N. V. Mitskiévic, 1 February 1973, PISGRG, folder 1.2.

[36] Georg Dautcourt to Møller, 29 October 1971, PISGRG, folder 1.3.

home country, the Soviet scientists abandoned the ad hoc assembly en masse—an action that was immediately followed by the majority of Eastern Bloc scientists.[37]

To solve the situation, those more sensible to the diplomatic character of this international endeavor negotiated to allow Eastern-Bloc scientists to continue to participate in discussions concerning this delicate phase of structural change, the transformation from a committee to a society. As the assembly work was partially disrupted by this politically charged conflict, it was decided to cast the vote about the replacement of the retiring ICGRG members the day after.[38] In order to come up with a common strategy approved by party authorities, the Soviet delegates went to the Embassy of the USSR in Copenhagen between the assembly and the vote. The directive was that Soviet scholars would vote for every other region but refrain from voting for the USSR region—a choice that was to be followed by the scientists of the other Eastern Bloc countries.[39] The official position of the Soviet Union with respect to the transformation of the ICGRG was communicated through a petition signed by all Soviet participants:

> Due to absence of normal conditions during discussions of various candidates at yesterday meeting (8th July), the Soviet participants of GR6 Conference are abstaining from voting any of the proposed new candidates from USSR. We cannot recognize the results of voting for this Region No. 3 (USSR) and believe that the question of the new Soviet members [of the ICGRG] must be settled by GRG Committee which will be informed of the opinion of the Academy of Science USSR and of the Soviet Gravitational Commission.[40]

The protests of the Soviet delegations did not prevent the election of two new Soviet ICGRG members, V. Braginsky and I. D. Novikov, through the vote of all attendees at the Copenhagen conference. In addition, it was confirmed that the next conference would be held in Israel, where the novel institution, the ISGRG, would be formally established.[41] Immediately after their return in the Soviet Union, Fock and other Soviet ICGRG members were questioned about their inability to avoid a situation that had become politically dramatic for them and about having accepted decisions they should have turned down (Khalatnikov 2012). With Fock under attack, the main Soviet exponent in the ICGRG became Dmitri Ivanenko. In the months following the conference he vociferously protested that it was not possible for Soviet scientists to accept the decisions of the ad hoc assembly concerning the new Soviet delegates, less so the place for the 1974 meeting because they jeopardized the chances of scholars from Soviet Bloc countries of being able to join the discussions on the new international institutional body. One of the most dramatic consequences was that

---

[37] Ivanenko to Møller, 22 July 1971, PGR6, Box A-L. Dautcourt, "Bericht über die 6. Internationale Gravitationskonferenz in Kopenhagen," DAUT.

[38] "Minutes of the 1st Meetings of a General Assembly towards the foundation of Society for General Relativity and Gravitation GRG," p. 4, DAUT; and Kip Thorne, personal communication.

[39] Dautcourt, "Bericht über die 6. Internationale Gravitationskonferenz in Kopenhagen," DAUT.

[40] Soviet participants of the GR6 Conference to the Chairman of the Plenary GR6, 9 July 1971, GR6P, Box A-L.

[41] "Minutes of the 1st Meetings of a General Assembly towards the foundation of Society for General Relativity and Gravitation GRG," p. 4, DAUT.

the establishment of the ISGRG in Israel might have meant the complete exclusion of scholars working in the Soviet Union and in other Eastern Bloc countries.[42]

Ivanenko pursued a number of strategies to avoid having the establishment of the society occur in Israel. The proposal of an alternative country for hosting the international 1974 conference or the idea to organize a second international meeting in a different country shortly before or after the official international conference met the firm opposition of many Western scholars.[43] When an official directive of IUPAP resolved that Israel had all the credentials to host international scientific meetings, many ICGRG members defended the view that the resolutions of the more official institutions regulating scientific cooperation should be followed in these politically sensitive matters.[44] Eventually, however, it was decided to find a diplomatic solution acceptable for Eastern-bloc colleagues and, more subtly, for Soviet authorities. The new ICGRG President, Møller, proposed to formally found the new society by mail. Once 150 written consents from at least ten different nations arrived by mail, the ISGRG could be declared as established.[45]

Another matter of political contention concerned the rules about the membership of the envisaged international society. The initial idea was to establish a society open to all relativity experts who wished to partake. This scheme was unacceptable to political authorities in the Soviet-Bloc countries. As officially stated in the petition of Soviet scientists at the Copenhagen meeting, delegates had to be selected by recognized national scientific organizations, as it occurred for ICSU and the international unions. American theoretical physicists Kip Thorne was very aware of this problem and suggested a solution:

> [T]o be viable, our organization must be composed of individual scientists and not of national delegations. On the other hand, the Soviet bureaucracy insists on regarding all such organizations as composed of national delegations. Somehow, a constitution must be drafted which makes it perfectly clear that ours is an organization of individual scientists; but the phraseology should probably be such that Soviet bureaucrats can misinterpret it if they wish.[46]

To negotiate between completely different views of the structure of the international society, a small subcommittee on the draft constitution was established. The final proposal made by the subcommittee was to admit two kinds of membership, individual and corporate, which, according to Mercier, met the wishes expressed by Thorne.[47] This regulation was the most explicit outcome of the negotiations between

---

[42] Ivanenko to Møller, 22 July 1971, GR6P, Box A-L; Ivanenko to Mercier, undated, probably June 1972, attached to Mercier to sub-committee, 14 July 72, PISGRG, folder 1.3.

[43] Wheeler to Møller, 4 April 1972, John Archibald Wheeler Papers, 1880–2008, Mss.B.W564, American Philosophical Society, Philadelphia, PA, Box 18; Thorne to Mercier, 25 July 1972, PISGRG, folder 1.3.

[44] Mercier to the members of the ICGRG, 30 May 1972, PISGRG, folder 1.2; Mercier to Rosen, 10 October 1972, PISGRG, folder 1.3.

[45] Møller to Mercier, 5 February 1973, CMP, folder GR7 1974.

[46] Thorne to Sciama, 1 November 1972, PISGRG, folder 1.3.

[47] Draft Constitution of The International Society for General Relativity and Gravitation, attached to the letter from Mercier to the members of the ICGRG, 18 November 1972, PISGRG, folder 1.2.

the different needs expressed by scientists working in different political systems during the Cold War. The authors of the constitution institutionalized the political nature of the enterprise by giving the society a hybrid character—a feature that was quite rare in the landscape of existing international scientific institutions (for exceptions, see Tromp 1960; Bernardini 1968; Bockris 1991; Tannenberger 2000).

After a few months of negotiations with American colleagues, who had proposed major changes in the constitution, and Soviet scholars, who insisted in maintaining the old structure, the ICGRG approved a final version that confirmed hybrid membership, essential in order to allow the participation of Eastern-Bloc scientists.[48] Mercier sent the final constitution as an attachment to the application form to his list of scientists in October 1973.[49] Less than three months later, on January 7, 1974, Mercier informed Møller that as many as 166 membership fees, including both individual and corporate, from more than twenty-three nations had been paid. The ISGRG could then "be declared into existence."[50] Among the society's initial members, there were three individuals from the Soviet Union whose membership fees were paid by the Academy of Sciences of the USSR.[51] Despite the very limited number of Eastern Bloc members in the new society, the subscription by three Soviet scientists constituted a clear sign that Soviet scientific and political authorities had approved the ISGRG. This implied that the ISGRG was born maintaining one of the initial elements that shaped the ICGRG: its being a vehicle for East-West scientific cooperation.

Conclusion: A Multi-voice Biography of the International Committee on General Relativity and Gravitation

The International Society on General Relativity and Gravitation had a strange birth: it came to life on January 7, 1974, with a letter from the Secretary of the ICGRG to its President. In fact, it cannot really be understood as a birth, but rather as a metamorphosis from a self-appointed group of national elite figures in general relativity research to a semi-democratic society. During the metamorphosis, the society maintained many of the facets that had characterized the previous institution, notably its diplomatic mission in furthering East-West collaboration and its dedication to a new field, general relativity and gravitation, the committee had helped establish in the previous fifteen years.

---

[48] Constitution of the International Society for General Relativity and Gravitation (as finalized by the ICGRG at its meeting in Paris on 22 June 1973), International Society on General Relativity and Gravitation Records, 1961–1982, AR235, American Institute of Physics, Niels Bohr Library and Archives, College Park, MD.

[49] Mercier to Relativists throughout the World, 10 October 1973, PISGRG, folder 1.4.

[50] Mercier to Christian Møller, 7 January 1974, PISGRG, folder 1.4.

[51] Mercier to Members at large of the International Society on GRG, 28 March 1974. International Society on General Relativity and Gravitation Records, 1961–1982, AR235, American Institute of Physics, Niels Bohr Library and Archives, College Park, MD.

The biographical approach was used in this essay both as a metaphor and an analytical tool to shed light on those most relevant passages that characterized the history of community-building activities in the tumultuous epistemic and political passages in the period between the Bern conference and the establishment of the society. By understanding the ICGRG as a living agent, the early conferences have been characterized as the generative events of both a shared research framework and a scientific community dedicated to it. This framework has also enabled the interpretation of debates within the ICGRG as a problem of institutional self-determination of a scientific identity, which was related to the determination of general relativity as a defined research domain. But besides this 'scientific life,' the ICGRG came to fulfill a political, diplomatic mission that was inscribed in its DNA since the Bern conference in 1955: the attempt to further East-West cooperation. This parallel 'political life' was as important as its scientific counterpart and became more and more relevant as real wars prompted the participants to explicitly address the political implications of building an international scientific institution during the Cold War.

The various tensions characterizing the work of the ICGRG and its relation with the relativity community at large made it impossible for the committee to really develop into an individual entity, however. In many respects, the ICGRG can more easily be understood as an interaction of individual biographies. While the ICGRG served the important function to establish a scientific field, it did so in an uncertain institutional terrain, where the positions of individuals were often irreconcilable. It provided unity at the organizational and institutional level, but remained disunited at the deeper level of planning, with its many cultural, generational, epistemic and political divergences. This is evident in some of the most important activities of the committee, which were strongly dependent on individual agendas, as showed by Mercier's unquestionable role in the publication means of the ICGRG. At the political level, only the transformation into a society eventually disentangled these strong individual agendas from the institutionalization of community-building activities in general relativity.

This leads to a paradoxical conclusion. The biographical approach serves the purpose to address in a unique narrative the different aspects, in this case scientific and political ones, of something that certainly existed: a committee that was generated, was born, changed, grew and finally transformed into something else. On the other hand, it never developed what we might call an institutional identity that was unequivocally superior to the sum of individual identities who composed it. Its activities were deeply conditioned by the scientific and political agendas of the involved scientists who tried to model the structure according to their own agendas, interests, views, ideologies and so on.

As some of the ICGRG members stated, it was a spontaneous "club" based on a gentlemen's agreement.[52] Even if the ICGRG nominated a Secretary who played the role of the coordinating figure, the committee was a platform in which all members were supposed to hold equal status. The President certainly had an important role, but those who held this role did so only because they organized a previous conference,

---

[52]H. -J. Treder to Møller, 17 June 1971, GR6P, Box M-Z.

and they could scarcely impose their visions concerning the future of the committee, even less of the field. The structure of the committee, therefore, resembled that of a chorus without a conductor. This structure poses several challenges to the historian wishing to use biographical approaches. The multi-voice character of such a biographical account implies that the history of the ICGRG might be understood as a set of several individual biographies in dynamical relation between each other, which poses problems both at the level of the narrative and of the historical sources. As for the first, it is highly problematic to provide a clear chronological narrative of the said institution, when one has to give space to so many actors, as there are no general principles to achieve a narrative balance between the institution and the individuals who made it. At the level of the primary sources, it implies a search into the different archives of the ICGRG members, as there was no centralized committee archive until the establishment of the ISGRG. The research results, therefore, significantly depend on random factors, such as the existence of personal archival collections and the presence of relevant materials related to the ICGRG in them. The views and activities of scholars who managed to preserve related documents appear more prominently in the final narrative. Thus, the ideal attempt to provide a faithful multi-voice representation of the activities of the committee has profound and so far unsolved limits.

References

Aspelmeyer, M., Brukner, Č., Giulini, D., & Milburn, G. (2017). Focus on gravitational quantum physics. *New Journal of Physics, 19,* 050401. https://doi.org/10.1088/1367-2630/aa6fdc.
Bartusiak, M. (2015). *Black hole: How an idea abandoned by Newtonians, hated by Einstein, and gambled on by Hawking became loved.* New Haven: Yale University Press.
Bergmann, P. G. (1956). Fifty years of relativity. *Science, 123,* 486–494.
Bernardini, G. (1968). The origin of the European physical society. *Physics Today, 25,* 34–38. https://doi.org/10.1063/1.3070996.
Blum, A. S., & Rickles, D. (Eds.). (2018). *Quantum gravity in the first half of the twentieth century.* Berlin: Edition Open Sources.
Blum, A., Giulini, D., Lalli, R., & Renn, J. (2017). Editorial introduction to the special issue "The renaissance of Einstein's theory of gravitation." *The European Physical Journal H,* 1–11. https://doi.org/10.1140/epjh/e2017-80023-3.
Blum, A. S., Lalli, R., & Renn, J. (2015). The reinvention of general relativity: A historiographical framework for assessing one hundred years of curved space-time. *Isis, 106,* 598–620.
Blum, A. S., Lalli, R., & Renn, J. (2016). The renaissance of general relativity: How and why it happened. *Annalen der Physik, 528,* 344–349. https://doi.org/10.1002/andp.201600105.
Blum, A. S., Lalli, R., & Renn, J. (2018). Gravitational waves and the 'long relativity revolution'. *Nature Astronomy, 2,* 534–543.
Bockris, J. O'M. (1991). The founding of the international society for electrochemistry. *Electrochimica Acta, 36,* 1–4. https://doi.org/10.1016/0013-4686(91)85171-3.
Bonolis, L. (2017). Stellar structure and compact objects before 1940: Towards relativistic astrophysics. *The European Physical Journal H, 42,* 311–393. https://doi.org/10.1140/epjh/e2017-80014-4.
Cassidy, D. (1986). Understanding the history of special relativity. *Historical Studies in the Physical and Biological Sciences, 16,* 177–188. https://doi.org/10.2307/27757561.

DeWitt, C. M., & Rickles, D. (Eds.). (2011). *The role of gravitation in physics: Report from the 1957 Chapel Hill conference*. Berlin: Edition Open Sources.

DeWitt-Morette, C. (2011). *The pursuit of quantum gravity: Memoirs of Bryce DeWitt from 1946 to 2004*. Heidelberg: Springer.

Earman, J., & Glymour, C. (1980). Relativity and eclipses: The British eclipse expeditions of 1919 and their predecessors. *Historical Studies in the Physical Sciences, 11*, 49–85. https://doi.org/10.2307/27757471.

Eisenstaedt, J. (1986). La relativité générale à l'étiage: 1925–1955. *Archive for History of Exact Sciences, 35*, 115–185. https://doi.org/10.1007/BF00357624.

Eisenstaedt, J. (1989). The low water mark of general relativity, 1925–1955. In D. Howard & J. Stachel (Eds.), *Einstein and the history of general relativity* (pp. 277–292). Boston: Birkhäuser.

Fennell, R. (1994). *History of IUPAC, 1919–1987*. Oxford: Blackwell Science Ltd.

Gieryn, T. F. (1999). *Cultural boundaries of science: credibility on the line*. Chicago: University of Chicago Press.

Glick, T. F. (Ed.). (1987). *The comparative reception of relativity*. Dordrecht: Reidel.

Goenner, H. F. M. (2004). On the history of unified field theories. *Living Reviews in Relativity, 7*, 2. https://doi.org/10.12942/lrr-2004-2.

Goenner, H. F. M. (2014). On the history of unified field theories. Part II. (ca. 1930–ca. 1965). *Living Reviews in Relativity, 17*, 5. https://doi.org/10.12942/lrr-2014-5.

Goldberg, J. N. (1992). US Air Force support of general relativity 1956–72. In J. Eisenstaedt & A. J. Kox (Eds.), *Studies in the history of general relativity* (pp. 89–102). Boston: Birkhäuser.

Greenaway, F. (1996). *Science international: A history of the International Council of Scientific Unions*. Cambridge: Cambridge University Press.

Greene, M. T. (2007). Writing scientific biography. *Journal of the History of Biology, 40*, 727–759.

Gutfreund, H., & Renn, J. (2017). *The formative years of relativity: The history and meaning of Einstein's Princeton lectures*. Princeton: Princeton University Press.

Held, A., Leutwyler, H., & Bergmann, P. G. (1978). To André Mercier on the occasion of his retirement. *General Relativity and Gravitation, 9*, 759–762. https://doi.org/10.1007/BF00760862.

Hentschel, K. (1990). *Interpretationen und Fehlinterpretationen der speziellen und der allgemeinen Relativitätstheorie durch Zeitgenossen Albert Einsteins*. Boston: Birkhäuser.

Hermann, A., Belloni, L., Mersits, U., Pestre, D., & Krige, J. (1987). *History of CERN, Volume I: Launching the European Organization for nuclear research*. Amsterdam: North Holland.

Hollings, C. D. (2016). *Scientific communication across the iron curtain*. Cham: Springer International Publishing.

Holton, G. (1969). Einstein, Michelson, and the "Crucial" experiment. *Isis, 60*, 133–197.

Ivanov, K. (2002). Science after stalin: Forging a new image of soviet science. *Science in Context, 15*, 317–338.

Kaiser, D. (1998). A ψ is just a ψ? Pedagogy, practice, and the reconstitution of general relativity, 1942–1975. *Studies in History and Philosophy of Science Part B: Studies in History and Philosophy of Modern Physics, 29*, 321–338. https://doi.org/10.1016/S1355-2198(98)00010-0.

Kaiser, D. (2000). *Making theory: Producing physics and physicists in postwar America*. Ph.D. dissertation. Cambridge, MA: Harvard University.

Kaiser, D. (2002). Cold war requisitions, scientific manpower, and the production of American physicists after World War II. *Historical Studies in the Physical and Biological Sciences, 33*, 131–159. https://doi.org/10.1525/hsps.2002.33.1.131.

Kaiser, D. (2012). Booms, busts, and the world of ideas: Enrollment pressures and the challenge of specialization. *Osiris, 27*, 276–302. https://doi.org/10.1086/667831.

Kaiser, D., & Rickles, D. (2018). The price of gravity: Private patronage and the transformation of gravitational physics after World War II. *Historical Studies in the Natural Sciences, 48*, 338–379.

Kennefick, D. (2007). *Traveling at the speed of thought: Einstein and the quest for gravitational waves*. Princeton: Princeton University Press.

Kennefick, D. (2012). Not only because of theory: Dyson, Eddington, and the competing myths of the 1919 eclipse expedition. In C. Lehner, J. Renn, & M. Schemmel (Eds.), *Einstein and the changing worldviews of physics* (pp. 201–232). Boston: Birkhäuser.

Khalatnikov, I. M. (2012). *From the Atomic Bomb to the Landau Institute: Autobiography. Top Non-Secret*. Berlin: Springer.

Kragh, H. (1996). *Cosmology and controversy: The historical development of two theories of the universe*. Princeton, NJ: Princeton University Press.

Kragh, H. (1999). *Quantum generations: a history of physics in the twentieth century*. Princeton, N.J.: Princeton University Press.

Lalli, R. (2016). 'Dirty work', but someone has to do it: Howard P. Robertson and the refereeing practices of physical review in the 1930s. *Notes and Records: The Royal Society Journal of the History of Science, 70,* 151–174. https://doi.org/10.1098/rsnr.2015.0022.

Lalli, R. (2017). *Building the general relativity and gravitation community during the cold war*. Cham: Springer.

Lichnerowicz, A. (1992). Mathematics and general relativity: A recollection. In J. Eisenstaedt & A. J. Kox (Eds.), *Studies in the history of general relativity* (pp. 103–108). Boston: Birkhäuser.

Lichnerowicz, A., & Tonnelat, M.-A. (Eds.). (1962). *Les théories relativistes de la gravitation*. Paris: Éd. du Centre national de la recherche scientifique.

Longair, M. S. (2006). *The cosmic century: A history of astrophysics and cosmology*. Cambridge: Cambridge University Press.

Martinez, J.-P. (2019). Soviet science as cultural diplomacy during the Tbilisi conference on general relativity. *Vestnik of Saint Petersburg University History, 64,* 120–135.

McCrea, W. H. (1955). Jubilee of relativity theory: Conference at Berne. *Nature, 176,* 330. https://doi.org/10.1038/176330a0.

Mercier, A. (1950). *De la science à l'art et à la morale*. Paris: Neuchatel Éditions du Griffon.

Mercier, A. (1959). *De l'amour et de l'etre*. Paris: Louvain.

Mercier, A. (1968). On the foundation of man's rights and duties. *Man and World, 1,* 524–539.

Mercier, A. (1970). Editorial. *General Relativity and Gravitation, 1,* 1–7. https://doi.org/10.1007/BF00759197.

Mercier, A. (1979). Birth and rôle of the GRG-organization and the cultivation of international relations among scientists in the field. In P. C. Aichelburg & R. U. Sexl (Eds.), *Albert Einstein: His influence on physics, philosophy and politics* (pp. 177–188). Braunschweig: Vieweg Verlag. https://doi.org/10.1007/978-3-322-91080-6_13.

Mercier, A. (1983). *André Mercier, physicien et métaphysicien*. Berne: Institut des sciences exactes de l'Université de Berne.

Mercier, A. (1992). General relativity at the turning point of its renewal. In J. Eisenstaedt & Anne J. Kox (Eds.), *Studies in the history of general relativity* (pp. 109–121). Boston: Birkhäuser.

Mercier, A., & Kervaire, M. (Eds.). (1956). *Fünfzig Jahre Relativitätstheorie: Jubilee of relativity theory*. Helvetica Physica Acta. Supplementum IV. Basel: Birkhäuser.

Mercier, A., & Schaer, J. (1962). General information. *Bulletin on General Relativity and Gravitation, 1,* 1–2. https://doi.org/10.1007/BF02983127.

Misner, C. W. (2010). John Wheeler and the recertification of general relativity as true physics. In I. Ciufolini & R. A. Matzner (Eds.), *General relativity and John Archibald Wheeler* (pp. 9–27). Dordrecht: Springer. https://doi.org/10.1007/978-90-481-3735-0_2.

Newman, E. T. (2005). A biased and personal description of GR at Syracuse University, 1951–61. In A. J. Kox & J. Eisenstaedt (Eds.), *The universe of general relativity* (pp. 373–383). Boston: Birkhäuser.

North, J. D. (1965). *The measure of the universe: A history of modern cosmology*. Oxford: Oxford University Press.

Pauli, W. (1956). Schlußwort den Präsidenten der Konferenz. In A. Mercier & M. Kervaire (Eds.), *Fünfzig Jahre Relativitätstheorie: Jubilee of relativity theory* (pp. 261–267). Basel: Birkhäuser.

Pauli, W. (1999). Wissenschaftlicher Briefwechsel mit Bohr, Einstein, Heisenberg u.a. Band IV, Teil II: 1953–1954. In K. Meyenn (Ed.). Berlin: Springer.

Peebles, P. J. E. (2017). Robert Dicke and the naissance of experimental gravity physics, 1957–1967. *The European Physical Journal H, 42*, 177–259. https://doi.org/10.1140/epjh/e2016-70034-0.

Renn, J. (2007). *The genesis of general relativity*. Dordrecht: Springer.

Rickles, D. (2011). The chapel hill conference in context. In C. M. DeWitt & D. Rickles (Eds.), *The role of gravitation in physics: Report from the 1957 Chapel Hill conference* (pp. 7–21). Berlin: Edition Open Sources.

Robinson, I., Schild, A., & Schucking, E. L. (Eds.). (1965). *Quasi-stellar sources and gravitational collapse, including the proceedings of the First Texas Symposium on Relativistic Astrophysics*. Chicago: University of Chicago Press.

Ryckman, T. (2005). *The reign of relativity: Philosophy in physics 1915–1925*. Oxford: Oxford University Press.

Schucking, E. L. (2008). The first texas symposium on relativistic astrophysics. *Physics Today, 42*, 46–52. https://doi.org/10.1063/1.881214.

Schweber, S. S. (2006). Einstein and Oppenheimer: Interactions and intersections. *Science in Context, 19*, 513–559.

Shortland, M., & Yeo, R. (1996). Introduction. In M. Shortland & R. Yeo (Eds.), *Telling lives in science: Essays on scientific biography* (pp. 1–44). Cambridge: Cambridge University Press.

Sponsel, A. (2002). Constructing a 'revolution in science': The campaign to promote a favourable reception for the 1919 solar eclipse experiments. *The British Journal for the History of Science, 35*, 439–467.

Stanley, M. (2003). "An expedition to heal the wounds of war": The 1919 Eclipse and Eddington as quaker adventurer. *Isis, 94*, 57–89. https://doi.org/10.1086/376099.

Strasser, B. J. (2009). The coproduction of neutral science and neutral state in cold war Europe: Switzerland and international scientific cooperation, 1951–69. *Osiris, 24*, 165–187. https://doi.org/10.1086/605974.

Tannenberg, H. (2000). From CITCE to ISE. *Electrochimica Acta, 45*, xxvii–xxviii.

Thorne, K. S. (1994). *Black holes and time warps: Einstein's outrageous legacy*. New York: WW Norton.

Tromp, S. (1960). Report of the secretary-treasurer. *International Journal of Bioclimatology and Biometeorology, 4*, 213–214.

van Dongen, J. (2010). *Einstein's unification*. Cambridge: Cambridge University Press.

Wheeler, J. A., & Ford, Kenneth. (2000). *Geons, Black Holes, and quantum foam: A life in physics*. New York: WW Norton.

Will, C. M. (1989). The renaissance of general relativity. In P. Davies (Ed.), *The new physics* (pp. 7–33). Cambridge: Cambridge University Press.

Will, C. M. (1993). *Was Einstein right? Putting general relativity to the test* (2nd ed.). New York, NY: BasicBooks.

Wilson, A. (2017). Science's imagined pasts. *Isis, 108*, 814–826. https://doi.org/10.1086/695603.

Wright, A. S. (2014). The advantages of bringing infinity to a finite place. *Historical Studies in the Natural Sciences, 44*, 99–139. https://doi.org/10.1525/hsns.2014.44.2.99.

Chapter 12
Whose Biography Is It Anyway? Shared Biographies of Institutions, Leaders, Instruments, and Self

Catherine Westfall

My life-long career preoccupation has been to better understand how large-scale science develops in its most obvious natural habitat, that is, the laboratory, in particular the U.S. national laboratories. Thus, in addition to exploring the dramas and drama queens of high-energy physics, I have also investigated more exotic nooks, kooks, and crannies of these laboratories. As a result my research has focused on nuclear physics, materials science, and engineering projects as well as a variety of laboratory actors: scientists, engineers, elite advisors, funding officials, and upon occasion, laboratory directors. One of the joys of a long career as a laboratory historian is the opportunity to consider what I have missed so far so that I can find promising new ways to view familiar territory. Recently I have been determined to pluck episodes from previously told stories—and from experiences inside and outside the laboratory that never found their way into articles—to discover new combinations, patterns, and avenues of inquiry. The resulting new perspectives, in turn, suggest fresh insights into how laboratories are formed and function and how they can be studied. Along the way I have also gained a deeper appreciation of my own life in the laboratory and how it has affected my work.

My desire for fresh perspectives had led me most recently to contemplate in a more rigorous way the analytical possibilities of biographical studies. As a start I drew inspiration from Theodore Arabatzis's notion that the electron and other theoretical entities "have a certain independence from the intentions of their makers; that is, they have a life of their own" (Arabatzis 2006, p. 36). I was intrigued by the possibility of conceptualizing nonhuman aspects of laboratory life as having characteristics similar to human life. Could this metaphorical use of biography highlight the intertwined development, the shared "lives", of institutions, actors, objects? And

C. Westfall (✉)
Michigan State University, East Lansing, MI, USA
e-mail: westfa12@msu.edu

could this conceptualization facilitate consideration of how autobiography fits into the mix? And finally, could this expanded view of shared biography, in turn, offer other interesting angles to explore in the search for an understanding of laboratory life?

Sizing up Laboratories, Directors, and Instruments

This is partially explored territory: I have long focused on individual laboratories as the unit of study, considering that a laboratory develops in stages similar to those in human life. Indeed, the subtitle of my 1988 dissertation is "The Birth of Fermilab" (Westfall 1988). At that same time, my dissertation and subsequent studies of U.S. laboratory life make clear that the agency of laboratories is limited, that is, they do not really have an independent life of their own. In fact, although there are a variety of influences, laboratory actors, in particular founding directors, shape the life of the laboratory in ways that are profound. As a result I have dared upon occasion to dabble in scientific biography, using the laboratory director as a unit of study, at least in relation to the biography of the laboratory. A prime example is a 2002 article that considers how E. O. Lawrence's launch of his Berkeley laboratory influenced how Robert Wilson founded Fermilab for high-energy physics in Illinois and how Hermann Grunder founded Jefferson Laboratory for nuclear physics in Virginia (Westfall 2002).

In past decades granting a starring role to directors has left me open to the criticism that I am indulging in non-scholarly "Great Man" history, that is ascribing agency simplistically to one gallant hero whose single-handedly orchestrates construction. Both historians and scientists were uncomfortable with the very concept of the director as a Great Man. I remember when I was gathering material for my dissertation on the founding of Fermilab, John Heilbron warned me not to be fooled by the charm of the grandfather-like men I was interviewing in the 1980s since they were quite different characters in the 1960s when they fought tooth and nail over the privilege of founding Fermilab. And at Jefferson laboratory, which was still being built when I began writing its history, nuclear physicists involved with the project were aghast to learn I was comparing Grunder with Wilson and E. O. Lawrence. One physicist who had spearheaded the construction of a Jefferson Laboratory experimental hall stood in the hall outside my office, wringing his hands, insisting that I would destroy my credibility by publishing such a comparison. Another, who had been an elite reviewer of the Jefferson Laboratory experimental program, seemed to invoke the unspoken pecking order in American physics to explain the underlying problem. Using his hands to measure three sizes, he noted that E. O. Lawrence was the giant who revolutionized accelerator-based physics in the post World War II glory days, Robert Wilson was a big, world-class high-energy physicist, and Hermann Grunder was this little itty bitty guy, an engineer turned nuclear physicist turned bureaucrat. (From pictures Lawrence, who I never met, looks on the tall side, but I towered over both Wilson and Grunder, which added a comic element to the hand measurements).

Whatever their comparative sizes, literally or figuratively, both Wilson and Grunder played a key role in setting the parameters of laboratory instruments, including the centerpiece of the laboratory, the particle accelerator. Both men made bold choices. In the mid-1960s, as rising costs threatened initial funding of the laboratory, Wilson rejected the expensive, carefully engineered risk-adverse original accelerator design, insisting instead on plans for an inexpensively spare (and therefore risky) machine that could be expanded from its original 200 GeV energy range to 1000 GeV. With a nod towards further cost savings, he arranged for spartan experimental areas with cheap, cobbled together equipment set in minimally developed experimental areas. For his part, in the mid-1980s Grunder rejected already approved plans for an accelerator based on standard, reliable technology and instead successfully convinced the funding agency to re-do the approval process for a completely different type of accelerator using superconducting radiofrequency technology. Although this technology promised greatly improved capabilities, it was enormously tricky, and in fact, had doomed a previous nuclear physics project. And Grunder didn't stop there. He continued to successfully badger the funding agency until they approved the construction of three fully equipped experimental halls, even though by any reasonable measure such expansive plans meant greatly exceeding the originally allocated budget.

So founding laboratory directors are important actors in their laboratories. Therefore, it is important to get a clear picture of how they function—to size them up so to speak—within the context of the life of the laboratory. But getting a clear picture can be challenging.

For one thing: I'll let you into a little secret known to those who do laboratory history in situ: laboratory directors are crazy people! And founding directors are especially crazy. I remember Wilson explaining that creating Fermilab took him out of himself. Certainly, the stories he and others told made him sound out of his mind. For example, he reportedly took the architect-created sketches for the Fermilab administration building, tore them up and danced on the pieces. And I can personally attest to the maniacal hyperactivity and volatility Grunder was known for since I experienced it first hand when Jefferson Laboratory was being built. I thus found it easy to believe the stories of how he would roam the halls of Department of Energy officials, staging temper tantrums—waving his arms, stomping his feet—as he went from office to office when funding news was bad. One official whispered that it was as if he thought he could scare money into his budget.

Of course, these acts were largely stagecraft and purposeful. Wilson needed to recruit users to work in unpleasant conditions—the users always mentioned the mud and lack of rest rooms—with an accelerator so under-built that it might not work. It was therefore important to have a welcoming, attractive beacon, the iconic "High Rise" administration building with its dramatic upswept walls that can be seen for miles, rising from the flat, undistinguished looking suburbs and farm land surrounding Chicago. And as for Grunder, he had to find the money to build a laboratory that would begin experiments in 1995, two years after the much larger, more impressive Super Conducting SuperCollider was defunded in the middle of construction.

Shared Biographies: Laboratories, Directors, and Instruments

Thanks to the drama of founding directors, I have cast the first director as the star and prime agent in the founding story of the laboratory since my dissertation days. But my current ruminations have suggested a new perspective for this and other elements of laboratory life that promises to provide deeper insights: visualizing shared biography, that is, the overlap between two or more life stories within the laboratory. My concept of shared biography is reminiscent of Peter Galison's "intercalated periodization," which in his words "drops the assumption of coperiodization and separates the subculture of physics" into parts, in his case "three quasi-independent groupings of theory, experiment, and instrument making." This idea is based on the observation that the efforts of these groups develop according to individual time spans so that the joint development of knowledge and artifacts (by way of his famous trading zone) is best conceptualized as a disjointed process. For me, shared biography is a way to conceptualize how the laboratory is created and functions in terms of disjointed biographies—in this case I consider the shared lives of founding directors, instruments, and the laboratory. In the process I consider disjointed time periods, as well as other shared disjointed frames of reference in particular the unequal importance of a development in shared biographies. The key point, in both conceptualizations, is in Galison's words "that breaks in one [element] need not coincide with breaks in the others" (Galison 1997, p. 799).

Viewing the shared biography of founding directors and their laboratories suggests that the crazy behavior of Wilson and Grunder stems in part from the disjointed relationship of the founding years in the life of the laboratory compared with the founding years in the life of the first director. It is true that Wilson and Grunder had a deeper influence on laboratories than subsequent directors; as one Jefferson Lab administrator put it, decisions made during construction are literally "cast into concrete." (Domingo 1998) Nonetheless, the stars in the lives of laboratories are results and the scientists who provide them, not directors or buildings, or instruments, all of which are simply the means to set the stage for the recruitment of scientists and results.

By contrast, Fermilab and Jefferson Laboratory played the starring role in the lives of Wilson and Grunder. In fact, both men experienced a precipitous career fall after founding their respective laboratories. Wilson was pushed out as Fermilab director and therefore was not allowed to lead the 1000 GeV accelerator construction project that was his heart's desire. After a few undistinguished years as the director of Argonne National Laboratory Grunder was also forced into retirement. Tellingly, one of Grunder's lieutenants remarked that Grunder's push for three fully equipped experimental halls made no sense in terms of the construction era budget crunch since it threatened the completion of the laboratory as a whole. He judged that Grunder felt the move was nonetheless necessary because the laboratory was Grunder's mausoleum. (Domingo 1998) That was meant figuratively, of course. And yet Wilson really is buried at Fermilab. I have often spoken of the legacies that Wilson and

Grunder left at their laboratories. I now realize that for both men, the laboratory they founded was the biggest monument of their lives.

Seeing the lopsided relationship between directors and their laboratories made me understand why I had the gut feeling when writing about the early years of Fermilab and Jefferson Laboratory that both directors had been consumed by their jobs. And this realization, in turn, made me recalibrate my understanding of the founding years themselves. Of course in both cases I was examining boxes of documents and interviewing dozens of people both inside and outside of the laboratory. And yet because of the obvious importance of the influence of the founding directors—and the color and drama of their behavior—I tended to conceptualize the beginning of the Fermilab and Jefferson Laboratory stories in line with the perspective of two men desperate not merely because of the considerable pressures of multi-hundred billion dollar construction projects but also due to the internal pressure to maintain a sense of self. This made me wonder if to some degree the drama of founding Fermilab and Jefferson laboratory was at least in part not only staged, but also conjured by their founding directors.

Identifying the disjointed shared biography of founding directors and laboratories brought into focus another such biography: the biography of the instruments that were the raison d'être for founding the laboratories in the first place. For one thing, Fermilab and Jefferson Laboratory (which remain in operation today) have outlasted those accelerators and related equipment, which have been replaced by newer, more capable devices. In addition, although Wilson and Grunder set the parameters for experimental equipment at Fermilab and Jefferson Laboratory—and spearheaded laboratory construction so that they could be built—the equipment outlasted these directors. Indeed, these instruments—like the laboratories that housed them—loom larger in the lives of the directors than the directors do in the lives of the instruments, which were built, maintained, and improved by many others.

This disjunction in the lives of instruments and directors makes clear that it is limiting to cast the director as the star and prime agent in a consolidated founding story of the laboratory. And clearly, it is limiting to break laboratory stories into the periods of directors that for example begin with the "Wilson years" and the "Grunder years." An alternate approach is to use something akin to intercalated periodization, that is, to view how a laboratory is created and functions in terms of disjointed stories. Such an approach would allow an analysis of the staged and conjured drama of the Wilson and Grunder years alongside a consideration of their longer-term impact as measured by the working lifetime of initial equipment. What can emerge, in the process, is deeper insights into how laboratories develop from their beginnings and what founding directors—and other factors—have to do with the process.

The Autobiographies of Wilson and Grunder

A first step in analyzing the staged and conjured drama of a founding director is to take a close look at the drama itself and how it functioned in the development of the laboratory in question. In this regard, Wilson and Grunder provide an interesting contrast. Both founding directors drew on their colorful personalities when mounting the effort to found their laboratories. And yet, Wilson was the only one who conjured autobiography, that is, he used the story of his origins—his Bildungsroman—to help mobilize efforts to construct the laboratory. Realizing this difference between Wilson and Grunder opened several questions in my mind. For one thing, why did Wilson use autobiography and Grunder not use it? Did Grunder simply miss a trick? (This notion struck me as wrong, but initially I did not have an alternate answer). I also wondered what difference did it make that Wilson used autobiography and Grunder did not in the founding and functioning of their respective laboratories?

As Wilson noted in a 1977 interview, he grew up within "a great tradition of telling stories" with "a fire outside" with men "standing around it [...] in a circle [...] it was magnificent, always, a great occasion, like a seminar". And he certainly went on to tell an engaging story. He was born in Wyoming in a town called Frontier. He remembered "riding all alone with a packhorse many miles away—sometimes days away—from other people, days without seeing anybody," which gave him "the impression of being completely independent". These early experiences helped form his identity. As he explained:

> I worked hard at ranching and wanted to be a cowboy, perhaps even was. I mean, I would ride after cattle. I learned how to throw a lariat. (Wilson 1977)

Life on the ranch also helped pave the way for his future. He worked alongside his uncle, an experience that taught important lessons in hands-on problem solving and self-confidence.

> We had to make everything in the blacksmith shop, all the haying equipment, except the rakes and the mowers, were made, but you had to keep repairing those yourself. You couldn't take them 30 miles away and get them repaired when they broke. You had to make a part in the blacksmith shop and fashion it and put it on the mower or the rake. (Wilson 1977)

He would sometimes "work in the blacksmith shop just for the hell of it, and learned how to use my hands and make things," which gave him confidence that he could "build large contraptions and make them work" (Wilson 1977).

As he tells it, this upbringing proved useful when Wilson left Wyoming to pursue a career in science. After enrolling in the prestigious University of California at Berkeley in 1932 in engineering, Wilson met Ernest Lawrence, joined the Berkeley Cyclotron-building research team, and obtained a Ph.D. in physics in 1940. Wilson became one of Lawrence's "boys," joining in the hard work and camaraderie of the Berkeley Laboratory (Fig. 1). Experiences on the ranch helped him excel there—for example, Wilson remembers battling the difficulty of obtaining a vacuum on one occasion by temporarily using a broom handle and then later machining his own seal based on the properties of a bicycle pump. In turn, early lessons about confidence were

Fig. 1 Robert Wilson and his wife Jane at a party during the Berkeley years

reinforced and extended. As he explained, from working with Lawrence he learned "if you want something to come true, you can make it come true just by pushing like hell." That meant "you twisted nature if necessary until you made things happen the way you wanted them to be" (Wilson 1977).

His early experiences also helped him thrive at wartime Los Alamos after a short stint at Princeton. In the midst of rustic conditions Wilson found it easy to cobble together equipment to find innovative ways to measure the properties of little known fissionable materials in the crash effort to design the first atomic bombs, also demonstrating his horseback riding prowess in the austere, New Mexican environment (Hoddeson et al. 1993, p. 80, pp. 184–186).

When Wilson was chosen as Fermilab's first director in 1967 he had been a professor, experimenter, and accelerator builder at Cornell University for twenty years. To rise to the task of founding Fermilab, however, he reached back to his Bildungsroman to conjure the drama he employed to help found the laboratory. As described in our 2008 Fermilab history: "Wilson would convene a meeting of the entire staff roughly every six months... At the first meeting, about two dozen employees gathered in an open area" on the as yet undeveloped site. "A campfire" as in Wilson's native Wyoming "would have been impractical." Instead, in the words of one staff member "people took benches and chairs. Some even sat on the hoods of their cars," to listen to Wilson's inspiring rhetoric. Wilson's story telling relied heavily on a frontier motif. After admitting the uncertainty of funding, he noted: "Your coming here is a kind of adventuresome thing." He would then exhort them "to be pioneers to come out and go through a period of inconvenience." Later, he would address researchers struggling with their cheap equipment and minimally developed experimental areas, explaining that nature's mysteries were "just as much of a challenge" at this point "as they were when our pioneer forebears started at the beginning of the century" (Hoddeson et al. 2008, p. 112).

Wilson's story telling with its autobiographical flavor was decidedly effective. I remember at lunch asking an European émigré how he could leave home to come work at the site with the flies and mud and other inconveniences he had just recounted, especially considering the uncertainty of funding. He replied with another memory: at the staff gatherings Wilson would sometimes scoop up the ground, pop it in his mouth, explaining that he could tell of the soil's possibilities by its very taste. How could he not stick with this man who was the quintessential American cowboy, so at one with the land? Leon Lederman, later Fermilab's second director noted as an experimenter at the time that his group "didn't bargain on the frogs, or the ditches, or the roof leaking, but its members accepted the primitive conditions" as part of "the pioneering spirit of doing experiments at early Fermilab" (as quoted, Hoddeson et al. 2008, p. 168).

Wilson's story that the laboratory set on small-town farm land in the greater Chicago area was a frontier settled by cattle ranchers stuck. It was, if you will, a very effective branding device, shaping how Fermilab users and personnel viewed themselves and the way that others viewed them for decades to come. When I was writing my dissertation in the 1980s I often heard Fermilab experimenters proudly refer to themselves as cowboys, by which they meant rambunctious, risk-taking conquerors of nature. And later I heard Fermilab competitors—and a funding agency official—refer to Fermilab's cowboys, by which they meant rule-breaking, uncooperative, and reckless, and yet somehow engaging and inspiring researchers.

Laboratory life proceeded differently in Virginia in the building years. Grunder's did use story telling as well as other techniques to conjure drama and motivate workers. For example, he periodically gathered the staff to give pep talks, particularly when funding prospects looked dim. One staff member remembered a repeated, emphatic declaration: "We will not fail," noting that this made such talks "a stunning demonstration of the intensity of his willpower." Another signature tactic was to inspire staff to work long hours by modelling what one staff member characterized as abnormal energy. He was at the laboratory day and night, never appearing to sleep. I repeatedly heard one particular story. One afternoon Grunder decided, on a whim, to clear the parking lot after a rare Virginia snow fall. Within minutes, armed with a shovel, this sixty-something director cajoled his bewildered scientific and administrative staff members to follow him outside in the cold. They tried to keep up, but he moved, as one of them later put it, in "fast animation, like a cartoon character" (Westfall 2002, 397).

However, Grunder did not use a Bildungsroman to help mobilize the staff and shape Jefferson Laboratory's identity. I found this odd when I left the Fermilab history project and began working at Jefferson Laboratory during the final years of its construction in the early 1990s. After all, one of my first impressions at Jefferson Laboratory was that Grunder in most respects closely followed in the footsteps of Wilson who had founded Fermilab just 25 years earlier. This was the observation that later led to the article comparing Wilson and Grunder as well as E. O. Lawrence. Why would he not want to put his stamp on Jefferson Laboratory, his mausoleum, with a personal story the way Grunder had with Fermilab? I cast around for possible explanations. Maybe Grunder did not have a personal story that lent itself to conjuring the necessary drama. Or perhaps he did not have a story he wished to share. Or maybe he simply lacked the story telling skills to mold early experiences into a compelling narrative that would motivate staff and provide a brand for Jefferson Laboratory.

By 1999 I no longer could believe any of these explanations. In that year I conducted an interview with Grunder that was recorded and made public that showed Grunder could conjure a Bildungsroman that was potentially every bit as serviceable for motivating and branding as the story Wilson had told. In the interview Grunder quite spontaneously described being a teenager in Basel during World War II and how he handled being caught in the cross fire when the fighting strayed into Swiss airspace. There was danger—for example several floors of his grandfather's house were accidentally obliterated by a British bomb. But there was also opportunity. As he explained: when airplanes went down they would "typically blow up, but there was a short time" allowed for the crew to get to safety and "you could go in and loosen up a little bit of the ammunition and carry it away on your bicycle." He thought it was fun "to make experiments in the back yard," although "the windows would go because you weren't too good [at] figuring out the dose." In the wreckage he found business opportunities as well: k rations, with their chewing gum and chocolate, could be sold (Grunder 1999).

Grunder acknowledged that growing up in Switzerland meant he was not subject to the deprivations, danger, and devastation of the war faced by other Europeans. But he and other Swiss teens were "useful observers" of the horrors of the war that affected him only peripherally. It meant he learned the lesson that "life is dangerous" so risks were worth taking, and "if you don't look out for yourself, forget it". Another part of his wartime experience was meeting American G.I.s, "guys from Iowa, from Arkansas" who risked and gave their lives to save Europe from the Nazis. He found these young men and by extension their country "impressive," and like many of his generation, he was also drawn to the America's booming economy and job possibilities. And at the same time, he felt pushed away from home. As he explained, due to his wartime experience, he felt a "deep resentment towards a generation who got us into World War II." Although he had a "love-hate relationship with Europe" he felt a misfit there (Grunder 1999).

He first established himself in Europe, earning a M.S. in mechanical engineering from the Karlsruher Institut für Technologie, marrying, and having three children. He then wrote 82 letters in search of a job in the U.S., including one to Lawrence Livermore Laboratory, a defense facility. He got no response from Livermore (and indeed Grunder could not have worked there as a non-citizen). But then he happened to run into the head of a mechanical engineering group from Lawrence Berkeley Laboratory, Livermore's non-defense sister laboratory. Discussion revealed that Berkeley was hiring engineers, and Grunder made clear he would jump at the chance to emigrate. A few months later Grunder got a call from the man, who had in the meantime been forwarded Grunder's letter from Livermore. There was no money for him to make the move and he would be given no contract beforehand. But if he showed up, there was a job for him.

Without a job in hand, getting a visa required a financial guarantee, which he obtained from a brother-in-law who was living in the U.S. After getting the necessary visas, he flew to San Francisco in 1959 along with his wife, three kids, and $34. As he later explained: they took a taxi straight from the airport to the laboratory and "deposited the kids" at the housing office run by a "charming" lady who said "the first thing we need to do is find a motel for you". In the meantime, Grunder went off, found the guy who had offered him the job, and became duly employed. The next step was to get "a station wagon" and pack "the whole family in it … it was all downhill from there". Of course he meant that life in America only got easier from there, but it struck me as I listened that ironically, his career only went up from there. With considerable energy, Grunder worked his way up the career ladder in accelerator-building projects as an engineer, then returned to Europe and obtained a Ph.D. in experimental nuclear physics from the Universität Basel in 1967. This work positioned him to obtain more and more influential roles in accelerator building and laboratory administration. And along the way he became a naturalized citizen (Grunder 1999) (Fig. 2).

In retrospect, I find it amazing that Grunder never employed this story when founding Jefferson Laboratory. After all, he was neither shy, nor one to avoid self-aggrandizement. And Grunder had seen what Wilson had been able to accomplish with the story of his frontier youth. Why did Grunder not add the story of his Swiss beginning to his storytelling repertoire to provide another way to mobilize staff members? The story itself seems replete with opportunity to inspire loyalty and hard work—Grunder casts himself into the role of the worthy, talented, entrepreneurial America-loving emigrate who rolls up his shirt sleeves and wants nothing more than build something wonderful for the country of his heart if not his birth. Because of Jefferson Laboratory's funding problems such a story would have been useful—and it also would have been useful to have an organizing motif around which to develop an identity for a laboratory being created in the South, a region not previously known for scientific excellence. Jefferson Laboratory researchers could have been invited to see themselves as explorers in a New World, for example. And yet in the decades I have spent at Jefferson Laboratory—during Grunder's directorship and thereafter—I have never once heard a single person even mention Grunder's rather amazing Bildungsroman.

Fig. 2 Hermann Grunder in his Berkeley days

Autobiography Considered

In mulling over the question of why Wilson used autobiography and Grunder did not, I stumbled across a line in the introduction of one of my favorite biographies, *Uncertainty: The Life and Science of Werner Heisenberg* by David Cassidy. Cassidy notes that every biography "brings together three lives: the subject's, the author's, and the reader's" (Cassidy 1992, x). It occurred to me that this insight could be extended when trying to make sense of Wilson and Grunder's use of autobiography at their respective laboratories. In the case of autobiography, the subject is the author, and in the case of Wilson and Grunder they were conjuring the drama not just to readers but also to an audience of listeners.

Once I focused on the issue of audience, I was less inclined to worry whether Grunder had missed a trick, which was a bit of an intellectual relief. After all, Grunder may have acted crazy as Jefferson Laboratory director, but the man was as savvy as Wilson or any other crazy laboratory director. Although I would need more information to be sure, my suspicion, as one who has lived and worked in various laboratories throughout my adult life, is that Grunder had not missed something. Instead, he just pitched his Bildungsroman to me, the historian, a person he correctly assumed was positioned to preserve his story for posterity, both in an audio tape and in a transcript. By contrast, Wilson pitched his story both to historians who have similarly preserved it for posterity and to those who were helping him build Fermilab.

The difference in how these two founding directors pitched their stories, in turn, points to the difference I have observed between the audiences at their respective laboratories as they were being constructed. When I thought about the nuclear physics community at Jefferson Laboratory compared with the high energy physics community at Fermilab it made perfect sense that Wilson conjured drama with his Bildungsroman and Grunder did not. I have long noticed after decades working with both groups that high energy physicists are much more prone to flair and drama, including self-dramatization than nuclear physicists.

An appreciation of this difference comes from informal interactions, like the more dramatic way the high energy physicists interacted with each other and with me. (I once had a coffee cup thrown at me during a Fermilab interview with a high energy physicist; nothing like that ever happened during an interview at Jefferson Laboratory or when interviewing a nuclear physicist.) Other measures do come to mind. For example, in the early years of Fermilab researchers were talking about quarks that came in colors, that is, up, down, and strange. By contrast, in the early years of Jefferson Laboratory some twenty years later researchers talked in terms of the letters and numbers that were part of the ordinary nomenclature of the phenomena they studied, like (e,e'p) coincidence experiments or elastic scattering of free nucleons. And even when nuclear physicists addressed quark phenomenon they tended to tone down the flair. For them it was deep inelastic scattering from quarks to nucleons. Added to this is the difference in drama and flair is also apparent in the ways high energy physicists are portrayed in the public sphere. CERN particle physicists are featured in the reasonably well known documentary Particle Fever, (Particle Fever 2013) and one of the more colorful characters in the successful TV show *The Big Bang Theory*, is Sheldon Cooper, a high energy physicist. There are no widely distributed documentaries featuring nuclear physicists and *The Big Bang Theory* lacks a representative from that community as well (*The Big Bang Theory*).

So, it makes sense to me that Wilson and Grunder made choices about the use of their personal stories that made sense considering the individual research communities they drew on to build their respective laboratories. But what can we learn by assessing the staged and conjured drama of the two founding directors alongside their longer-term impact as measured by the working lifetime of the equipment built under their direction? And what, if any, difference did it make that Fermilab was shaped by a Bildungsroman and Jefferson Laboratory was not?

When I looked at the tenure of Wilson and Grunder alongside the life-long performance of the instruments they built I was struck by the realization that although both Wilson and Grunder made bold instrument choices, Grunder's choices were much more successful. As summarized by an administrator of the Jefferson Laboratory physics program, the laboratory's superconducting radiofrequency accelerator and three fully equipped experimental halls exceeded expectations allowing more and better experiments in the decades-long inquiry into nuclear structure as well as experiments probing the quark structure of nucleons and nuclei not originally envisioned (Westfall forthcoming). By contrast, as noted in my 2002 article, Fermilab was a research disappointment. Wilson's cheap, cobbled-together equipment proved inadequate for the large, sophisticated experiments needed to confirm the quark hypothesis and develop the Standard Model in the 1970s (Westfall 2002, pp. 386–387).

The disparity in instrument performance was surprising given that Wilson was the one with the reputation for being a big, world-class high-energy physicist and Grunder had very little experience doing nuclear physics experiments and no experience doing any of the experiments performed at Jefferson Laboratory. Why did Wilson's experimental talent and experience not translate into an instinct for building successful instruments? I then remembered the time when I got a surprise visit from Nobel Laureate Leon Lederman when he was Fermilab's second director. I was all alone in the History Room at the High Rise and Lederman came wandering in, looked at me, and said: always remember that luck has a lot to do with what happens in physics. I never figured out what specifically had led to this comment because he then wandered away. I do know that Lederman himself, who led the most successful experiment in the Wilson years to discover the bottom quark, would considerably enlarge and enhance Fermilab's experimental capabilities. So I assume he would agree that luck does not determine all outcomes in physics. And yet I think luck is part of the answer for why Grunder's equipment choices performed brilliantly and Wilson's did not. Nature, after all, is tricky and fickle. And bold choices are by their nature risky: I think in part the comparison of Wilson's failure and Grunder's success simply shows that the one who knows more can achieve less while the one who knows less can achieve more.

At the same time I think it is useful to consider what if any difference it made to Fermilab that it was shaped by Wilson's story and what if any difference it made to Jefferson Laboratory that it was not shaped by such a story. When Wilson cast Fermlab as a frontier settled by cowboys in line with the story of his youth he did get experts, including Lederman, to come to Illinois to build his bare bones laboratory and do experiments with skimpy equipment. Thus, on the one hand, that cowboy identity did very little to bring research success to the laboratory with its frontier-inspired under built instruments. But on the other hand, the laboratory was built and staffed so that Lederman could then direct the construction of a much larger accelerator, major detectors, and other more sophisticated equipment that brought more research success. In addition, even amidst the drama in the high energy physics community, Wilson's branding of Fermilab has given the laboratory undeniable flair, a distinct sense of community, and a robustness that Jefferson Laboratory and most other laboratories lack. Of course there are many factors that play into whether a laboratory perseveres or is closed. And yet I think that Fermilab's frontier identity has something to do with the fact that it is the U.S.'s last facility for high energy physics and that it remains open even though it operates no large accelerator and has not managed to secure funding for a future large project. Fermilab's frontier identity might not be everything. But Fermilab staff and experimenters have that going for them.

As I think back on the musings I have indulged in here about how founding directors and their autobiographies shape laboratories and their instruments I realize it is good to recall David Cassidy's words in a wider context. I have sought in this analysis to bring together three lives: the lives of the subject, the lives of my audience, and, yes, the life of the author—me. Indeed, much of what I have brought to this analysis has come from actually living in and being a part of laboratory life at Fermilab and Jefferson Laboratory, including being an enthusiastic supporter of each enterprise. So, in that sense, my analysis combines my own autobiography alongside the biography and autobiography of others. I think all historians need to be introspective about how our analysis is shaped by our experiences and values, which determine not only what we consider to be true, but also how we choose subjects, frame subject matter, and what counts as evidence. In other words, shared biography includes the historian.

Expanding Shared Biography

Finally, my musings about the insights gained by conceptualizing the shared lives of institutions, leaders, instruments, and self, lead me to wonder about the possibilities of extending the metaphor of shared biography. What other elements could be added to the mix to provide a more complete and vivid picture of laboratory life?

For one thing, it might be interesting to add the biography of key ideas. For example, interestingly enough, the notion of quarks played a big role in the development of both Fermilab and Jefferson Laboratory. As mentioned, Fermilab's equipment was conceived and built before quark physics became dominant in high energy physics in the 1970s, and therefore the laboratory's underbuilt original equipment was poorly suited for the first quark studies at high energy physics energies. Although Jefferson Laboratory equipment ultimately rose to the challenge of quark studies, it too was conceived before it was understood that quarks would be an important topic at nuclear physics energies in the 1980s. It would be interesting to trace the "birth," development, and influence of the notion of quarks, which involved institutions, leaders, researchers, instruments, and breakthroughs both outside and inside Fermilab and Jefferson Laboratory.

In fact, upon consideration, really understanding the life of a laboratory requires knowledge about the widest possible context of all that surrounds and interacts with the laboratory. In addition to ideas, this wider context also includes, for example, the scientific communities that provide standards, elite decision makers, and outside accelerator users; local communities that shape and are shaped by the presence of the laboratory; and political establishments that provide financial and cultural support. When it comes right down to it, large-scale science is a deeply and widely shared experience, and our investigations need to reflect that.

References

Arabatzis, T. (2006). *Representing electrons: A biographical approach to theoretical entities*. Chicago: University of Chicago Press.
Cassidy, D. C. (1992). *Uncertainty: The life and science of Werner Heisenberg*. New York: W. H. Freeman and Company.
Domingo, J. (1998). Interview by Catherine Westfall. Thomas Jefferson National Accelerator Facility Archive, November 9.
Galison, P. (1997). *Image and logic: A material culture of microphysics*. Chicago: University of Chicago Press.
Grunder, H. (1999). Interview by Catherine Westfall, Thomas Jefferson National Accelertor Facility Archive, October 7.
Hoddeson, L., Henriksen, P., Meade, R., Westfall, C., & Baym, G. (1993). *Critical assembly: A technical history of Los Alamos during the Oppenheimer years, 1943–1945*. Cambridge: Cambridge University Press.
Hoddeson, L., Kolb, A., & Westfall, C. (2008). *Fermilab: Physics, the frontier, and megascience*. Chicago: University of Chicago Press.
Particle Fever. (2013). https://particlefever.com.
The Big Bang Theory. Wikipedia. https://en.wikipedia.org/wiki/The_Big_Bang_Theory.
Westfall, C. (1988). The first "truly national laboratory": The birth of Fermilab. Ph.D. dissertation, Michigan State University.
Westfall, C. (2002). A tale of two more laboratories: Readying for research at Fermilab and Jefferson Laboratory. *Historical Studies in the Physical and Biological Sciences, 21*, 369–407. https://www.jstor.org/stable/10.1525/hsps.2002.32.2.369.
Westfall, C. (Forthcoming). From desire to data: How JLab's experimental program evolved. Part 3: From experimental plans to concrete reality, JLab gears up for research, Mid-1990 through 1997. *Physics in Perspective*.
Wilson, R. (1977). Interview by Spencer Weart. American Institute of Physics, May 19. http://www.aip.org/history-programs/niels-bohr-library/oral-histories/4972.

Part III
Objects

Chapter 13
Lost in the Production of Time and Space: The Transformation of the Airy Transit Circle from a Working Telescope to a Museum Object

Daniel Belteki

Introduction

The Royal Observatory at Greenwich is best known for its association with the Greenwich Prime Meridian. Visitors on a daily basis venture to climb the hill of Greenwich Park upon which the Observatory rests, just to snap a photo of themselves 'standing on both sides of the world' at the same time. The line crossing the courtyard and marking the meridian serves not only as a material manifestation of a geographical coordinate system, but also embodies Britain's historical relations to the world by being 'the Line' that aided the Empire's maritime power throughout centuries. During the middle of the nineteenth century, the Observatory transformed its stance to expand its regulation of space to include the regulation of time. With the advent of the telegraph and railways, it was possible to both control and distribute Greenwich Time at an unprecedented speed around England. Galvanic connections between the Observatory and several distribution points in London gradually appeared, forming the central connections for the galvanic 'nervous system of Britain' (Morus 2000). Yet, hidden behind Longitude Zero and the ticking of the clocks stood the Sisyphean astronomical labour that produced and defined time and space for the Empire on a daily basis.

The founding documents of the Observatory defined its aims as the production of astronomical tables and solving the problem of finding longitude at sea. Positional astronomy was the basis for both avenues of research. It focused on the ever more precise determination of the positions of celestial bodies and the publication of such astronomical data in catalogues and tables. In light of this, the Observatory focused most of its attention on procuring meridian instruments (Stott 1985). These instruments were specially designed to be placed along the local meridian. Until

---

D. Belteki (✉)
University of Kent, Canterbury, UK
e-mail: daniel.belteki@gmail.com

the middle of the nineteenth century the two most widely used meridian instruments were transit instruments and mural circles. The former (used in conjunction with timekeepers) was used to measure the time when a celestial body crossed the meridian (i.e. the point of transit), while the latter gave an indication of the angle at which the transit occurred (i.e. declination). Together these two measurements gave the position of a body on the sky. Towards the very end of the eighteenth-century, design and technical solutions allowed for combining the two instruments, thereby creating what today is known as a transit circle (or a meridian circle).

The first transit circle of the Observatory was installed at the site in 1850 (Satterthwaite 2001a). It was the result of the collaboration between George Airy (Astronomer Royal and director of the Observatory), and instrument makers Troughton and Simms, and Ransome and May. The Transit Circle as a combination of two instruments not only eliminated a few errors of the previous instruments, but by allowing one person to make the same observation that previously required two, also sped up the observing process. However, through the combination of two instruments, transit circles also made the process more complicated (Satterthwaite 2001a, p. 129). Besides making astronomical observations quicker and more reliable, it also reflected Airy's approach to the Observatory as a factory for astronomical labour (Schaffer 1988; Smith 1991; Chapman 1988). Under his directorship, members of staff were assigned to carry out work with one specific instrument or on only one set of measurements made with it. The mental labour of calculating the positions of celestial bodies underwent a similar transformation. Standardised forms were created that only required computers (individuals carrying out the computations) to carry out the most basic arithmetic calculations. Due to the reliance on mathematics, division of labour, and such rigorous methods, work with and related to the Transit Circle was boring, meticulous, and almost devoid of the stars themselves as opposed to being a romantic exploration of the universe (Donnelly 2014). Dickens (1854 [1920], p. 85) interpreted such astronomical work in a way that observatories no longer required "windows" to the sky, instead "the astronomer within should arrange the starry universe solely by pen, ink, and paper".

The applications of such work with the Transit Circle culminated in the practical results of the ever more accurate determination of longitude and measurement of time based on the motion of celestial bodies. However, underneath these practical applications, the work of the Observatory always remained grounded in astronomy. The annually published *Greenwich Observations* were widely distributed among observatories around the world in order to provide essential reference data for observations and theoretical calculations. Long term studies showing both the fluctuations and the regularities in the motions of selected celestial bodies were published in collated catalogues of stars (Satterthwaite 2001a, 136). Although taking place prior to the installation of the Transit Circle, the Neptune Controversy of the 1840s highlighted the extent to which astronomers relied on such positional data (Smith, 1989). Towards the end of the nineteenth century, the Carte du Ciel project revived the need for confirming positional data for which observations the Transit Circle played a crucial role (Daily Mail 1899). Unfortunately, within the stories of time and longitude, the underlying astronomical work was "black-boxed", and as a result its impact received less attention (Winner 1993; Bijker et al. 2012, p. xliii). This paper investigates why such contributions from the instrument received less attention, and in what other ways it contributed to astronomy beyond 'time and longitude'.

Material Dialogues and Object Biographies

There have been two different framings of the history of Observatory (and the Transit Circle): either as an astronomical or as a 'nautical' institution. When the Observatory became part of the National Maritime Museum in the 1950s, the nautical contributions of the site came to dominate the way in which its history was told. The aim of this paper is to take the first steps at reconstructing the astronomical framing of the history of the instrument with the help of the object biographies approach (Kopytoff 1986; Gosden and Marshall 1999; Daston 2000, 2004; Hoskins 2006). Object biographies allow us to reconstruct the "astronomical" framing of the Transit Circle through their focus on artefacts as (1) "talking things", (2) "emerging" things, and (3) thinking through as opposed to with things. Daston (2004, p. 9) raised the criticism that approaches to the history of objects often consider things as "mute" and "speechless". In consequence, such approaches resort to "ventriloquism or projection" of external voices on things. Instead, Daston (2004, p. 11) argued that the materiality itself also talks, and therefore, meanings can emerge not only through the context within which things are placed, but also through the materiality of things. Going one step further, this allows for things to enter into a "material dialogue" (Callen and Criado 2015, p. 25) with either the individual or other things they interact with. Such dialogue is present in Latour's (1994, p. 38) example of the speed bump or the "sleeping policeman", where he argued that the materiality not only mediates a message, but also transforms it (from morality and reflection to selfishness and reflex action). As we will see, nineteenth-century astronomers and observatory staff entered into such dialogues, which made them aware of the ways in which the materiality of instruments such as the Transit Circle embodied and transformed not only their intended uses but also their outputs (Herschel 1869, p. 78).

A similar active dialogue between the maker and the material employed was present in the 'functional adaptation of ax' throughout history. Leroi-Gourhan argued that "the problem of function [in the history of the ax] [...] assumed a series of forms dictated by the raw materials successively employed" (1993, p. 308). According to Ingold's interpretation, this demonstrates the active engagement and dialogue between the user and the material:

> The rhythmic repetitions of gesture entailed in handling tools and materials [...] are set up through the continual sensory attunement of the practitioner's movements to the inherent rhythmicity of those components of the environment with which he or she is engaged. (Ingold 2013, p. 115)

Such approaches to materiality allow for securing both the agency and the malleability of human and non-human entities. Furthermore, the focus on the dialogues between humans and things highlights the ongoing co-production between the various entities involved. This can be illustrated through the act of playing a musical instrument such as the violin. The sound produced by playing the violin is a co-production between the materiality of the instrument and the musician playing it. Even the slightest changes in the materiality of the instrument transforms the sound produced to be out of tune, and in turn require the musician to adjust their playing style. At the same time, different musicians produce different sounds based on how they use and respond to the materiality of the violin. In this light, the produced sounds become the manifestations of the dialogue in which both the thing and the individual actively participate. As it will be shown, the Transit Circle and its observers engaged in a similar performance, where both entities shaped the outcome of the performance—an action of which astronomers at the time were conscious of. However, instead of producing music, the Transit Circle produced large sets of data based on astronomical observations.

Approaching a thing as a 'talking thing' shifts our understanding to its relationship to temporality, and from the *being* of things to the *becoming* of things (Daston 2004, pp. 20–21). Such a shift questions the 'sharp edges' of objects, the distinction between artificial and natural, and the apparent separation of matters of morals and metaphysics. It places objects into a state where both the meanings attached to them and their material realities are constantly changing. In brief, 'talking things' are always in the process of becoming. As we will see in the case of the Transit Circle, the repeated daily and hourly measurements of its errors reflected such an approach towards the instrument by its users. Since instruments go through constant transformation, even though they 'gather' and 'mediate', they are also transforming their former selves. What instruments tell us or the way in which they tell their own stories can take different directions over time. Similarly, the materiality of the instrument also changes (or changed). As a result, the things are not only in a constant state of becoming, but also in a state of unbecoming, i.e. losing their ability to live up to the functions and meanings initially ascribed to them: a speed bump gradually degrades, the head of a hammer slowly loosens from the handle, over time the violin becomes out of tune, or the mechanical parts of a telescope wear out.

In relation to the history of the Transit Circle, this paper attempts to demonstrate the dialogue between the materiality of the instrument and the people who engaged with it. The paper argues that the material 'fixity' of the instrument in relation to the Greenwich Prime Meridian played a crucial role in how curators and historians have engaged with framing its history. However, the voices of astronomers and the staff of the Royal Observatory at Greenwich seem to be exempt from such histories. As a result, the paper also investigates how the astronomers entered into the material dialogue with the Transit Circle, and how such a dialogue frames the history of the instrument within its contribution to theoretical astronomy as opposed to nautical astronomy. By doing so, the paper demonstrates how the biography of an object changes its course depending on which one of the material dialogues we focus on.

So how did the Transit Circle 'talk'? During its working life (1851–1954), only a selected few were fortunate enough to engage in a material dialogue with the instrument. This was due to access to the Observatory being restricted to members of staff and invited visitors only. However, instrument makers, specialist labourers, observers, astronomical assistants, and astronomers were able to engage in such dialogues. For the instrument makers, the materiality spoke of the wellbeing of its parts and embodied the relationships between the Observatory and the artisans it employed externally (Bennett 1980; Mennim 1992); for the specialist labourers carrying out the weekly maintenance of the instrument, the creaks of the mechanical parts and the dust gathering on the instrument spoke of the intensity with which the Transit Circle was used (Belteki 2017); for the astronomical assistants measuring the errors of the instrument told about the extent to which the material parts gradually shifted their positions (Herschel 1869, p. 78). Astronomers (and the First Assistant of the Observatory) engaged in all of the above mentioned practices, which meant that they had to be proficient in participating in any material dialogue taking place between individuals and the Transit Circle (Ibid.).

This paper argues that the material dialogue between the Transit Circle and its users is also represented through the astronomical contributions of the instrument. In fact, such contributions preserve more of the material dialogue between 'the thing' and its users than the practical applications can (e.g. time, geographical coordinate system). As a result, through the astronomical contributions, the biography of the object can be understood as its own history written through the instrument as opposed to with it.

Communicating the Transit Circle

Before venturing into the astronomical contributions of the Transit Circle, it is important to highlight how such practices have been communicated to different audiences in the past. This was carried out through a content analysis of 26 articles in newspapers, magazines, journals, and books that discussed the instrument during its lifetime. Only publications that directly referenced the Transit Circle were used. Furthermore, publications that did not refer to the application of the work of Transit Circle were excluded. Overall, three major narrative categories were identified: (1) advancement of technical development (7 publications), (2) contributions to meridian observations (6 publications), (3) practical uses of the observations (8 publications), and 5 publications that mentioned all three topics. The analysis showed that during the first 20 years of the instrument's working life, the majority of the focus was on the technical novelties that it introduced into the design of meridian instruments (e.g. Morning Post 1850). Afterward, both meridian observations and practical uses were represented in publications, but not always together. Texts that mentioned all three categories of the instrument's contributions were most frequently written by the assistants of the Observatory or by people very closely familiar with the instrument. However, the most significant finding of the content analysis was the existence and recurring reference to the contributions of the instrument to the advancement of meridian observations. For example, the Inverness Courier (1854) described the Transit Circle being used for recording the transits of celestial bodies to ascertain their exact positions. Similarly, the Morning Post (1874) referenced the instrument's use in relation to the determination of the position of fixed stars. Finally, The Times (1922) reported on the repairs made to the instrument, and the beginning of its use for a new star catalogue. This demonstrates that an 'astronomical' framing of the Transit Circle was present throughout its working life as opposed to relating the use of the instrument exclusively to 'nautical astronomy'.

The existence of multiple framings did not mean that they were incompatible with each other. Instead, the articles written by the members of the Observatory staff demonstrated the ways in which they could be brought together. The account of the instrument written by Dunkin (1862) follows the narrative thread of the instrument's historical connection to similar instruments, its connection to maps and surveying, its technical features, and then to the practices within meridian observations. Carpenter (1866a, b) took a different narrative line that spun across two articles. He began by emphasizing the role of the Observatory in relation to practical astronomy (i.e. providing positional data for mathematical astronomers), then expanded on the technical details of the Transit Circle. However, rather than connecting the Transit Circle to navigation or timekeeping, it was through the Transit Clock "Hardy" that allusions to time dissemination was made. This highlighted the approach to the Transit Circle as a set of multiple instruments being used in conjunction with each other in order to attain its 'practical' uses. Forbes (1872), a close associate of the Observatory took a different path. The article itself began by relating the historical connection between the Observatory and the problem of longitude. Yet, the paragraphs containing descriptions of the Transit Circle did not make any direct connections between the two. Instead, the description began by comparing the Transit Circle to a child that inherited all the greatest (technical) features of its ancestors. This introductory segment was then followed by a short lesson on positional astronomy and the various methods with which observations were carried out with the instrument. The last description of the century was produced by another assistant of the Observatory, Walter Maunder. His account (1898) was written almost 15 years after the Transit Circle was recommended for marking the Prime Meridian of the World. As a result, Maunder introduced it as the instrument that marks Longitude Nought. When providing a technical description, all elements of the instrument were linked to the determination of longitude. This was then related to its use in connection to the expansion of the British Empire. In the second part of his description, he briefly outlines the 'history of time' and time reckoning, as well as the uses of the Transit Circle for time measurement and positional astronomy. In summary, the three astronomical assistants of the Observatory, and a close associate of the institution, all provided various solutions with which both the astronomical and practical applications of the Transit Circle could be described.

Science History or Maritime History

The last observation with the Transit Circle was made in 1954 (Satterthwaite 2001a, p. 132). The same year marked the transfer of operations of the Royal Greenwich Observatory from the Old Observatory situated at Greenwich Park to a new location at Herstmonceux, Sussex. The fate of the Old Observatory along with the future of the instruments that it housed (including the Transit Circle) became an important question. By the 1950s, the Observatory lost its architectural appeal as parts of the building undergoing significant decay due to lack of use and proper maintenance (Littlewood and Butler 1998, p. 145). Despite this, there was no doubt about the historical value of the site. In light of this, the possibility of converting the buildings into a museum was raised at the Ministry of Works, which department was in charge of its upkeep (Ibid., p. 146). Two museums were considered for taking over the buildings: the Science Museum and the recently opened National Maritime Museum (NMM). Both museums were able to make good cases for taking charge of the Observatory. The Science Museum was the leading institution in relation to displays of scientific and technological achievements of the British nation, while the NMM focused on the history of British naval power over the centuries that incorporated displays about its scientific and technological basis. By the 1940s the Science Museum had already had an extensive collection of astronomical instruments. On the other hand, the historical connection of the Observatory to the 'Quest for Longitude' brought forth the site's contribution to the development of navigation techniques, which made it a perfect fit for the NMM. Finally, the Old Observatory was located on the top of the hill at Greenwich Park, which brought it into close geographical connection with the NMM (the NMM being on the opposite side of Greenwich Park). By contrast, the Science Museum occupied a building in west London. In light of these details, the director of the NMM, Frank Carr, began an active effort for the annexation of the Observatory (Ibid. pp. 147–148). He framed its history as an Observatory devoting all of its attention to supporting the development of navigational techniques, thereby placing emphasis on the nautical and practical applications of astronomy. However, such an emphasis diminished the contributions of the Observatory to the theoretical side of positional astronomy. Carr's efforts were successful as the Ministry of Works ultimately agreed to the incorporation of the Observatory into the NMM.

Within the internal policy developments of the NMM, a similar debate around the type of astronomy to be exhibited arose. The initial aim of the NMM specified exclusive focus on maritime history, which fueled the need to frame the Observatory's history within such context (Ibid., p. 147). Once the Observatory buildings were officially promised to the NMM, the aims and scope of the museum were updated to include a more general scope of maritime history and astronomy. Within the initial proposal they were defined through '(a) British seafaring and shipbuilding, and (b) positional astronomy' (Ibid., p. 183). As it can be seen here, the work of the astronomy was highlighted with the term positional astronomy, which encompassed both the theoretical and practical parts of the field. However, the trustees of the NMM were critical of other parts of the draft, and they set up a sub-committee for the revision of the aims. Within the new aims, the term positional astronomy was abandoned and replaced by the concept of 'Nautical Astronomy' (Ibid.). In contrast to positional astronomy, nautical astronomy denoted the astronomical practices that aided in locating the positions of ships at sea. In the context of the nineteenth-century, this meant a focus on longitude and time measurements. These two aspects of astronomical works were only the practical applications of positional astronomy, as opposed encompassing its theoretical contributions more widely. As a result, the revised internal policy document limited the historical scope of the activities at the Observatory only to its nautical contributions. In brief, the association of the Observatory with the NMM shaped the way in which the displays were exhibited at the site, and the ways in which stories about its instruments were told.

The discordance between the various framings of the history of the Observatory and the Transit Circle also surfaced at the opening of the Meridian Building as a museum to the public. At the ceremony, Sir Richard Woolley (Astronomer Royal at the time) reflected upon the history of the site, and the contribution of George Airy to the development of nineteenth-century astronomy. Surprisingly, rather than highlighting Airy's connection to nautical astronomy, Woolley emphasized the theoretical contributions to nineteenth-century astronomy. For instance, the 'diligence and efficiency' implemented into the work of the Observatory by Airy allowed Simon Newcomb to carry out 'his great analysis of the motions for the solar system' (Wooley 1967, p. 4). Similarly, Airy's Observatory maintained the "principal interest of Victorian astronomers and mathematicians" in the extension of "Newton's law of gravitation to account for the motions of the objects in the solar system" (Ibid.). What Woolley's speech highlighted was a different framing of history of the Observatory that embedded it within the history of 'Victorian astronomers and mathematicians' instead of within the developments of nautical astronomy exclusively.

Star Catalogues

The most frequent task carried out with the Transit Circle was the making of observations for deriving the positions of celestial bodies. The output of such work was published in the Greenwich Observations and served as the basis for star catalogues. The production of catalogues of stars has been a practice exercised since antiquity (Eichhorn 1974, pp. 101–102). They have taken textual, tabular, and visual forms (Kanas 2009). Their main uses related to the determination of exact positions of celestial bodies in reference to each other, and to recording their historical motion. The history of star and fundamental catalogues was dealt with in greater length by other historians. Though offering a eurocentric view, Fricke (1985) traces the history of fundamental catalogues back to Hipparchus, and discusses how they led to the discoveries of the precession of equinoxes and the stellar motions with respect to each other. His historical account finishes with the work of Bessel (Fundamental Astronomie) as the first precursor of a fundamental catalogue by offering the standard of precision for future catalogues in the reduction of observations made by astronomers.

Bessel's novelty was the reduction of James Bradley's (third Astronomer Royal at the Royal Observatory, Greenwich) observations in such a way as to incorporate two astronomical constants (aberration and notation) and the instrumental errors into the final results. Such method set a standard for the production of future fundamental catalogues (Ibid., p. 213). When George Airy was appointed as the director of the Cambridge Observatory in 1828, he incorporated the Besselian novelties into the Cambridge Observations, thereby offering reduced results for the observations made at the site (Airy 1829, pp. iv–vii). Once appointed as director of the Royal Observatory at Greenwich, he continued the Besselian analysis. Later catalogues during the nineteenth century continued the incorporation of further astronomical constants, as well as instrumental and human factors into the reduction of the observations.

In relation to object biographies, the data published in catalogues are the outputs of the hybrid "inscription devices" composed of the human-transit-circle assemblage (Latour and Woolgar 1986, p. 51). As such, each data point tells us a micro story about the state of the instrument (and its parts) and the observer during a single observation. For example, the *Greenwich Observations* describe the instrumental errors of the Transit Circle, the state of the transit wires in the telescope tube, the personal equation of the observers, and the error of the clock used in conjunction with the telescope. In addition, the *Greenwich Observations* included the derived astronomical constants and the errors of the microscopes with which the declination of the celestial object was read off from the Transit Circle. Finally, they even recorded the weather conditions affecting the observations. In this light, they are the inscriptions of the shards and fragments of the materiality exhibited through a textual format (Schaffer 2011, p. 707). From these data points, the history of the instrument emerges not as one that passively gave into the wishes of its users, but rather, as a sensitive instrument that through its ever-changing character resisted the physical control of its users. In consequence, the measurements of its errors (i.e. character) serve as a basis for revealing the activity of the Transit Circle, and as such the biography of the object as told by itself, as opposed to through activity of its users.

The Transit Circle at Greenwich contributed to several catalogues that reduced observations of fundamental stars. Satterthwaite (1995), who made the last official observation with the instrument, identified twelve such different catalogues spanning from the year 1854–1954. Going beyond the catalogues, the Transit Circle helped in the "determination of the solar parallax from astrometric observations of the minor planet Eros at its close approach to Earth in 1931" (Ibid., p. 59.) Similarly, the published observations on the Sun, Moon and inner planets served as the basis for demonstrating the possible irregularities in the rotation of the Earth.

Big Data

Star catalogues and published observations were presented in the form of tables. Such a mode of presentation allowed for the management of large astronomical data

(Norberg 2003). Taking the Greenwich Observations as an example, the names of celestial bodies, their positions, and the factors that contributed to the errors of the final results were presented in a clear and simple way. The Royal Observatory at Greenwich can be seen as one of the historical headquarters for the gathering of large sized astronomical data, and as a result a 'centre of calculation' (Latour 1987; Jons 2011). At the beginning of the nineteenth century this was highlighted through Bessel's reliance on the observations made by Bradley (Fricke 1985). Delambre's memoir of Maskelyne awarded a compliment that highlighted the same strength of the Observatory:

> if by any great revolution the works of all other astronomers were lost, and [Maskelyne's catalogues of the stars were] preserved, it would contain sufficient materials to raise again, nearly entire, the edifice of modern astronomy. (Delambre 1813, p. 11)

Even though the work of the Observatory was praised in such a way, it also had its critics. One of those people was Airy himself. While still being the director of the Cambridge Observatory, he reflected upon the recurring inaccuracies found in the reductions of the positions of fundamental stars within the Greenwich Observations (Airy 1829). The attack on the Nautical Almanac by the 'business astronomers', which took place in the first decades of the nineteenth century and led to the formation of the Royal Astronomical Society, similarly criticised the work of the Royal Observatory at Greenwich on the basis of publishing inaccurate observations (Ashworth 1994). Under Airy's directorship, the focus on data collection and its publication continued. By 1881, positional astronomy at Greenwich gained such a reputation that Newcomb (1881, p. 198) repeated Delambre's praise of Maskelyne, but within the context of Airy's work: "if this branch of astronomy were entirely lost, it could be reconstructed from the Greenwich observations alone".

Airy faced the problem of managing large amounts of astronomical data. His answer to the problem relied on the introduction of 'factory techniques' (Smith, 1991). These were largely derived from the principles of division of labour applied at the factories that emerged during the industrialisation of Britain. It consisted of breaking down long manufacturing processes into smaller parts that could be carried out repeatedly, at a quicker pace, and with less reliance on specialist skills. The novelty of Airy's approach was the application of such division of labour into astronomical practices as well as the calculating work carried out by his staff. To achieve this, he introduced a new organisational hierarchy that consisted of the computers, the observers, the astronomical assistants, and the chief assistant. The computers were only required to carry out simple mathematical calculations, and as a result, only required basic arithmetic skills. Similarly, the positions of observers and the astronomical assistants did not require a solid background knowledge in astronomical matters. Instead, Airy looked for diligent workers who could carry out the observations and the related calculations reliably. The more competent computers, observers, and assistants furthered their knowledge in astronomy through spending time at similar or higher ranked positions at other observatories (and through their access to the library of the Observatory). It was mainly the Chief Assistant (later renamed First Assistant) who was educated to a degree.

Besides the new organisational structure, the practices involved in astronomical labour underwent similar transformations. Airy's re-instrumentation programme that began in the 1840s included specialist instruments that either focused on the observations of specific celestial bodies (e.g. Altazimuth and Reflex Zenith Tube) or incorporated multiple instruments in order to reduce the personnel and time needed for the observations (e.g. the Transit Circle) (Satterthwaite, 2001b). Besides the instruments, Airy implemented the use of standardised forms. These forms (along with slider scales) can also be interpreted as material manifestations of the mathematical tools employed at the Observatory (Aubin 2017). By arranging entries on the form underneath each other and following the logical direction of calculations, the computers simply had to insert the relevant numbers and carry out the calculations dictated by the forms.

As critics of modernity have shown, such an approach to division of labour transformed people's mindsets about what it means to be human. As in the case of the human computers and Babbage's difference engine, human actions were seen more and more as mechanical tasks. Similarly, as Schaffer (1988) has argued, observers were reduced to machines that erred, in order to account for the differences between the perceptions of individuals during observations. The instruments themselves were treated under the same close scrutiny. For instance, the errors of the Transit Circle were measured twice a day, in order to numerically repair them during calculations. In this light, both humans and instruments were thought to possess a certain of autonomy ('freedom') that could only be controlled through the standard procedures set by the foremost authority of the institution and then placed under close surveillance by members of the Observatory staff (Tresch 2010). Since the datasets published in the Greenwich Observations highlight the measured variations of the instrument, they become representations of the autonomy of the instrument, as well as the history of an autonomous instrument as recorded through constant surveillance.

Beyond Time and Space

Within both framings of the history of the Transit Circle (nautical and positional astronomy), the instrument emerges as an object that creates and defines the representation of time and space used by society. Through the use of the Transit Circle time does not appear as an entity existing independently, but rather as a measurement of the motion of celestial bodies derived by the assemblage of the instrument and the humans using it. The simultaneous coexistence of multiple "times" further complicated this issue. Until the standardisation of time across England, different locations derived their own local time. In addition, astronomers differentiated between solar time and sidereal time. Finally, clocks and watches were never perfect, which meant that their clock faces had to be constantly brought in synchronicity with another clock. In brief, we can derive two major characteristics of time in relation to the use of the Transit Circle. First, time was derived by human defined measurements. Second, clocks and watches were always inaccurate representations (or 'keepers') of such time. They kept a time, which was their own individual rate, but never *the (absolute) time.*

It is within such a scientific and technological context that Pearson (1900, pp. 186–192) reflected upon the issue of "conceptual time and its measurement". Within the passages, Pearson attempted to demonstrate that time is "a relative order of sense-impressions, and there is no such thing as absolute time". By using the clocks-Royal Observatory-stars sequence, Pearson argued that there is always a fixed reference point from which time is derived. However, the fixity (or regularity) of that point is always only conceptual to aid our understanding, as opposed to being realities. Or as Pearson put it:

> Absolute intervals of time are the conceptual means by which we describe the sequence of our sense-impressions, the frame into which we fit the successive stages of the sequences, but in the world of sense-impression itself they have no existence. (Ibid., p. 189)

Within this framing, the Transit Circle emerges as a crucial instrument within the sequence of time regulation. It was not only the instrument that guided the process of deriving Greenwich Time, but also the artefact that transformed and inscribed into it its own errors and precisions. In this way, the history of time as well as its relativity also became the history of the Transit Circle.

Besides time, the Transit Circle also contributed to the production of space itself through longitude determinations. Its contribution was acknowledged at an international level at the 1884 International Meridian Conference, where the longitude passing through the centre of the Transit Circle was recommended to define the Prime Meridian of the world (Howse 1980, pp. 138–151). With this step, the new standard was directly connected to the instrument itself. The current framing of the history of the instrument highlights how previous longitudes aided the navigation of ships, and the significance of the widespread use of such charts played in the decision to choose the Greenwich Observatory and its Transit Circle as the Prime Meridian. Within this framework, the 'shifting' of the longitude of Greenwich with new successive instruments is also present. Prior to the installation of the Transit Circle, two other instruments marked two different longitudes for the Observatory. First marked by the transit instrument commissioned by Edmond Halley and set up in 1721 (around 43 m to the west of the Transit Circle), and later marked by the transit instrument commissioned by James Bradley and installed in 1750 (around 6 m to the west of the Transit Circle). Within this framing of the history of longitude, it is not seen as absolutely fixed entity, but rather as one that is intertwined with the instrumentation.

Despite the primacy of the meridian defined by the Transit Circle, visitors to the Observatory who are eager to measure their longitude with GPS have always found themselves perplexed at the apparent error of their portable devices. In a research published in 2015, scientists called attention to GPS identifying zero longitude 102 m to the east of the currently marked meridian line inside the premises of the Observatory (Malys et al. 2015). As the researchers pointed out, this was due to the differences between the astronomical (1884) and geodetic (1984) methods being the basis/standards for determining geographical positions. While the history of longitude connected the meridian to the material instrumentation, this framing of the history connects the history of 'the Line' to the methods and standards used by scientific communities. In this light, the Greenwich Prime Meridian is disassociated from the Transit Circle, and Longitude Nought takes on an almost independent existence. At the same time, by comparing the materiality of the instrument with the conceptuality of the theoretical methods, the Transit Circle emerges as a materialisation of the nineteenth-century practice as opposed to an instrument on its own. Such a change in approach allows us to open up the black box of methods and observe whether it is empty or not.

Conclusion

An instrument's materiality is its own autobiography, and it enters into material dialogue with people who investigate and use it. However, different people will interpret what the materiality of an instrument tells them in different ways. For example, the assistants of the Observatory approached the Transit Circle differently from the makers of the instrument. The same differences appeared at a larger institutional level. For instance, the National Maritime Museum made a conscious decision to place the history of the Transit Circle within the 'Quest for Longitude', thereby highlighting its practical applications. Finally, as time passed and the Transit Circle became entangled in new settings, the narrative of its history also changed. As this paper has shown, throughout the working life of the instrument there existed multiple framings within which its history could be placed, among which the measurements of 'space' and 'time' were just two framings.

This paper also made preliminary explorations of what the history of the Transit Circle would look like beyond the narrative boundaries of 'space and time'. It demonstrated the existence of two other framings of the history of the Transit Circle: as a mechanical marvel, and as a contributor to theoretical astronomy. Within the scope of this paper, the latter framing was explored further. By approaching the history of the telescope through its contributions to positional astronomy, it was possible to show its contributions to the production of fundamental and other star catalogues.

Star catalogues and publications such as the Greenwich Observations showed the instrument not as a passive executor of the commands of the astronomers, but rather as one that actively behaved according to its own 'personality'. In consequence, similarly to the observers, the instruments themselves were put under the rigorous surveillance in order to control their 'behaviour'. This allows us to consider the related data on the errors of the instrument published in the Greenwich Observations as a collection of moments from the life of the Transit Circle.

References

Airy, G. B. (1829). *Astronomical observations made at the observatory of Cambridge*. Cambridge: J. Smith.
Ashworth, W. J. (1994). The calculating eye: Baily, Herschel, Babbage and the business of astronomy. *The British Journal for History of Science, 27*(4), 409–441.
Aubin, D. (2017). On the epistemic and social foundations of mathematics as tool and instrument in observatories, 1793–1846. In J. Lenhard & M. Carrier (Eds.), *Mathematics as a tool* (pp. 177–196). Cham: Springer.
Belteki, D. (2017). Caring for the circle: The maintenance of the airy transit circle, 1851–1861. The Maintainers. http://themaintainers.org/s/Daniel-Belteki-Caring-for-the-Circle-The-Maintenance-of-the-Airy-Transit-Circle-1851-1861.pdf. Accessed April 21, 2018.
Bennett, J. (1980). George Biddell Airy and horology. *Annals of Science, 37*(3), 269–285.
Bijker, W. E., Hughes, T. P., & Pinch, T. (2012). *The social construction of technological systems*. London; Cambridge, MA: The MIT Press.
Callen, B., & Criado, T. S. (2015). Vulnerability tests: Matters of "care for matter" in e-waste practices. *Technoscienza, 6*(2), 17–40.
Carpenter, J. (1866a, February). John Flamsteed and the Greenwich observatory. *The Gentleman's Magazine*, 239–252.
Carpenter, J. (1866b, March). John Flamsteed and the Greenwich observatory. *The Gentleman's Magazine*, 378–386.
Chapman, A. (1988). Science and the public good: George Biddell Airy (1801–92) and the concept of a scientific civil servant. In N. A. Rupke (Ed.), *Science, politics and the public good* (pp. 36–62). London: Palgrave Macmillan.
Daston, L. (Ed.). (2000). *Biographies of scientific objects*. Chicago and London: The University of Chicago Press.
Daston, L. (Ed.). (2004). *Things that talk: Object lessons from art and science*. New York: Zone Books.
Delambre, J. B. (1813). Memoirs of the life and works of the late Dr. Maskelyne. *The Philosophical Magazine, 42*(183), 3–14.
Dickens, C. (1854 [1920]). *Hard times*. London and Toronto: J.M. Dent & Sons Ltd. & E.P. Dutton & Co.
Donnelly, K. (2014). On the boredom of science: Positional astronomy in the nineteenth century. *The British Journal for the History of Science, 47*(3), 479–503.
Dunkin, E. (1862, January). The royal observatory, greenwich: A day at the observatory. *The Leisure Hour*.
Eichhorn, H. (1974). *Astronomy of star positions*. New York: Frederick Ungar Publishing Co.
Forbes, G. (1872, January). The royal observatory, greenwich. *Good Words*, 792–796 and 855–858.
Fricke, W. (1985). Fundamental catalogues: Past, present and future. *Celestial Mechanics, 36*, 207–239.
Gosden, C., & Marshall, Y. (1999). The cultural biography of objects. *World Archeology, 31*(2), 169–179.
Herschel, J. (1869). *Outlines of astronomy*. New York: Sheldon and Company.
Hoskins, J. (2006). Agency, biography and objects. In C. Tilley, W. Keane, S. Kuchler, M. Rowlands, & P. Spyer (Eds.), *Handbook of material culture* (pp. 74–84). London: Sage.
Howse, D. (1980). *Greenwich time and the discovery of longitude*. Oxford: Oxford University Press.
Ingold, T. (2013). *Making: Anthropology, archaeology, art and architecture*. Abingdon and New York, NY: Routledge.
Jons, H. (2011). Centre of calculation. In J. A. Agnew & D. N. Livingstone (Eds.), *The Sage handbook of geographical knowledge* (pp. 158–170). London: Sage.
Kanas, N. (2009). *Star maps: History, artistry, and cartography*. Chichester: Praxis.

Kopytoff, I. (1986). The cultural biography of things: Commodization as process. In A. Appadurai (Ed.), *The social life of things: Commodities in cultural perspective* (pp. 64–91). Cambridge: Cambridge University Press.
Latour, B. (1987). *Science in action: How to follow scientists and engineers through society*. Cambridge, MA: Harvard University Press.
Latour, B. (1994). On technical mediation—Philosophy, sociology, genealogy. *Common Knowledge, 3*(2), 29–64.
Latour, B., & Woolgar, S. (1986). *Laboratory life: The construction of scientific facts*. Princeton: Princeton University Press.
Leroi-Gourham, A. (1993). *Gesture and speech*. Cambridge, MA and London: The MIT Press.
Littlewood, K., & Butler, B. (1998). *Of ships and stars: Maritime heritage and the founding of the National Maritime Museum, Greenwich*. London: The Athlone Press and the National Maritime Museum.
Malys, S., Seago, J. H., Pavlis, N. K., Seidelmann, P. K., & Kaplan, G. H. (2015). Why the Greenwich meridian moved. *Journal of Geodesy, 89*(12), 1263–1272.
Maunder, W. (1898, February). Greenwich observatory. *The Leisure Hour*, 228–238.
Mennim, E. (1992). *Transit circle: The story of William Simms 1793–1860*. York: W. Sessions.
Morus, I. R. (2000). 'The nervous system of Britain': space, time and the electric telegraph in the Victorian age. *The British Journal for the History of Science, 33*(4), 455–475.
Newcomb, S. (1881). Astronomical observatories. *The North American Review, 133*(297), 196–203.
Norberg, A. L. (2003). Table making in astronomy. In M. Campbell-Kelly, M. Croaken, R. Flood, & E. Robson (Eds.), *The history of mathematical tables: From sumer to spreadsheets* (pp. 177–208). Oxford; New York: Oxford University Press.
Pearson, K. (1900). *The grammar of science*. London: Adam and Charles Black.
Satterthwaite, G. E. (1995). *The history of the airy transit circle at the royal observatory, greenwich*. Unpublished MA dissertation, University of London.
Satterthwaite, G. E. (2001a). Airy's transit circle. *Journal of Astronomical History and Heritage, 4*, 115–141.
Satterthwaite, G. E. (2001b). Airy and positional astronomy. *Journal of Astronomical History and Heritage, 4*(8), 101–113.
Schaffer, S. (1988). Astronomers mark time: Discipline and the personal equation. *Science in Context, 2*(1), 115–145.
Schaffer, S. (2011). Easily cracked: Scientific instruments in states of disrepair. *Isis, 102*, 706–717.
Smith, R. W. (1989). The Cambridge network in action: The discovery of Neptune. *Isis, 80*(3), 395–422.
Smith, R. W. (1991). A national observatory transformed: Greenwich in the nineteenth century. *Journal for the History of Astronomy, 22*(1), 5–20.
Stott, C. (1985). The Greenwich meridional instruments: (Up to and including the Airy Transit Circle). *Vistas in Astronomy, 28*, 133–145.
Tresch, J. (2010). Even the tools will be free: Humboldt's romantic technologies. In D. Aubin, C. Bigg, & H. Otto Sibum (Eds.), *The heavens on Earth: Observatories and astronomy in nineteenth-century science and culture* (pp. 253–284). Durham; London: Duke University Press.
Unknown. (1850, August 9). British association for the advancement of science. *Morning Post*.
Unknown. (1854, June 22). Annual visitation of the royal observatory, greenwich. *Inverness Courier*.
Unknown. (1874, June 8). Greenwich observatory. *Morning Post*.
Unknown. (1899, April 16). New greenwich observatory. *Daily Mail*.
Unknown. (1922, June 6). The greenwich observatory. *The Times*.
Winner, L. (1993). Upon opening the black box and finding it empty: Social constructivism and the philosophy of technology. *Science, Technology and Human Values, 18*(3), 362–378.
Wooley, R. (1967). Opening of meridian building at greenwich, 19th July 1957. *Caird Library, 069*(26:421)7, 520.1.

Chapter 14
Scientific Instruments Turning into Toys: From Franklin's Pulse Glass to Dipping Birds

Panagiotis Lazos

Not only experiments have 'a life of their own', rather also scientific instruments. They have shaped the evolution of physics due to their design, construction, and use. Until the early twentieth century, these instruments were often and almost instantly transformed into supervisory instruments for the teaching of physics, becoming a basic pillar of education at all levels. In this paper I analyze the historical evolution of an object that has been on both sides of the borderline between scientific instruments and toys, namely the pulse glass. I follow the development of the objects for several generations to study more evolutionary changes than would be possibly during one 'lifetime' of a single object.

In an almost circular path, it started its life as a peculiar type of toy made by artisans and earned the title of philosophical instrument in 1768 thanks to Benjamin Franklin. Three decades later, Sir John Leslie and Benjamin Thomson (Count Rumford) almost simultaneously built two similar instruments (differential thermometers) that had many similarities with the pulse glass. They were widely used, along with Leslie cube, at the frontier of research. Specifically, they played a significant role in the study of the propagation of 'thermal' radiation in the first decades of the nineteenth century. Later, when they were not used in research, because they had been replaced by newer instruments, they continued to be used in education until the twentieth century, moving into school science laboratories due to their low cost, ease of operation and ease of supervision.

In 1812 Wollaston adopted the pulse glass to use it in a quite different way, namely to freeze water at a distance, inventing the cryophorus. An interesting and quite unexpected application of the pulse glass was presented by Clegg in 1831 in a new type of gas-meter, while during the last quarter of nineteenth century there were some ideas for making thermal machines using pulse glasses, but they had no

---

P. Lazos (✉)
PhD Candidate, Department of Primary Education, National and Kapodistrian University of Athens, 52, Pontou Str, 11527 Athens, Greece
e-mail: taklazos@gmail.com

© The Editor(s) (if applicable) and The Author(s), under exclusive licence to Springer Nature Switzerland AG 2020
C. Forstner and M. Walker (eds.), *Biographies in the History of Physics*,
https://doi.org/10.1007/978-3-030-48509-2_14

particular success. During the twentieth century, the basic idea behind pulse glasses came full circle and reappeared in the form of toys like the hand boiler and dipping birds. The seeming simplicity and the elegance of these toys can be a strong tool in physics education.

It is worth noting that in the course of the described evolution there are a few gaps with regard to whether and how a stage has been directly or indirectly affected by a precedent. It is not known, for example, whether the unknown first maker of the dipping bird was aware of the principles of the differential thermometers or the pulse glass. It would be useful, therefore, to investigate not only the development of a device, but also the successive but possibly unrelated applications of an initial idea about physics.

Differential Thermometers in the Greek Schools of Istanbul

The National Hellenic Research Foundation established the Hellenic Archives of Scientific Instruments in 1997. The archive is hosted on www.hasi.gr and includes photographs and data from scientific instruments built up to 1970 which belong to universities, schools and private collections in the greater Greek region. The author of this paper has recorded the relevant collections in the Greek schools in Constantinople (Phanar Greek Orthodox College, Zografeion High School and Zappeion High School) and in the Theological School of Halki Island.

It is particularly interesting to have found differential thermometers or accompanying instruments in two of the four institutions. Specifically, the Phanar Greek Orthodox College has a Leslie Differential Thermometer manufactured by J. Salleron (Fig. 14.1), a Leslie cube and two large-scale parabolic mirrors. In Zappeion, the scale and the base of a Leslie differential thermometer have survived along with a Leslie cube and two paraboloid mirrors. The above findings are an indication of the role played by differential thermometers in the experimental teaching of physics, in particular the propagation of radiation in the nineteenth century, and at the same time a starting point for our research.

Franklin's Pulse or Palm Glass

The pulse glass is attributed to Benjamin Franklin (1706–1790), but he just introduced it as a "philosophical" instrument. Franklin writes in a letter to John Winthrop, professor of mathematics and natural philosophy at Harvard, dated July 2, 1768, that he first saw such an instrument, made by artisans, on his trip to Germany in 1767 (Franklin 1769, pp. 489–492).

His description includes all the elements of the instrument this paper studies:

14 Scientific Instruments Turning into Toys …

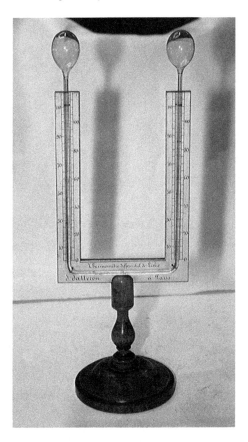

Fig. 14.1 A Leslie differential thermometer from the collection in the Phanar Greek Orthodox College (Μεγάλη του Γένους Σχολή) in Istanbul. Manufacturer: J. Salleron, France. Photo by Panagiotis Lazos

> I met with a glass, being a tube about eight inches long, half an inch in diameter, with a hollow ball of near an inch diameter at one end, and one of near an inch and a half at the other, hermetically sealed, and half filled with water. (Fig. 14.2)

Fig. 14.2 Franklin pulse glass (Black 1803, Plate 1)

Fig. 14.3 A Franklin pulse glass and how it is used (Salleron 1864, p. 225)

Franklin was particularly impressed, even puzzled by the function of this device therefore he acquired some. In his own words:

> If one end is held in the hand, and the other a little elevated above the level, a constant succession of large bubbles proceeds from the end in the hand to the other hand. (Fig. 14.3)

He contacted 'an ingenious artist', the well-known instrument manufacturer Edward Nairne (1726–1806) to build a number of such instruments, which even functioned better than his own, since they were more sensitive. Franklin subsequently cooperated with Nairne again (Franklin 1972, p. 65). Unfortunately, no details are given about how Nairne improved the pulse glass.

Franklin describes a number of experiments with the pulse glass, including some critical points related to later applications of this idea. He experimented with a version with bigger balls and a narrower tube that was bent in right angles. In that case "the water will be depressed in that (ball) which is held in the hand, and rise in the other as a jet or fountain". This characteristic shape is the basic contribution of Franklin in what is now called the Franklin pulse glass and it was adopted later in the differential thermometers and thermoscopes.

He also observed that, if the surface of one ball becomes wet with alcohol, then the water will move towards this ball, because of the evaporation of the alcohol and the temperature reduction inside the ball. Moreover Franklin proposed that "the power of easily moving water from one end to the other of a moveable beam, suspended in the middle" could be used for mechanical applications. The dipping bird works based on these two observations.

Finally, he used it to detect currents of air in a room by placing one of the two bulbs inside a cold current, while the second one rests in a lower and hotter point of the room. This results in the formation of bubbles from the lower bulb to the upper one. The bubbles "were continually passing day and night, to the no small surprise of philosophical spectators". The word surprise best describes this layout and it is the reason why pulse glasses have found a place in most nineteenth century school and university physics cabinets. Such instruments were found in all three schools in Istanbul, which highlights how popular and impressive they were (and are).

The structure of the instrument often found in schools' and universities' labs consists of a closed and airtight U-shaped glass tube ending in two bulbs (Jones 1832, p. 79). Its interior is filled about 1/3 to 1/2 with a quantity of a volatile liquid such as an alcohol or ether. If one of the bulbs is placed in a palm, then the liquid in it is heated and evaporates rapidly, resulting in increased vapor pressure in the bulb. This results in a portion of the vapors being transferred to the other bulb and turbulence appearing in its liquid that resembles boiling, but is not.

A practically identical instrument is the Wollaston cryophorus, even if Wollaston does not mention its similarity with Franklin's pulse glass (Fig. 14.4). Cryophorus is used to demonstrate how a liquid can be frozen fast by evaporation. The one ball of the instruments must be empty and in contact with a freezing mixture of snow and salt. The condensation of water vapors inside this ball is accompanied by evaporation of water in the other ball. As a result the water remaining in liquid phase finally freezes in a few minutes.

It is worth noting that Wollaston gives an interesting detailed description of how a cryophorus–and consequently a pulse glass–can be made:

> Let a glass tube be taken, having its internal diameter about ½ of an inch, with a ball at each extremity of about one inch diameter; and let the tube be bent to a right angle at the distance of half an inch from each ball. One of these balls should contain a little water, and the remaining cavity should be as perfect a vacuum as can readily be obtained. The mode of effecting this is well known to those who are accustomed to blow glass. One of the balls is made to terminate in a capillary tube, and when water admitted into the other has been boiled over a lamp for a considerable time, till all the air is expelled, the capillary extremity, through which the steam is still issuing with violence, is held in the flame of the lamp till the force of the vapour is so far reduced, that the heat of the flame has the power to seal it hermetically. (Wollaston 1813)

The shape of the cryophorus made later by instrument makers is often different, e.g. with the tube being vertical and the one bulb higher than the other. The function remains the same, but these other forms may be more convenient for the set-up of the experiment.

The pulse glass became very popular, partly because it was relatively inexpensive, and it could even be found in private homes. Robison writes:

> We have seen this little toy suspended by the middle of the tube like a balance, and thus placed in the inside of a window, having two holes a and b cut in the pane, in such a situation than when A is full of water and preponderates, B is opposite the hole b. Whenever the room

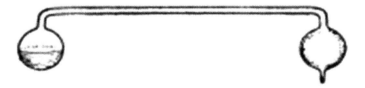

Fig. 14.4 The cryophorus of Wollaston, practically a pulse glass

became sufficiently warm, the vapour was formed in A, and immediately drove the water into B, which was kept cool by the air coming into the room through the hole b. By this means B was made to preponderate in its turn, and A was then opposite to the hole a, and the process was now repeated in the opposite direction and this amusement continued as the room was warm enough. (Robison 1822, p. 15)

The description implies that this toy was quite common, verifies that it is amusing and perhaps demonstrates the first application of the pulse glass as a kind of thermal machine.

Rumford's Differential Thermoscope and Leslie's Differential Thermometer

The next step in this course was particularly episodic. Sir John Leslie (1766–1832) in Scotland and Benjamin Thomson (Count Rumford) (1753–1814) in Bavaria almost simultaneously (1803) constructed two similar instruments. These are the first differential thermometers. The controversy that erupted between the two scientists and people like Sir Humphrey Davy or the widow of Lavoisier (and later wife of Rumford) about the paternity of the instrument was intense (Brown 1967, pp. 192–198). Moreover, Leslie was more or less accused of copying the differential thermometer from Van Helmont (Davy 1812, p. 76 and Plate 1) and from John Christopher Sturmius (Brewster 1824, pp. 144–147), which led to an exchange of articles.

The structures of the two thermometers are similar and very much reminiscent of the Franklin pulse glass. They consist of a U-shaped glass tube, with the two ends forming glass bulbs. The bulbs contain air while the tube contains some kind of liquid. The whole layout is closed and does not communicate with the atmosphere. One difference between the two instruments is that on the Leslie thermometer the horizontal part of the U is small and the temperature scale is located on one vertical while in the Rumford thermoscope the horizontal part of U is larger and bears the scale. Moreover, the Rumford thermoscope has only a bubble of liquid, whereas the Leslie thermometer contains enough liquid to fill the lower part of the U tube.

The structural and operational similarity between these instruments and the pulse glass, plus the time proximity of their appearance, make probable that Leslie and Rumford were inspired by the pulse glass when designing them. This is quite obvious when Leslie describes that the instrument is adjusted: "…by a little dexterity, […] by forcing with the heat of the hand a few minute globes of air form the one ball into the other" (Leslie 1804, pp. 10–11).

In 1803 Rumford presented a series of significant experiments about the nature of heat, based on a new instrument he had invented; the thermoscope, as he called it. It is interesting to follow his detailed description:

> Like the hygrometer of Mr. Leslie (as he has chosen to call his instrument) it is composed of two glass balls, attached to the two ends of a bent glass tube; but the balls, instead of being near together, are placed at a considerable distance from each other; and the tube which connects them, instead of being bent to the middle, and its two extremities turned upwards,

is quite straight in the middle, and its two extremities, to which the two balls are attached, are turned perpendicularly upwards, so as to form each a right angle with the middle part of the tube, which remains in a horizontal position. (Brown 1967, p. 174)

Rumford introduced "a very small quantity of spirit of wine, tinged of a red colour" through a small tube inserted at one of the elbows and after that the instrument was sealed.

By taking appropriate action a small "bubble" of this quantity would be introduced in the long horizontal tube until it rests in the middle of it, when the temperature in the two bulbs is the same. If the temperature was higher in one of the bulbs, then the increased pressure of the air inside it would move the bubble towards the other bulb, until the two pressures equalize. The horizontal tube bears "a scale of equal parts" with which the "movement of the bubble can be observed". An interesting fact is that Rumford did not try to use this scale in order to measure the temperature difference between the two bulbs, rather just to observe the movement of the bubble. This is probably why he chose to call his instrument a thermoscope and not a thermometer.

Rumford made quite a few trials to find the best tube's diameter. This was a critical point for the proper function for the instrument, since a larger diameter meant that the bubble was spread quite far and was difficult to observe precisely, while a smaller one divided the bubble in two parts. Rumford evidently ended up with a tube able to contain 15–18 grains troy of mercury (about 0.971–1.166 g) in a length of 1 in. This means a diameter of 0.37–0.41 cm.

The version of the instrument used in the experiments described by Rumford consisted of a horizontal tube 17 in. in length, two vertical tubes 10 in. in length and two bulbs of 1.625 in. in diameter. The set was supported on a wooden board 27 in. in length, 9 in. in width and 1 in. in thickness. Between the two bulbs there was a circular thermal shield to protect each bulb from the radiated heat from objects presented to the other ball. The shield was supported on a wooden pillar on the board (Fig. 14.5). The instrument was so sensitive that even the radiated heat from the body of the experimenter was enough to put the bubble in motion. Rumford used

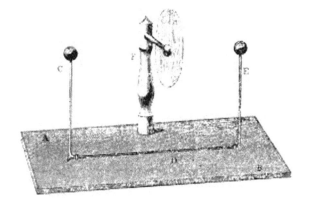

Fig. 14.5 Rumford's thermoscope (Rumford 1804, Fig. 2)

an appropriate type of screens and mounted them around the instrument (Rumford 1804, pp. 103–104).

Leslie published his important book "An experimental inquiry into the nature and propagation of heat" in 1804. The first pages are devoted to the description of the instruments he used and especially to the one "most essential in this research"; namely the differential thermometer. Leslie mentions that "its general construction is the same as that of the hygrometer"; this is an instrument he had presented some years earlier than the photometer (Leslie 1800) and with minor changes could be adapted either as a hygrometer or photometer (Chen 2005, pp. 166–170).

> Nothing indeed could be simpler….Two glass of unequal lengths, each terminating in a hollow ball and having their bores somewhat widened at the other ends, a small portion of sulphuric acid tinged with carmine being introduced into the ball with the longer tube, are joined together with the flame of a blow-pipe, and afterwards bent into nearly the shape of the letter U, the one flexure being made just below the joining, where the small cavity facilitates the adjustment of the instrument, which, by a little dexterity, is performed by forcing with the heat of the hand a few minute globes of air from the one ball into the other. The balls are blown as equal as the eye can judge, and from four-tenths to seven-tenths of an inch in diameter. The tubes are such as are drawn for mercurial thermometers, only with wider bores; that of the short one, and to which the scale is affixed, must have an exact caliber of a fiftieth or a sixtieth of an inch; the bore of the long tube need not be so regular, but should be visibly larger, as the coloured liquor will then moving quicker under any impression. Each leg of the instrument is from three to six inches in height, and the balls are from two to four inches apart. The lower portion of the syphon is cemented at its middle to a slender wooden pillar inserted into a round or square bottom, and such that the balls stand on a level with the centre of the speculum. (Leslie 1804, p. 9)

The intention of Leslie to take measurements on heat is obvious when he states that the instrument "is calculated to measure difference (of heat) with peculiar nicety".

The described form and structure of the instrument must not be taken as granted, because as Leslie points out: "…I had a variety of those differential thermometers, of different sizes, and of some diversity of forms, adapted for particular occasions". Actually, some types of his differential thermometers are illustrated in the very first page of his later book *A short account of experiments and instruments depending on the relations of air to heat and moisture.* The form of the instrument in Fig. 14.1 of this page was adopted by later instrument makers as the more typical differential thermometer.

Apart from the obvious similarities among Leslie's differential thermometer, Rumford's differential thermoscope and Franklin's pulse glass, there is a main difference; namely the air in their tube is not exhausted. The presence of air does not affect the functional ability of the instruments and makes their construction simpler. Moreover, the Rumford thermoscope contains a significantly smaller quantity of wine spirit than the pulse glass, since Rumford used the liquid's bubble just as an indicator and not for producing vapors. It is mainly the difference of the pressure of the air inside the bulbs and the tube that moves the bubble, not of its vapours. Despite these differences the operational principle of the instruments is the same as with the pulse glass.

Experiments with Differential Thermometer and Thermoscope

Although it is beyond the scope of this article to present analytically the experiments made by Rumford and Leslie with these instruments, it is useful to give some examples, especially in cooperation with other instruments.

In experimental arrangements, the differential thermometer was often associated with a parabolic mirror and a Leslie cube. The latter is a hollow metal cube rested on a wooden base (Fig. 14.6). The top of the cube is removable so it can be filled with hot water, while its four vertical sides were coated with different materials or painted in different colors, usually white, black and red. After the water has been inside the cube for some time, all sides reach the same temperature. The cube serves as a source of infrared radiation, or thermal radiation as it was called. Similar but simpler vessels were also used by Rumford in his experiments (i.e. the vessels were cylindrical and each one had a surface of different material).

Below is a description of three of the experiments that could be carried out with the above equipment as they were described and illustrated in a physics textbook:

1. A screen is placed between Leslie cube (A) and differential thermometer (D) as in Fig. 14.7 (Peck 1869, pp. 198–199). The screen has a small hole. A reflective surface (B) is placed between the screen and the differential thermometer. Radiation from the cube passes through the hole, reflects on the surface and directs toward the differential thermometer. A suitable position of the thermometer is

Fig. 14.6 A Leslie cube from the collection of the Phanar Greek Orthodox College in Istanbul (Photo by Panagiotis Lazos)

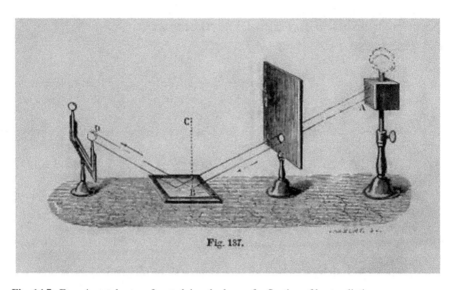

Fig. 14.7 Experimental set up for studying the laws of reflection of heat radiation

sought for one bulb to be heated by the reflected radiation. It can then be confirmed that the CBD and CBA levels coincide and that the angle of incidence is equal to the reflectance angle.

2. The Leslie cube is positioned opposite a parabolic mirror and at a fairly large distance to the focal length of the mirror, which has the property of concentrating the parallel beams on its surface (Fig. 14.8) (Peck 1869, pp. 202–203). A small flat piece of some material is placed at a very short distance before the focus

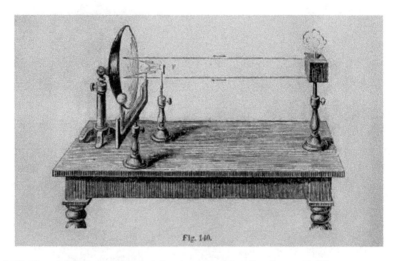

Fig. 14.8 Experimental set up for studying the reflectivity of various materials

and in such a way that the radiation is reflected and directed to one of the bulbs of the differential thermometer. Comparing the temperature values for surfaces of different materials gives a measure of the reflectivity of these materials for infrared radiation.

3. The Leslie cube is positioned opposite a parabolic mirror and at a fairly long distance with respect to the focal length of the mirror, which has the property of focusing the parallel rays hitting it (Peck 1869, pp. 204–205). One bulb of the differential thermometer is placed in the focus. Then we turn different sides of Leslie cube to the mirror and record the temperature values on the thermometer. The material to which the highest temperature corresponds is the one with the highest thermal radiation emission factor. It is reported that the coefficient decreases with the alternation of the surfaces as follows: black, paper-covered, glass-covered, and tin.

Differential thermometers were used in the study of heat and heat radiation during the first decades of the nineteenth century. The invention of the thermopile (Fig. 14.9) by the Italian physicist Leopoldo Nobili in 1829 and the adoption of it in the Melloni optical bench in 1835 gave scientists a more accurate and handy instrument (Melloni 1835). After 1841, when the Melloni bench was commercialized, it gradually pushed the differential thermometers to the margins of research because it lacked one of their main disadvantages; the fact that the glass of the bulbs absorbed radiant heat. The Melloni bench was used until around 1880 when in turn it was replaced by the bolometer (Colombi et al. 2017). As a popular physics textbook explained in 1877: "The instrument (differential thermometer) is now only used as a thermoscope; that

Fig. 14.9 A Nobili thermopile on a Melloni bench, made by Breton Frères, from the collection of the Phanar Greek Orthodox College in Istanbul (Photo by Panagiotis Lazos)

is, to indicate a difference of temperature between the two bulbs, and not to measure its amount" (Atkinson 1877, p. 243).

Nevertheless, they continued to be used extensively in education due to their simplicity of supervision, relatively low cost and ease of operation. The wide use (or at least tendency to be used) of the differential thermometers in education is demonstrated by three factors. First, they are found in the collections of scientific instruments in most physics laboratories at schools or universities. Second, they are illustrated and presented in many physics textbooks of the nineteenth century (E.g. Arnott 1856, p. 302, Atkinson 1877, pp. 242–243, Thresh 1880, p. 85), which shows, and perhaps vice versa caused, the spread of these instruments in the educational institutions. Finally, they were made by several instrument makers and presented in their commercial catalogues as late as in the 1920s (E.g. CENCO 1912, p. 131, Deleuil 1848, p. 20, Griffin John and sons 1910, pp. 482–483, Max Kohl 1905, p. 588) (Fig. 14.10).

Fig. 14.10 A Rumford and a Leslie differential thermometer (Max Kohl 1905, p. 588)

In some cases many different types of differential thermometers were offered, like the one designed by August Matthiessen around 1867, having the bulbs pendent so they could be easily immersed in vessels. By this modification the differential thermometer could be used in experiments with liquids, something not possible for Leslie's version. Matthiessen himself presented an experiment to the Chemical Society about the specific heats of metals (Matthiessen 1867, p. 9). In another type of differential thermometers a tube with a stopcock is placed exactly under the bulbs connecting the two vertical tubes. This allows the immediate adjustment of the level of the liquid inside the thermometer before an experiment.

The Pulse Glass in Measuring Gas

Samuel Clegg (1781–1861) was one of the pioneers of the gas industry. Having worked for fifteen years in companies in that field, he realized that paying for gas according to the elapsed time of use was not the best solution. On December 9th 1815 he got a patent for a gas apparatus that included a rotative gas meter (Clegg 1831, p. 322). It is thought to be the first wet gas-meter since water is essential to the measurement (Webster 1847, p. 106). The details of that gas meter are not important for this essay but it should be mentioned that, despite its success, there were some disadvantages, so that Clegg invented a gas-meter without water (a "dry" one) in 1831. The advantages of the new gas-meter in his own words were:

> It does not work in water, and hence is not subject to the inconveniences and uncertainty of action that may occasioned by the waste of water from evaporating or other causes [and] it operates without that resistance to the passage of the gas through it, which has heretofore prevented the use of gas-meters in situations where the pressure of the gas in main pipes has been very slight. (Clegg 1831, pp. 322–323)

The instrument consists of an airtight container with two openings in the lower and upper part of it, from which the gas enters and exits. The upper opening is a bit smaller. In the middle of the container there is a pulse glass containing "spirits of wine" and it is able to rotate around a horizontal axis. The gas is heated by a burner outside the instrument as it enters the lower opening, so that the gas is always hotter in the lower part (Fig. 14.11). As a result the liquid in the lower bulb of the pulse glass moves towards the upper one, eventually rotating the pulse glass a half turn. By an appropriate mechanism this rotation moves an index in the front side of the instrument, thereby the number of these oscillations can be counted and the volume of the gas passed can be measured. The instrument had success in France but not in England. Some years later Clegg stressed that his dry gas-meter could be used not only to measure gas but also: "…to register the average pressure of high pressure steam, the average temperature of heated air, or the average of any variable temperature for any period" (Clegg 1839, p. 286).

Fig. 14.11 Back side of the "Clegg's improved gas-meter" (Clegg 1831, Fig. 1)

Thermal Machines and Toys (Once More)

A modern variation of the Franklin pulse glass, which is a particularly interesting toy, is the so-called hand boiler or love meter (Fig. 14.12). In fact it is just a lightly restructured pulse glass. The tube is vertical but usually not straight as it includes some curves for aesthetic purposes and it descends almost to the bottom of the lower bulb, immersed in the liquid.

Nowadays the liquid used in the toy is not water, as in the pulse glass, but methanol (Ucke & Schlichting 1995). This is because the liquid must have a low boiling point but also be safe. For the latter reason diethyl ether–the theoretically best choice–has never been used and the methyl chloride (dichloromethane) is not used any more, because it is considered to be toxic (OSHA site). If we put the lower bulb into our palm, then the liquid evaporates rapidly and the vapor pressure increases. As a result the liquid rises through the tube into the upper bulb, given that there is quite a large temperature differential between the two bulbs. As soon as the whole liquid rises, a quantity of its vapors comes up and ends up in the upper bulb creating a strong turbulence resembling boiling.

The demonstration of the device is a wonderful trigger for discussion in the context of teaching the relationship between temperature and pressure of a gas in a vessel, the evaporation of liquids and the energy transformations taking place in the phenomenon. Especially for young children the layout can also be used as a toy,

Fig. 14.12 Hand boiler
(Photo by Panagiotis Lazos)

whereby the students try to raise the liquid in the upper bulb as fast as they can. Compared to the pulse glass, the hand boiler has two advantages. It can stand upright alone and it has an even greater degree of surprise because the liquid not only moves from one bulb to another, but at the same time defies gravity.

A slightly more complicated toy is the so-called dipping bird (Fig. 14.13). This is virtually a hand boiler with two significant differences. First, the top bulb is covered with felt, which, for reasons of appearance, resembles a bird's head. Moreover, the tube is straight and is supported by a metal ring with protrusions on a plastic base so that the whole device can rotate freely around a horizontal axis. The position of the ring is adjusted so that the device balances with the "head" leaning over.

If the felt is wet then the liquid gradually rises in the upper bulb and the center of mass of the device elevates. As soon as the center of mass is higher than the axis the device rotates towards a horizontal position (where there is a stopper). Simultaneously the liquid drains back to the lower bulb and part of the tube. The device then returns to its upright position and the phenomenon is repeated as long as the felt is wet. The period of the oscillations is inversely proportional to the rate of evaporation of the water (or other liquid) in the felt (Guemez et al. 2003). If a container of water is placed in such a position that the head gets wet every time it falls, the phenomenon takes days until the level of water in the glass drops due to evaporation and the felt gets completely dry. The evaporation of water in the felt causes a decrease in the temperature in the upper bulb and consequently a reduction of the vapor pressure within the tube (Bohren 2001, pp. 15–19; Crane 1989, p. 470; Jargodzki and Potter 2001). This results in the rise of liquid in the upper bulb, since the pressure in the lower bulb is greater. In order to raise the liquid to the upper bulb in both toys the

Fig. 14.13 Dipping bird
(Photo by Panagiotis Lazos)

temperature of the top bulb must be lower than that of the bottom. This is achieved either by increasing the temperature in the lower bulb (in hand boiler) or by lowering the temperature in the upper bulb (in dipping bird).

The dipping bird was patented in the USA by Miles Sullivan (US patent 2,402,463) in 1946 but it seems to have been first made by unknown artisans in China during the first decades of the twentieth century (Perelman 2008, pp. 226–228. The book was first published in Russian in 1913 and the edition translated in English was printed in 1936). This pattern is similar to what happened with Franklin and pulse glass. It is true, however, that there were at least three similar inventions in the last quarter of nineteenth century that used pulse glasses to create a motor (thermal machines), giving life to Franklin's idea of using the pulse glass to mechanical application.

In 1871 the Italian physicist Enrico Bernardi proposed a motor using three pulse glasses, containing sulfuric acid, able to rotate around a common horizontal axis. Every moment the bulbs of one or two pulse glasses are immersed in water (Bernardi 1872, pp. 297–300). The motor starts to rotate at the direction of the arrow at (Fig. 14.14) because of evaporation of the water. In this sense this motor is the closest relative to the dipping bird, with the difference that the latter is an oscillating motor.

In 1881 Israel Landis (US patent 250,821) (Fig. 14.15) and one year later the Iske brothers (US patent 253,868) also proposed oscillating motors using pulse glasses, but instead of taking advantage of evaporation, they used a source of heat in order to raise the temperature in one of the bulbs. All of these thermal machines were unsuccessful, mainly because of their low efficiency. In any case it appears difficult to connect the aforementioned inventions directly with the invention of the dipping

Fig. 14.14 The Bernardi motor (Bernardi 1872)

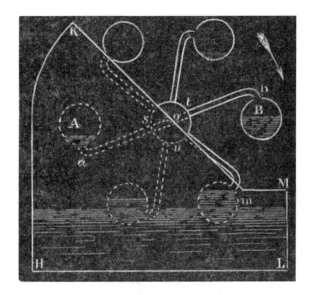

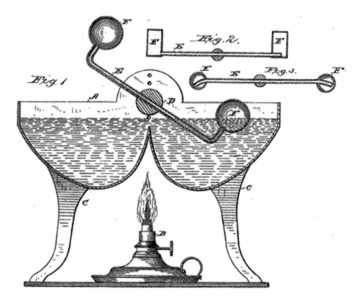

Fig. 14.15 The Landis oscillating motor. The vessel contains water, the pulse glass is "charged with volatile liquid—such as alcohol, ether, or the like"

bird in China. Moreover, there is an obvious difference. The research on the motors is an effort to find practical applications of the pulse glass and a step towards technology, whereas the dipping bird is an amusing–and potentially educating–artwork.

A thermal machine similar to the dipping bird, but with an obvious educational purpose, is the pulse-glass engine (Fig. 14.16) mentioned in "Demonstration Experiments in Physics" (Sutton 1938).

This consists of a pulse glass able to rotate vertically around a horizontal axis. At a short distance above each bulb pieces of cotton dipped in ice water or ether are attached. If one of the bulbs is in contact with the cotton, then it is cooled, the pressure decreases and it enters the liquid from the other bulb. The displacement of the fluid causes the device to rotate, causing the other bulb to come in contact with the other piece of cotton. The phenomenon is repeated until the cotton pieces reach ambient temperature.

A slightly different and more recent version of this thermal machine is presented in Fig. 14.17 (UC Berkeley Physics Lectures Demonstrations). The repetition of the movement of the pulse glass is due to an electric heater, an idea very close to Landis. The educational purpose is once again evident.

Obviously, the dipping bird is not just a toy but a thermal machine as well (Mentzer 1993, pp. 126–127); it can also be used in teaching the relevant concept of thermodynamics. The efficiency of this thermal machine is about 2% (Ng and Ng 1993), which makes its application in practical machines quite improbable. Nevertheless, the conservation of energy, combined with the periodic motion of the device, can be the starting point for a creative debate on the operating principles of thermal machines. All these are supported by the rich literature and the fact that Franklin wrote about the surprise the pulse glass caused to everyone.

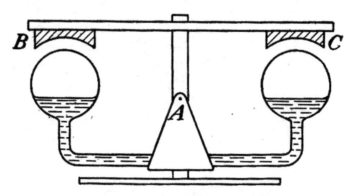

Fig. 14.16 Pulse glass engine. Demonstration Experiments in Physics (Sutton 1938, p. 216, Fig. 179)

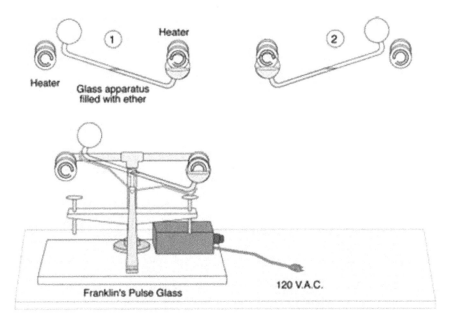

Fig. 14.17 Thermal pulse-glass machine with an electric heater (Courtesy of the U.C. Berkeley Physics Lecture Demonstrations)

Conclusions

We have followed the multiple transformations of the Franklin pulse glass over several generations for more than 200 years. It is clear that a wide range of technological, scientific, educational and recreational applications is based on the structural and operational principles of the pulse glass. This is due to its far-reaching distribution, both as a real object in workshops or homes, as well as descriptions in texts. It is also due to the impact it has on those who learn about it and especially on those who use it. It would probably be worthwhile to investigate whether a similar evolutionary pattern applies to other physics instruments (e.g. in devices for the study of the impacts which have common characteristics with Newton cradle) and, if so, what are the common features between these cases.

The pulse glass combines some rare attributes, which few scientific instruments have to this degree: simplicity, grace, supervision and, above all, surprise. It could be said that surprise is embedded in this object. Perhaps it is this property that has given such strong impetus to human imagination and resourcefulness and has led to such versatile constructions.

References

Arnott, N. (1856). *Elements of physics or natural philosophy, general and medical*. Philadelphia: Blanchard and Lea.

Atkinson, E. (1877). *Elementary treatise on physics experimental and applied. Translated and edited from Ganot's elements de physique*. New York: William Wood and Co.

Bernardi, E. (1872). Modo di utilizzare il calorico dell' ambiente per produrre un piccolo lavoro. *Rivista scientifico-industriale delle principali scoperte ed invenzioni, 4*, 297–300.

Black, J. (1803). *Lectures on the elements of chemistry*. Edinburgh: Mundell and Son.

Bohren, C. (2001). *Clouds in a glass of beer*. New York: Dover Publications.

Brewster, D. (1824). Professor Leslie's differential thermometer invented by professor Sturmius. *The Edinburgh Journal of Science, 1*, 144–147.

Brown, S. (1967). *Men of physics: Benjamin Thompson-Count Rumford*. Oxford: Pergamon Press.

CENCO. (1912). *Physical and chemical apparatus*. Chicago: Catalogue M.

Chen, X. (2005). Visual photometry in the early 19th century: A "good" science with "wrong" measurement. In J. Buchwald & A. Franklin (Eds.), *Wrong for the right reasons* (pp. 161–183). The Netherlands: Springer.

Clegg, S. (1831). Clegg's Improved Gas Meter, *Mechanics' Magazine, Museum, Register, Journal and Gazette 15*(415).

Clegg, S. (1839). On the dry meter. *Journal of the Franklin Institute, 23*, 286.

Colombi, E., Leone, M., & Robotti, N. (2017). The emergence of Melloni optical bench. *European Journal of Physics*. https://doi.org/10.1088/0143-0807/38/1/015802.

Crane, H. R. (1989). What does the drinking bird know about jet lag. *The Physics Teacher, 27*, 470.

Davy, H. (1812). *Elements of chemical philosophy* (Vol. 1). London: J. Johnson and Co.

Deleuil, J.A. (1848). *Catalogue d'instruments de physique, de chimie, d'optique, de mathématiques, de chirurgie, d'hygiène et d'économie domestique*. Paris: d'A. René.

Franklin, B. (1769). *Experiments and observations on Electricity, Made At Philadelphia in America: To which are Added, Letters and Papers On Philosophical Subjects*. London: David Henry.

Franklin, B. (1972). The Papers of Benjamin Franklin, Vol. 16: January 1, 1769, through December 31, 1769, ed. Willcox W. Yale University.

Griffin John and sons. (1910). *Scientific handicraft*. London: An illustrated and descriptive catalogue of scientific apparatus.

Guemez, J. V., Fiolhais, R., & Fiolhais, M. (2003). Experiments with the drinking bird. *American Journal of Physics, 71*, 1257–1263.

Jargodzki, C. & Potter, F. (2001). *Mad about physics. Braintwisters, paradoxes and curiosities*. New York: John Wiley and Sons.

Jones, T. (1832). *New conversations on chemistry*. Philadelphia: Hohn Crigg.

Iske Brothers. (1882). Motor. https://www.google.com/patents/US253868. Accessed December 18, 2018.

Lantis, I. (1881). *Oscillating motor*. https://www.google.com/patents/US250821. Accessed December 23, 2018.

Leslie, J. (1800). Description of an hygrometer and photometer. *A journal of Natural Philosophy, Chemistry and the Arts, 3*, 461–467.

Leslie, J. (1804). *An experimental inquiry into the nature and propagation of heat*. London: J. Mawman.

Leslie, J. (1813). *A short account of experiments and instruments depending on the relations of air to heat and moisture*. Edinburgh: Blackwood and Ballantyne.

Max Kohl, A.G. (1905). *Price List, Vol II & III*. Chemnitz.

Mentzer, R. (1993). The drinking bird. The little heat engine that could. *The Physics Teacher, 31*, 126–127.

Matthiessen, A. (1867). *Improved differential thermometer* (p. 1). Laboratory: A Weekly Record of Scientific Research.

Melloni, M. (1835). Description d'un appareil propre à répéter toutes les expériences relatives à la science du calorique rayonnant contenant l'exposes de quelque faits nouveaux sur les sources calorifiques et le rayons qui en émanant. *L'Institut, 3,* 22–26.

Ng, L., & Ng, Y. (1993). The thermodynamics of the drinking bird toy. *Physics Education, 28,* 320–324.

OSHA—Occupational Safety and Health Administration. Methylene chloride. https://www.osha.gov/SLTC/methylenechloride/index.html Accessed November 17, 2018.

Peck, W.G. (1869). *Introductory course of natural philosophy for the use of schools and academies. Edited form Ganot's Popular Physics.* New York: A. S.Barnes.

Perelman, Y. (2008). *Physics for entertainment, book 2.* New York: Hyperion.

Robison, J. (1822). *System of mechanical philosophy* (Vol. 2). Edinburgh: Thoemmes Continuum.

Rumford, B. T. (1804). An inquiry concerning the Nature of Heat, and the Mode of its Communications. *Philosophical Transactions, 94,* 77–182.

Salleron, J. (1864). *Notice sur les instruments de précision.* Paris: Marais.

Sullivan, M.V. (1946). *Novelty device.* https://www.google.com/patents/US2402463. Accessed December 16, 2018.

Sutton, R. M. (1938). *Demonstration experiments in physics.* New York: McGraw-Hill.

Thresh, J. (1880). *Physics, Experimental and Mathematical: A Handbook for the Physical Laboratory, and for Students Preparing for the Science Examinations of the London University.* London: W. Stewart and Co.

U.C. Berkeley Physics Lecture Demonstrations. http://berkeleyphysicsdemos.net/node/350. Accessed October 28, 2018.

Ucke, Christian, & Schlichting, Hans-Joachim. (1995). Der Kaffeekugelschreiber oder das Liebesthermometer. *Physik in Unserer Zeit, 26*(4), 192–193.

Webster, Thomas. (1847). *The principles of hydrostatics: An elementary treatise on the laws of fluids, and their practical applications.* London: John Parker.

Wollaston, W. H. (1813). On a method of freezing at a distance. *Philosophical Transactions of the Royal Society of London, 103,* 71–74.

Part IV
Limitations

Chapter 15
I'm Not There. Or: Was the Virtual Particle Ever Born?

Markus Ehberger

Introduction

The virtual particle is an integral part of the conceptual framework of modern quantum electrodynamics (QED) and quantum field theory (QFT). Although these particles are in principle unobservable and, according to a popular narrative, violate energy conservation for the short time of their existence, their centrality in the conceptual framework is unquestionable. Through their representation in Feynman diagrams, virtual particles were not only brought to the masses of the working physics community [for a historical account of the dispersion of Feynman diagrams, see Kaiser (2005)], but spread into the realm of popular culture and were even included in the redesign of the windows of the St. Nicolai Church in Kalkar (North-Rhine-Westphalia, Germany) in 2000 (Pawlak 2004).

The centrality of this, at first sight, rather strange concept is at least partly due to the close connection between QED and perturbation theory. It is through this latter method of approximation that virtual particles show up in modern day QED. The link between virtual particles and such a technique, as well as the strange attributes of virtual particles, has prompted philosophical discussions concerning their ontological status in the framework of quantum field theory [the most important arguments brought forward until 2008 have been evaluated by Fox (2008), resulting in the denial of any realistic interpretation of virtual particles; this view has been contested by Bacelar Valente (2011)]. As is common for the philosophy of physics, these works take the modern description as a starting point and seldom discuss historical connections. But even within historical literature, the formation of the concept is only touched upon, and the modern language of QED at times overshadows the description of the past physicists' calculations and conceptions.

M. Ehberger (✉)
Technical University of Berlin, Berlin, Germany
e-mail: markus.ehberger@tu-berlin.de

The collection of papers and developments between roughly 1927 and 1949 presented in this article were chosen to provide an overview of the formation of the virtual particle concept. Although some of the strands of the historical development, in which the virtual particle and related concepts played a crucial role (like the diverging self-energy of the electron or the infrared divergence) will not be considered, and many of the intricacies of the calculations and the conceptualizations will be brushed over in broad strokes, the given selection will allow us to engage with the overall theme of this volume: biographical approaches in the history of physics. In the literature on concept formation and development, biographical metaphors were applied not only in popular works (one can find such colourful examples as Charles Seife's (2000) *Zero: The Biography of a Dangerous Idea*) but also by some scholars in the history and philosophy of science, most notably by Arabatzis (2006). The development sketched in this contribution will shed a rather pessimistic light on the application of such metaphors. In particular, I want to call the appropriateness of the term 'birth' in connection with concept formation into question.

Therefore, I will focus on a selection of physical processes analysed by the historical actors mainly using time-dependent perturbation theory. This was one of the standard ways of doing QED calculations in the 1930s until the late 1940s. The description of the physical processes as multi-step phenomena is conceptually close to the modern description and was integrated into the mathematical treatment early on. Moreover, some of the papers discussed here were described by historians of science explicitly using the notion of virtual particles.

We will start out with Paul Dirac's introduction of the intermediate state in QED in 1927; from a modern vantage point he identified the most important attributes of those states. This section will also include some technical details to render the notions accessible to the reader. Afterwards I will outline how the connection with the term 'virtual' and the intermediate states became established in discussions around the Raman effect. The use of perturbation theory and the state transition model—still in play in QED until the late 1940s—will be exemplified by the work of Bethe and Fermi (1932) as well as Bernhard Kockel and Hans Euler (1936), in which particles only present in intermediate states were necessary for the theoretical evaluation of the processes of electron-electron and light-by-light scattering. I will conclude the historical outline by briefly considering the form the virtual particle took in the late 1940s, especially in the hands of Richard Feynman.

Examples from the History of the Virtual Particle

In the following sections some important steps and developments in the formation of the concept of the virtual particle will be presented. These sections can be read in their own right as a sketch of the development of the concept studied. I still want to warn the reader that this will not provide a concise history of the virtual particle concept. As already mentioned, the length of the article only allows a small selection of papers to be studied in detail and important strands of the concept formation process, such as the divergency difficulties of the 1930s and 40s, will not be considered. Nonetheless, the following sections will provide enough material to judge the applicability of the birth-metaphor in the concept formation process, which will be done in the conclusion.

Introduction of the Intermediate State and First Applications

Although there is a considerable pre-history to the concept of an intermediate state and to the connected conception of scattering as a two-step process of emission and absorption of radiation, we will start out with Dirac's work in the mid-to-late 1920s. The focus on his papers is especially motivated by the observation that they

> set the basic language and concepts characteristic of the modern conception of the elementary processes as a sequence of absorption/emission of light quanta, passing through a sequence of intermediate states. (Lacki et al 1999, p. 484)

Most importantly, Dirac's perturbative scheme and the corresponding conceptual picture became prominent in quantum field theoretic work of the 1930s and 1940s and was included in the influential didactic works on QED by Fermi and Heitler (Fermi 1932; Heitler 1936).

By 1926, Dirac had already devised a way of doing time-dependent perturbation theory in quantum mechanics (Dirac 1926, §5; for a historical contextualization of the work, best see Bromberg 1977 or Kragh 1990, Chap. 2). First, Dirac expanded the general solution ψ of the unperturbed Hamilton equation $(H - W)\psi = 0$ into its eigenfunctions $\psi = \sum_n c_n \psi_n$ and then showed that the solution to the perturbed problem $(H - W + V)\psi = 0$, where the perturbation V sets in at some time t = 0, can be written as $\psi = \sum_n a_n(t)\psi_n$, where the coefficients $a_n(t)$ are now time-dependent. The evolution of the coefficients was given by:

$$i\hbar \dot{a}_n = \sum_m V_{mn} a_m$$

where the V_{mn} are the matrix elements of the perturbation. This constitutes the basic equation of time-dependent perturbation theory. It is important to note that the a_n, although they change over time, are connected to the eigenfunctions of the unperturbed system. As Dirac considered an assembly of atomic systems, they represented the change in the number of atoms in a specific unperturbed state. The perturbation, therefore, does not alter the stationary states of the system but induces transitions between them.

The connection between this perturbation theory, Dirac's transformation theory and the quantization of the electromagnetic field in his famous "On the Quantum Theory of Emission and Absorption of Radiation" (Dirac 1927a) and "On the Quantum Theory of Dispersion" (Dirac 1927b) has been the subject of many studies [see for example Jost 1972; Bromberg 1977; Darrigol 1986 (Sect. 2); Schweber 1994 (pp. 23–32), Kragh 1990 (Chap. 6); Cao 1997 (Chap. 7.3) or Lacki et al. 1999 (Sect. 4.2)], and since we are mainly interested in the introduction of the intermediate state, we will move on directly to Dirac's paper on dispersion theory (Dirac 1927b), referring to specific interpretations of the physical processes along the way.

Dirac had already derived first order terms in his perturbative treatment of the radiation field, which provided a mathematical description of emission and absorption processes. To describe dispersion, he had to push his method one step further. The quantization of the radiation field brought Dirac to a Hamiltonian which included, besides the terms for the "proper energy" (this is what Dirac called the energy of the system minus the interaction energy), two terms containing the interaction between the radiation field and the material part of the system: these were treated as perturbations (let's call them V_1 and V_2). V_1 was already derived in (Dirac 1927a) and was linear in creation and annihilation operators of field excitations: he therefore interpreted it as responsible for emission and absorption processes of photons in the first order of the approximative calculation. The second term was proportional to the product of such operators (arising from the quadratic term of the vector potential in the classical Hamiltonian). The processes due to this term were called "direct scattering" by Dirac, a phrase that will become clear in a second.

The first order terms had the following structure:

$$a_m = a_{m0} + \sum_n V_{1mn} a_{n0} \frac{1 - e^{i(W_m - W_n)t/\hbar}}{W_m - W_n}$$

where a_{m0} and a_{n0} are the initial values of the state of the system (Dirac 1927b, p. 712). Since the coefficients described the state of the whole system, one should expect that the probability of a transition into a state with a different energy ($W_m - W_n \neq 0$) should, due to energy conservation, be zero. From this equation one sees however,

> that when two states m and n have appreciably different proper energies, the amplitude a_m gets changed only by a small extent, varying periodically with the time, on account of transitions from state n. (ibid., p. 712)

Dirac could show, nevertheless, that only for those states m' with (approximately) the same energy as the initial state m would the amplitude grow linearly with time. Consequently, only those transitions should be considered observable [this was already derived in (Dirac 1927a)].

To second order, Dirac arrived at the following terms (Dirac 1927b, p. 721):

$$a_m = \left(V_{2mk} - \sum_n \frac{V_{1mn} V_{1nk}}{W_n - W_k} \right) \frac{1 - e^{i(W_m - W_k)t/\hbar}}{W_m - W_k}$$
$$+ \sum_n \frac{V_{1mn} V_{1nk}}{W_n - W_k} \frac{1 - e^{i(W_m - W_n)t/\hbar}}{W_m - W_n}$$

Here, n denotes the intermediate state of the system; k and m correspond to the initial and final state respectively. The summation goes over all states of the total system, and hence also over the ones that do not conserve the proper energy. Again, the solution fluctuates when the initial and final state have considerably different energies. Only for initial and final states having the same total energy will the first term increase linearly with the time. The second term, in Dirac's interpretation, only becomes important for resonances and shall not be discussed here any further. v_{2mk} represents the contribution from direct scattering as it couples initial and final state directly. The other contribution to the scattering amplitude—the sum—is interpreted by Dirac in the following way:

> The scattered radiation thus appears as the result of the two processes k → n and n → m [v_{1nk} and v_{1mn} in the formula above], one of which must be an absorption and the other an emission, in neither of which is the total proper energy even approximately conserved. (ibid., p. 712)

This interpretation, highly suggestive from the structure of the theory, is a straightforward extrapolation from the interpretation of the matrix elements v_{1nk}. Since these correspond to emission and absorption, the terms proportional to $v_{1mn}v_{1nk}$ give rise to a combination of these processes. The only constraint for evaluating the formulas was energy conservation for the whole process. It should be noted that the probability of finding the system in the final state is actually $|a_m|^2$ (ibid., p. 722, especially Eq. (22)); consequently only the absolute square of the term in brackets has observable consequences, and, as Gregor Wentzel noted in his *Handbuch* article of 1933, "there can be no possibility of physically distinguishing the two terms [scattering through intermediate states and direct scattering]" (Wentzel 1933, p. 747).

Dirac notes one further peculiarity of the intermediate states (ibid., p. 722), when inserting the perturbation as described by the Hamiltonian into the perturbative scheme: the sequence of emission and absorption can be interchanged and both terms must be taken into account in order to derive the Kramers-Heisenberg formula (Kramers and Heisenberg 1925). This formula was devised by Hendrik Anthony Kramers and Werner Heisenberg in 1925, even before the establishment of a coherent mathematical framework for quantum mechanics. Its derivation became *"un passage obligé"* (Lacki et al. 1999, p. 462) for approaches to quantum mechanics during its infancy. Although not validated experimentally until 1928, it reassured Dirac that his procedure was a sensible one.

In his dispersion paper Dirac consequently identified the most important attributes of the intermediate states: the proper energy is not conserved in the transitions; they are unobservable in principle; all possible combinations of emission and absorption of light quanta have to be taken into account; and only the absolute square of the sum over all possible processes through intermediate states corresponds to an observable quantity. Still, the conceptualization of the perturbative scheme suggested a means of explaining the seemingly non-conservation of energy, as exemplified by two papers published by Yakov Illich Frenkel in 1929.

Although Frenkel insisted that "the usual assumption […] does not regard the state *n* as 'really' occurring" (Frenkel 1929b, p. 758) he argued that the intermediate states have a "definite duration, although a much shorter one than that of the two end states". (ibid.) On the one hand, he says that one can consider this a special case of the Heisenberg energy-time uncertainty. On the other (and he credited Dirac for this insight), he points out that, if one considers the process of scattering in Dirac's theory of radiation

> one has to take into account the interaction energy [gegenseitigen Energie] between the two [photon and atom], where the difference $h\nu_{no} - h\nu$ [i.e. energy needed for the transition to the intermediate state minus the energy of the incoming photon] has to be compensated at the expense of this interaction energy.[1] (Frenkel 1929a, p. 802)

In this way, the fact that the interaction energy was taken to induce transitions between the states of the system and was not part of the proper energy, supplied an explanation "in harmony with the conservation of energy" (ibid.).

Frenkel's reasoning constitutes a rare example of a direct defence of the reality of the transitions through intermediate states. Most physicists took a rather pragmatic attitude towards the method proposed by Dirac. They applied it to diverse problems and showed how higher order terms could be established.

In Göttingen, where Dirac had finished his second paper on radiation theory, two students of Max Born applied Dirac's methods. Maria Göppert-Mayer (1929, 1931) worked out processes of second order in the interaction between the radiation field and an atom. These included, besides the Raman effect (see also the next section), double emission and double absorption. The corresponding processes are shown in Fig. 15.1 and were interpreted by Göppert-Mayer as follows:

> [...] both of the processes discussed [double emission and double absorption], as well as the Raman effect, behave as if two processes, of which neither satisfies energy conservation, occur in one act.[2] (ibid., p. 284)

She described these processes as "the synergy of two light quanta in one elementary act"[3] (Göppert 1929, p. 932).

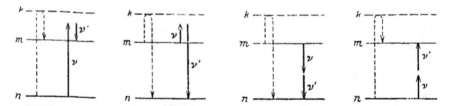

Fig. 15.1 Second order processes in Dirac's radiation theory [taken from Göppert (1929)]. From left to right: Stokes Raman-effect; Anti-Stokes Raman-effect; double emission; double absorption. The dashed lines refer to the transition of the atom (k is an intermediate state), solid lines pointing upwards (downwards) refer to absorbed (emitted) photons

---

[1] "[...] so muß man noch ihre gegenseitige Energie in Betracht ziehen, wobei die Differenz $h\nu_{no} - h\nu$ auf Kosten dieser gegenseitigen Energie zu kompensieren ist".

[2] "[...] beide besprochenen Prozesse ebenso wie der Ramaneffekt verhalten [sich so], als ob zwei Vorgänge, von denen jeder nicht dem Energiesatz genügt, in einem Akt geschehen".

[3] "[das] Zusammenwirken zweier Lichtquanten in einem Elementarakt".

Victor Weisskopf, also a student at Göttingen, applied Dirac's method in his work with Eugen Wigner on the breadth of spectral lines (Weisskopf and Wigner, 1930) and in his dissertation on resonance scattering (Weisskopf 1931). He eventually showed (Weisskopf 1933) that the perturbative expansion to an arbitrary order (z) is given by:

$$a_n^{(z)} = \sum_{n_1} \cdots \sum_{n_{z-1}} \frac{V_{0n_1} V_{n_1 n_2} \cdots V_{n_{z-1} 0}}{(E_{n_1} - E_0)(E_{n_2} - E_0) \cdots (E_{n_{z-1}} - E_0)} \frac{e^{2\pi i (E_n - E_0)t/h} - 1}{E_n - E_0}$$

Here, all n_i correspond to intermediate states in the sense given by Dirac.

Consequently, soon after the introduction of the intermediate state in Dirac's radiation theory, this notion was successfully applied in the prediction of new phenomena and the perturbative treatment from which it emerged was formally pushed to an arbitrary order. The interpretation of the formalism was pushed into the background, while the application of the calculational scheme gathered prominence. However, in the following section we will see how the theoretical interpretation of an experimentally established effect was responsible for the connection of the intermediate state to the term 'virtual'.

Naming the Child: How Transitions to Intermediate States Became 'Virtual'

As I mentioned in the previous section, when Dirac published his papers on radiation theory, the Kramers-Heisenberg formula (which he had rederived) had not yet gathered empirical support. This changed in 1928 through two different lines of experimental enquiry: the work of Rudolf Ladenburg and Hans Kopfermann on anomalous dispersion of excited gases, and the discovery of "a new type of secondary radiation" (Raman and Krishnan 1928) by C.V. Raman and his group, which interests us here [for a history of the discovery of the Raman effect, see Singh (2002). For the priority dispute because of the simultaneous discovery by Mandelstam and Landsberg, see Singh and Riess (2001). The following paragraphs also draw from Brand (1989)].

Raman experimentally observed that visible light scattered from liquids or gases did not only contain the frequency components of the incident light. He also found lines in the secondary radiation, which were discretely shifted from the original frequency. Such an effect was proposed for the first time theoretically by Smekal in 1923 and was put on firmer theoretical ground in the dispersion theory of Kramers and Heisenberg (1925). Theory predicted a frequency shift in the scattered radiation proportional to the energy difference between two stationary states of the scattering atom or molecule: $\Delta v = \frac{1}{h}(E_n - E_m)$.

Although Raman's discovery was made independently of the theoretical predictions, he nonetheless referred to the Kramers-Heisenberg paper in his first full-length publication on this effect in the *Indian Journal of Physics* (Raman 1928) and proposed an interpretation that rested on the idea of a direct energy exchange between a light quantum and the molecule:

> As a tentative explanation, we adopt the language of quantum theory, and say that the incident quantum of radiation is partially absorbed by the molecule, and that the unabsorbed part is scattered. This suggestion does not seem to be altogether absurd and indeed such a possibility is already contemplated in the Kramers-Heisenberg theory of dispersion. If we accept the idea indicated above, then the difference between incident and absorbed quantum would correspond to a quantum of absorption by the molecule. (ibid., p. 373)

As we can see, Raman's interpretation entails that the energy difference would correspond to absorption and emission lines of the scattering material, which he found to be the case in his experimental studies.

The set-up was easy to replicate and the effect not only provided a test of quantum mechanics, but also promised a new means of studying atomic spectra. This immediately prompted many physicists to investigate its occurrence in diverse materials. By July 1929, at least 160 papers had been published with reference to the newly found effect (Ganesan 1929).

The increasing amount of data and the study of simpler molecules put the above interpretation under increasing pressure. The observed frequency shifts did not correspond to emission and absorption lines of the scattering material or, at least, raised serious doubts about Raman's interpretation (see e.g. Langer 1929; Dieke 1929; or Rasetti 1929). On the contrary, the findings pointed to a proportionality of the observed frequency shift with the frequency difference between two of the lines, which would be explained by modelling the phenomena according to Dirac's theory as a two-step process (that the selection rules would change for incoherent scattering was already pointed out by Kramers and Heisenberg but was seemingly ignored by Raman). Therefore, the differing interpretations of the effect actually had empirically testable consequences and the data suggested that the intermediate states ought to be taken into account.

Raman accepted the empirical material but, in his capacity as keynote speaker at the symposium "The Raman effect", held by the Faraday Society in August 1929, he defended the "simple and easily understood picture of the process involved in this new phenomenon" (Raman 1929, p. 789); in other words, direct transitions between the energy levels of the scattering material. Raman explained that the energy of the visible light used for the experiments was not sufficient to cause transitions to the intermediate state involved in the calculation. One should therefore think of this third level only as a "mathematical device" (ibid., p. 789). He conjectured that "the transition of the molecule for the purpose of calculation is a purely virtual one which cannot actually occur" (ibid., p. 790).

It was apparently Raman's speech that led to the usage of the term 'virtual' in connection with transitions or processes. Robert Wood, who was present at the symposium, and Gerhard Dieke defended the picture implied by Dirac's radiation theory, but at the same time invoked "virtual transitions", "virtual absorption act" or "virtual reemission act" (Wood and Diecke 1930, p. 1357). In a context different from the Raman effect, it appears that this terminology was first used by Hulme (1932), who at the time was probably working in Leipzig with Heisenberg. He reminded the reader "that the transitions which occur in the dispersion formula are virtual ones only" (ibid., p. 238). In connection with problems of full quantum electrodynamics Heisenberg wrote to Pauli about the "infinitely many virtually possible transitions" (Heisenberg to Pauli, 5th of February 1934, cited in von Meyenn 1985, p. 273) with reference to the self-energy of the electron. Peierls (1934) used the term "virtual transitions" (ibid., p. 439) to describe the same phenomenon. In the subsequent years, also through the work of Euler and Kockel on light-by-light scattering, the terminology became rather widespread.

To be sure, I am well aware of the fact that the Kramers-Heisenberg formula was first proposed in the framework of the BKS-theory of virtual oscillators and the virtual radiation field (Bohr et al. 1924), although in the Heisenberg-Kramers paper the terminology of virtual oscillators had already been dropped. I am also aware that, although the BKS-theory was experimentally disproven quite quickly, the notion of virtual oscillators was still applied as a substitute concept for matrix elements in semi-classical approaches in quantum mechanics. Nevertheless, I want to argue that we need to take Raman's speech into account when we want to understand the connection between the terminology and transitions in perturbative approaches.

First of all, we have conceptually very different approaches. The BKS-theory rested explicitly on the notion of classical radiation (as did later semi-classical approaches), whereas Dirac quantized the electromagnetic field and thus the physical picture implied was based on the absorption and emission of light quanta. Physicists consequently tried to account for the phenomena with terminologies, which were closer to considerations using the light quantum hypothesis in the old quantum theory than to the BKS paper. Furthermore, Raman's speech was actually the first instance I could find in which the terminology and the concepts introduced in Dirac's paper merged. The continuity of the term 'virtual' mostly referring to processes (emission, absorption, pair creation, ...) in the perturbative treatments of the 1930s and 40s is especially striking.

It is probable that the BKS theory was Raman's source when connecting the terminology and the concepts. Nevertheless, I want to emphasize that, on the one hand, this connection was not a necessary one insofar as physicists used different terminologies to account for the same concepts and "virtual" was not a very popular choice at least until the late 1930s. On the other hand, I think it is at least noteworthy that the earliest connection between the terminology and Dirac's concept of transitions to intermediate states I could find was proposed by a physicist who initially disregarded their observable consequences and denied them any realistic interpretation.

QED, Hole Theory and Light-by-Light Scattering

The years following Dirac's papers on radiation theory saw some radical changes in theory. Methods were developed to quantize a fermionic as well as the full electromagnetic field (Dirac had only quantized the transversal part), culminating in the monumental work of Heisenberg and Pauli on a full QED (Heisenberg and Pauli 1929). They also considered longitudinal and scalar polarized photons in their perturbative treatment of the (electrostatic) self-energy of the electron and the (Coulomb) interaction, based on an expansion of the energy and the wave function. Although one might read, from a modern point of view, an exchange of photons into their mathematical scheme, there is no interpretation of the intermediate states in terms of processes in their paper. Through the theoretical investigation of Rosenfeld (1929) it became clear that the Coulomb interaction was entirely due to the non-transversal part of the electromagnetic field (see Carson 1996, especially Sects. 3.1 and 3.2).

Fermi (1932), in a very popular paper for learning QED in those days, showed how those longitudinal and scalar modes could be subsumed in the Hamiltonian to give the Coulomb interaction. This description, which reestablished a close connection to Dirac's radiation theory, became standard between the 1930s and the late 1940s. The work of Bethe and Fermi (1932) on the quantum electrodynamical description of the interaction between two electrons will serve as an example for this line of reasoning.

Carson (1996) pointed out that the emission and absorption of a quantum in this work was only taken to give the relativistic corrections to electron scattering, which problematizes David Kaiser's assessment that Bethe and Fermi "had put virtual particles to work" (Kaiser 2005, p. 30). Moreover, not only did Bethe and Fermi simply not use the term 'virtual' in any sense, but the calculation and conception of the process was greatly influenced by the state transition model implied in QED at the time:

> In first approximation, the coupling between matter and radiation field only causes such transitions, in which the quantum state of the electrons changes and a light quantum is emitted or absorbed. But we are interested in the matrix elements of the interaction energy of the electrons which cause a change in the electron states alone without altering the state of the radiation field [...]. One only gets such transitions of the electrons alone through a double process, where first a quantum is emitted and then the same quantum is absorbed.
>
> (Bethe and Fermi 1932, pp. 296–297; translation by the author)

Energy did not have to be conserved in the transitions to the intermediate state, but the photons in such a state were 'regular' transversal ones.

Carson (1996) has made a convincing argument that the notion of exchange forces as a particle exchange was developed in nuclear theory first and then taken over into the realm of QED, rather than the other way around. For the sake of the present argument, it should suffice that early calculations in Fermi and meson field theory rested as well on a perturbative treatment. The concepts of intermediate states and 'virtual transitions' also took centre stage here (see for example Fröhlich et al. 1938, or Wick 1938). Nevertheless, we also find strong visual aspects in the work of Gregor Wentzel, as represented in Fig. 15.2. Although one might argue that those figures were taken from a rather popular account of the theory, it is to note that such representation can also be found in his influential textbook (Wentzel 1943).

Returning to QED, we will consider the process of light-by-light scattering. Therefore, we shortly have to introduce the notion of the Dirac sea. In 1928 Dirac gave a relativistic description of a single electron in the famous "Dirac equation" (Dirac

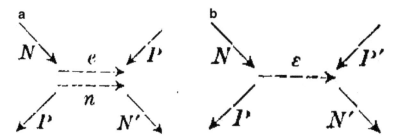

Fig. 15.2 a–b (from left to right): Representation of nuclear interactions, **a** in Fermi's theory of beta-decay, **b** in Yukawa's theory. [Both taken from Wentzel 1938, p. 274]

1928). In 1930, he tried to solve the apparent unphysicality of negative energy solutions included in his one-particle theory by introducing the so-called Dirac sea, an infinity of electrons occupying all the negative energy states. Due to the structure of the Hamiltonian of his relativistic theory the term responsible for "direct scattering" (see section "Introduction of the intermediate state and first applications") disappeared from the formulas. At the same time, the combination of Dirac's radiation theory with the relativistic theory of the electron made it necessary, as Werner Heisenberg (Heisenberg to Pauli, July 31st 1928, cited in von Meyenn et al. 1979, p. 466–469), Tamm (1930), and Waller (1930) realized independently, to take intermediate states of negative energy into account to arrive at the scattering formulas known at the time. Since Dirac had already filled his sea this only led to "formal changes" (ibid., p. 837), which he anticipated when introducing his idea of the infinitely many negative energy electrons (Dirac 1930).

At first, Dirac interpreted 'holes' in this sea as protons, but after serious criticism due to a much too high annihilation rate (Tamm 1930; Oppenheimer 1930) and, more importantly, because of a symmetry argument (Weyl 1931) he revised this interpretation and identified the holes with "a new kind of particle, unknown to experimental physics, having the same mass and opposite charge of the electron. We may call such a particle an anti-electron" (Dirac 1931, p. 61). Following the discovery of this particle in 1932–33 and its theoretical identification with a hole in the Dirac sea, the acceptance of the existence of anti-particles, pair creation and annihilation, and the heuristic use of Dirac's hypothesis became widespread (see for example Roqué 1997).

Physicists soon began to grapple with how such a filled vacuum should be treated theoretically and investigated and which in principle observable consequences were to be expected. The specific process that I want to discuss here—light-by-light scattering—was first proposed by Peter Debye in private communication with Heisenberg (Heisenberg to Pauli, June 16th 1934, cited in von Meyenn (1985), pp. 331–332), and in print by Halpern (1933). Although Halpern did not develop a quantitative account, his qualitative description of the process of light-quantum splitting (in Halpern's view analogous to light-by-light scattering) shows the physical picture he had in mind. Assuming the sum of the energy of the two scattering light waves to be smaller than

$2mc^2$ so that no real pair production could occur, Halpern looked for the "scattering properties of the 'vacuum'" (Halpern 1933, p. 856):

> In the language of Dirac's theory of radiation such splitting of the incident quantum occurs in processes of the following type: An electron in a negative energy state passes by absorption of the incident quantum into a state of positive energy; the electron then returns in several steps under the emission of $h\nu$ in total to its original state. (ibid., p. 856)

The scattering of two light quanta from each other in this stepwise process was calculated by Hans Euler and Bernhard Kockel in Leipzig under the supervision of Werner Heisenberg (for an analysis of the paper, with specific emphasis on the diagrammatic representation, see Wüthrich (2010), Chap. 2. Kaiser (2005), pp. 34–35, mainly stresses the cumbersome calculations that had to be performed to get the results).

In his dissertation, in which the inclusion of the term 'virtual' was clearly due to Heisenberg (e.g., "virtual possibility of pair creation"), Euler (1936) distinguishes two levels of theory. On the "intuitive [anschaulich]" level, he applied "general considerations" (ibid., p. 399) to arrive at the form the Lagrangian should have in the approximation of slowly varying fields. He found that he had to replace the current and sources that ordinarily featured in the Maxwell equations by functions of the field strength, "which one can call 'the matter virtually created by the field'" (ibid., p. 403). By imposing general physical constraints such as Lorentz invariance, Euler arrived at the following effective Lagrangian:

$$\frac{L}{4\pi} = \frac{E^2 - B^2}{8\pi} + \frac{hc}{e^2}\frac{1}{E_0^2}\left[-\alpha(E^2 - B^2)^2 - \beta(EB)^2\right] = \frac{L_0 + L_1}{4\pi}$$

where α and β are dimensionless coefficients which had to be deduced by applying perturbation theory.

On this level, the process of light-by-light scattering was of fourth order (four emission/absorption processes were involved) and the corresponding matrix element was given by:

$$H_{in}^4 = \sum_{klm} \frac{V_{ik}^1 V_{kl}^1 V_{lm}^1 V_{mn}^1}{(E_i - E_k)(E_i - E_l)(E_i - E_m)} + V_{in}^4$$

The term V^4 arises due to the so-called 'subtraction physics' of Dirac and Heisenberg (Dirac 1934; Heisenberg 1934), which provided a means of handling the divergencies of vacuum polarization. Again, energy is not conserved for the transitions to the 'virtual intermediate state' (n → m, m → l, l → k, k → i), in contrast to the whole process (n → i). The six possible ways for connecting the initial and the final state by intermediate states are shown in Fig. 15.3. Each has 24 possible permutations according to which of the four light quanta is involved in the individual subprocesses. Euler needed to calculate two special cases of incident and outgoing photons in order to determine the numerical coefficients α and β. This was a tedious task, involving

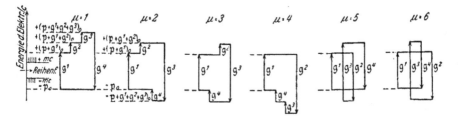

Fig. 15.3 Euler's visualization of the possible ways through which light-by-light scattering could occur, taken from Euler (1936), p. 419

pages of tables that displayed the results of his and Kockel's calculations, but which in the end led to specific numerical values of α and β.

The two levels of theoretical investigation were both essential to arrive at the final effective Lagrangian. It is noteworthy that, in the course of applying perturbation theory, Euler used the term 'virtual' to refer to the intermediate states, transitions to and from them, pair creation, and the possibility of pair creation, while the term 'virtual matter' only arises once, on the intuitive level. This terminology was definitely motivated by the perturbative calculations, which, according to the announcement of the paper, set the starting point of the considerations (Euler and Kockel 1935). I emphasize this point, because even in the more careful considerations of Wüthrich (2010) he states that "the pairs are only 'virtual'" (ibid., p. 25).

Feynman, Diagrams and Virtual Particles

The essentially non-covariant methods of perturbation theory and the state transitions model in QED were still in play in the middle of the 1940s. There were attempts to modify this method, one of the more promising ones being the theory of radiation damping of Heitler and Peng, in order to get rid of the divergencies of QED. But, in effect, this approach turned out to reintegrate the infra-red divergency and therefore proved to be unsuccessful. On the contrary, through the experiments of Lamb and Retherford announced publicly in 1947 it had become clear that one could not simply discard the self-energy of the electron (a second order effect in perturbation theory), which programs such as Heisenberg's S-Matrix or the Wheeler-Feynman absorber theory were aiming at. Indeed, most of the reactions to the divergency difficulties encountered in QED during the 1930s were rather radical (see e.g. Rueger 1992), and it was often pointed out that, in contrast, no revolution in a Kuhnian sense took place in the late 1940s, since the physical basis of the theory was left unaltered. (see for example Weinberg 1977). Blum (2017) has recently contested this view by elaborating how scattering became the paradigmatic problem of QED at the end of the 1940s).

Instead of going into the details of the programs that would eventually lead to a consistent means of regularization and renormalization, or into the calculations which tried to reconstruct the experimental finding of the Lamb shift, I will skip directly to the work of Richard Feynman. This focus stems from the observation that theoretical works in connection with the Lamb shift "still used the computational methods of the 1930s" (Schweber 1983, p. 169), and that we find in Feynman's work not only a covariant perturbative treatment, but also the diagrammatical method with which the virtual particles are still so closely connected.

The story of Feynman's path to his "Space-Time Approach to Quantum Electrodynamics" (Feynman 1949), resulting in a covariant theory, in which the radiation oscillators had been eliminated, positrons were conceived of as electrons travelling backwards in time, a strong diagrammatical or visual component had evolved, and only asymptotically free states of the particles were considered, has been told elsewhere. For the subsequent connection of this method to the more traditional ways of doing QED of Tomonaga and Schwinger and the S-Matrix of Heisenberg, the reader is also referred to the corresponding secondary literature (see e.g. Schweber 1986, 1994, Chaps. 8 and 9, for detailed overviews; Galison 1998, for war related work; Wüthrich 2010, for the diagrammatical aspect; Blum 2017, for the shift in perspective from a state transition model to a theory which only involved asymptotically free states). As we are only interested here in the impact of Feynman's work upon the virtual particle concept, we will go directly to his 1949 paper.

Figure 15.4 depicts what Feynman called, "the fundamental interaction of QED" (Feynman 1949, p. 772). The first order correction to two electrons moving freely in space described by the propagator K(3, 4; 1, 2) was given by Feynman as:

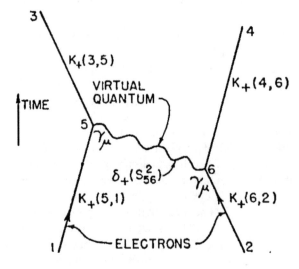

Fig. 15.4 The "fundamental interaction" of QED: the exchange of a photon between two electrons. [Taken from Feynman 1949, p. 772]

15 I'm Not There. Or: Was the Virtual Particle Ever Born?

$$K^{(1)}(3, 4; 1, 2) = -ie^2 \iint K_{+a}(3, 5)K_{+b}(4, 6)\gamma_{a\mu}\gamma_{b\mu}$$
$$\delta_+\left(s_{56}^2\right)K_{+a}(5, 1)K_{+b}(6, 2)d\tau_5 d\tau_6$$

"This is our fundamental equation for electrodynamics. It describes the effect of exchange of one quantum (therefore first order in e^2) between two electrons" (ibid., p. 772). The corresponding interpretation was given by Feynman as follows:

> It can be understood (see Fig. [15.4]) as saying that the amplitude for [particle] 'a' to go from 1 to 3 and [particle] 'b' to go from 2 to 4 is altered to first order because they can exchange a quantum. Thus, 'a' can go to 5 (amplitude $K_+(5, 1)$) emit a quantum (longitudinal, transverse, or scalar Ya_μ) and then proceed to 3 ($K_+(3, 5)$). Meantime 'b' goes to 6 ($K_+(4, 6)$), absorbs the quantum (Yb_μ) and proceeds to 4 ($K_+(4, 6)$). The quantum meanwhile proceeds from 5 to 6, which it does with amplitude $\delta_+\left(s_{56}^2\right)$. (ibid., pp. 772–773)

Interestingly, the term 'virtual' is connected in Feynman's work in far more regular intervals to quanta or particles than in, for example, the major work of Julian Schwinger of the time, where mostly the transitions or processes were termed 'virtual'. The connection between those more conventional theories of Schwinger or, for that matter, Tomonaga and Feynman's approach was made by Dyson (1949), who also gave explicit rules for drawing Feynman diagrams, which could then be connected to the perturbative expansion of the S-Matrix.

The distinction between the older methods of perturbation theory and the new method of Dyson and Feynman can best be exemplified by a quote from the textbook of Jauch and Rohrlich (1955). Introducing the "Feynman-Dyson diagrams" to their readers, they explained that

> in the older form of perturbation theory the time variable played a role distinguished from the space variables. The resulting unsymmetrical formulas were much more complicated than the corresponding formulas in the invariant perturbation treatment. Corresponding to this lack of symmetry, one had in the older formulation conservation of momentum but not conservation of energy for the intermediate states. In the present formulation one has conservation of energy and momentum at each corner of the diagram. On the other hand, the energy-momentum relation which is valid for free particles is not true for the particles in the intermediate states. (Jauch and Rohrlich 1955, p. 158)

In the covariant formulation, the energy is conserved at the vertices, but the particles present in the intermediate state are now, to use the modern term, 'off-shell'. Whether it was this realization, the graphical representation, or the emphasis on the particle picture in Feynman's work that led to the shift from the transitions to the particles being more regularly referred to as virtual, has still to be evaluated. Most likely, a monocausal explanation will not suffice and the historical reconstruction will have to take into account the combination of all of these aspects.

Was the Virtual Particle Ever Born?

We are now in a position to ask again the question posed in the beginning: can we sensibly speak of the birth of the virtual particle? Theodore Arabatzis has defended the notion of birth as "the emergence of an explanatory entity, in response to various problem situations, empirical and/or conceptual" (Arabatzis 2006, p. 43) in his biographical approach to theoretical representations of hidden entities, or in short: theoretical entities. I will follow some of the commentaries by using the terms 'concept' and 'theoretical entity' synonymously, an impression also borrowed from the concluding sections of Arabazis' book (see e.g. pp. 254–255). Arabatzis explicitly defended his biographical approach by referring to the stability of the experimentally established attributes of his protagonist, the electron (foremost its charge-to-mass ratio). For the case of the virtual particle, no such numerical values are at hand. Yet the coherence in the history can be created by referring to the mathematical techniques involved (perturbation theory), some stable attributes (unobservability in principle and energy non-conservation), and the effects explained by the corresponding concepts. The notion of birth, "usually associated with an event" and "a beginning in time" (Arabatzis 2006, p. 43), on the other hand, shall be seriously questioned in the following paragraphs by taking the historical development described above as our case study. To this end, let's modify our question a bit and ask: When was the virtual particle born?

Let's first opt for the temporal extremes. Looking at Dirac's introduction of the intermediate state, there is the problem that all of the particles in the description are real ones (albeit inhabiting an intermediate state for a short time). The photons involved are either present before or after the scattering takes place, and the atom (or the electron) makes transitions to and from intermediate states, but was not created or destroyed in between.

The other extreme would be Feynman's work in the late 1940s, where the particles are 'off-shell' and the virtual particle became an integral part of the diagrammatical method named after him. But this would run counter to the observation that physicists had already a clear idea that particles, which were in principle unobservable and only present because of the apparent violation of energy conservation, could be made responsible for (again in principle) observable effects, such as, for example, (the corrections to) the Coulomb interaction, light by light scattering, or the interaction of nuclear particles. Indeed, the terms 'virtual pairs' or 'virtual quanta' were at times applied, if only rarely.

The picturesque representations given by Göppert-Mayer, Euler or Wentzel, although they are in contrast to Feynman diagrams only post hoc visualizations of the formalism, clearly show that physicists were thinking in terms of particles being exchanged or only present 'in between'. Nevertheless, the calculational techniques for the papers presented in this article were always non-covariant and violated energy conservation in the transitions ('virtual transitions'), whereas the particles in the intermediate states were physical ones. Furthermore, even in the covariant work of Julian Schwinger of the late 1940s, the term 'virtual' is more often connected to processes than to the particles themselves.

From just these few examples, which did not even take into account such important developments as considerations of the self-energy of the electron, vacuum polarization, or radiative corrections to scattering, we see that the notion of birth is hardly applicable in the case of the virtual particle. One could make matters even worse, by incorporating examples as Ernst Stueckelberg's 1934 formulation of a covariant perturbation theory, which was not taken up by the theoretical community. As Lacki et al. (1999, p. 496) pointed out, in this calculational scheme energy and momentum are conserved at the interaction points while the particles in intermediate states are off-shell. Those particles would in this respect be analogous to the ones in the calculations of the late 1940s. If we would connect the notion of the virtual particle to the technique of a covariant perturbation theory and the characteristics of the particles in intermediate states that arise from it, we would deal in pseudo-biographical terms with a "zombie": an entity that had already emerged, was disregarded by the community, consequently faded away, only to show up in a different setting in the late 1940s.

The problem with the term 'birth' is, in my view, that it entails—if we take the metaphor seriously—that something completely new is formed. It thereby highlights breaks and puts less emphasis on continuity. But continuities and breaks do not necessarily exclude each other in concept formation processes. In the history of the virtual particle, as I have endeavored to describe above, we find mathematical techniques, physical ideas and processes, diagrammatical methods, empirical evidence, and terminological issues at play, which did not evolve at the same pace. Furthermore, a concept is always embedded in a theoretical and a wider conceptual framework that regulates its meaning and utility. Drawing from Hans Peter Hahn's critique of the "object biography" (Hahn 2015), one might say that not only are objects "fragments and assemblages", but that this attribution also holds for concepts. Applying the notion of birth would force us to make a distinct judgement when a specific conception or theoretical representation could be called a virtual particle, at the expense of treating the framework in which the actors worked and thought on their own terms. The possibility of falling into anachronism and radical present-centeredness should be evident. Therefore, my answer must be: no, the virtual particle may have evolved, but was never born.

Acknowledgements This work was supported by the research unit "The Epistemology of the LHC", funded by the Deutsche Forschungsgemeinschaft (DFG) under the grant number FOR 2063. I want to thank the members of the Research-Unit for reading a preprint of this paper. Especially, I want to thank the members of the subproject A1 "The Formation and Development of the Concept of the Virtual Particles", Robert Harlander, Daniel Mitchel, Friedrich Steinle, and Adrian Wüthrich, for valuable input concerning the structure and content of the paper and for help with the language. I also want to thank Simon Rebohm for pointing me to the paper on object biographies by Hans Peter Hahn and, last but not least, the organizers of the workshop for making this publication possible and the participants for input on the presentation of a preliminary version of this paper.

References

Arabatzis, T. (2006). *Representing electrons. A biographical approach to theoretical entities.* Chicago and London: University of Chicago Press.

Bacelar Valente, M. (2011). Are virtual quanta nothing but formal tools? *International Studies in the Philosophy of Science, 25*(1), 39–53.

Bethe, H., & Fermi, E. (1932). Über die Wechselwirkung von zwei Elektronen. *Zeitschrift für Physik, 77*(5–6), 296–306.

Blum, A. (2017). The state is not abolished, it withers away: How quantum field theory became a theory of scattering. *Studies in History and Philosophy of Science Part B: Studies in History and Philosophy of Modern Physics, 60,* 46–80.

Bohr, N., Kramers, H. A., & Slater, J. C. (1924). Über die Quantentheorie der Strahlung. *Zeitschrift für Physik, 24*(1), 69–87.

Brand, J. C. D. (1989). The discovery of the Raman effect. *Notes and Records: The Royal Society Journal of the History of Science, 43,* 1–23.

Bromberg, Joan. (1977). Dirac's quantum electrodynamics and the wave-particle equivalence. In Charles Wiener (Ed.), *History of twentieth century physics* (pp. 147–157). New York and London: Academic Press.

Cao, T. Y. (1997). *Conceptual developments of 20th century field theory.* Cambridge and New York: Cambridge University Press.

Carson, Cathryn. (1996). The peculiar notion of exchange forces II: From nuclear forces to QED, 1929–1950. *Studies in History and Philosophy of Science Part B: Studies in History and Philosophy of Modern Physics, 27*(2), 99–131.

Darrigol, O. (1986). The origin of quantized matter waves. *Historical Studies in the Physical Sciences, 16*(2), 197–253.

Dieke, G. H. (1929). Difference between the absorption and the Raman spectrum. *Nature, 123*(3102), 564.

Dirac, P. A. M. (1926). On the theory of quantum mechanics. *Proceedings of the Royal Society of London A: Mathematical, Physical and Engineering Sciences, 112*(762), 661–677.

Dirac, P. A. M. (1927a). The quantum theory of the emission and absorption of radiation. *Proceedings of the Royal Society of London A: Mathematical, Physical, and Engineering Sciences, 114*(767), 243–265.

Dirac, P. A. M. (1927b). The quantum theory of dispersion. *Proceedings of the Royal Society of London A: Mathematical, Physical, and Engineering Sciences, 114*(769), 710–728.

Dirac, P. A. M. (1928). The quantum theory of the electron. *Proceedings of the Royal Society of London A: Mathematical, Physical, and Engineering Sciences, 117*(778), 610–624.

Dirac, P. A. M. (1930). A theory of electrons and protons. *Proceedings of the Royal Society of London A: Mathematical, Physical, and Engineering Sciences, 126*(801), 360–365.

Dirac, P. A. M. (1931). Quantised singularities in the electromagnetic field. *Proceedings of the Royal Society of London A: Mathematical, Physical, and Engineering Sciences, 133*(821), 60–72.

Dirac, P. A. M. (1934). Discussion of the infinite distribution of electrons in the theory of the positron. *Proceedings of the Cambridge Philosophical Society, 30,* 150–163.

Dyson, F. J. (1949). The radiation theories of Tomonaga, Schwinger, and Feynman. *Physical Review, 75*(3), 486–502.

Euler, H., & Kockel, B. (1935). Über die Streuung von Licht an Licht nach der Diracschen Theorie. *Die Naturwissenschaften, 23*(15), 246–247.

Euler, H. (1936). Über die Streuung von Licht und Licht nach der Diracschen Theorie. *Annalen der Physik, 418*(5), 398–448.

Fermi, E. (1932). Quantum theory of radiation. *Reviews of Modern Physics, 4*(1), 87–132.

Feynman, R. (1949). Space-time approach to quantum electrodynamics. *Physical Review, 76*(6), 769–789.

Fox, T. (2008). Haunted by the spectre of virtual particles: A philosophical reconsideration. *Journal for General Philosophy of Science, 39*(1), 35–51.

Frenkel, Y. I. (1929a). Über die quantenmechanische Energieübertragung zwischen atomaren Systemen. *Zeitschrift für Physik, 58*(11–12), 794–804.

Frenkel, Y. I. (1929b). The quantum theory of the absorption of light. *Nature, 124,* 758–759.

Fröhlich, H., Heitler, W., & Kemmer, N. (1938). On the nuclear forces and the magnetic moments of the neutron and the proton. *Proceedings of the Royal Society of London A: Mathematical, Physical, and Engineering Sciences, 166*(924), 154–177.

Galison, P. (1998). Feynman's war: modelling weapons, modelling nature. *Studies in History and Philosophy of Science Part B: Studies in History and Philosophy of Modern Physics, 29*(3), 391–434.

Ganesan, A. S. (1929). Bibliography of 150 papers on the Raman effect. *Indian Journal of Physics, 4,* 281–346.

Göppert, M. (1929). Über die Wahrscheinlichkeit des Zusammenwirkens zweier Lichtquanten in einem Elementarakt. *Die Naturwissenschaften, 17*(48), 932.

Göppert-Mayer, M. (1931). Über Elementarakte mit zwei Quantensprüngen. *Annalen der Physik, 401*(3), 273–294.

Hahn, H.P. (2015). Dinge sind Fragmente und Assemblagen. Kritische Anmerkungen zur Metapher der ‚Objektbiographie'. In D. Boschung, P.-A. Kreuz, T. Klienlin (Eds.), *Biography of Objects. Aspekte eines kulturhistorischen Konzepts* (pp. 11–33). Paderborn: Wilhelm Fink Verlag.

Halpern, O. (1933). Scattering processes produced by electrons in negative energy states. *Physical Review, 44*(10), 855–856.

Heisenberg, W. (1934). Bemerkungen zur Diracschen Theorie des Positrons. *Zeitschrift für Physik, 90*(3–4), 209–231.

Heisenberg, W., & Pauli, W. (1929). Zur Quantendynamik der Wellenfelder. *Zeitschrift für Physik, 56*(1–2), 1–61.

Heitler, W. (1936). *The quantum theory of radiation*. Oxford: Clarendon Press.

Hulme, H. R. (1932). The Faraday effect in ferromagnetics. *Proceedings of the Royal Society of London A: Mathematical, Physical, and Engineering Sciences, 135*(826), 237–257.

Jauch, J. & Rohrlich, F. (1955). *The theory of photons and electrons* (Second Printing 1959). Reading, Massachusetts, and London: Addison-Wesley Publishing Company.

Jost, R. (1972). Foundation of quantum field theory. In Abdus Salam, E.P. Wigner (Eds.), *Aspects of quantum theory* (pp. 61–78). Cambridge: Cambridge University Press.

Kaiser, D. (2005). *Drawing theories apart: The dispersion of Feynman diagrams in postwar physics*. Chicago and London: University of Chicago Press.

Kragh, H. (1990). *Dirac: A scientific biography*. Cambridge et al.: Cambridge University Press.

Kramers, A., & Heisenberg, W. (1925). Über die Streuung von Strahlung durch Atome. *Zeitschrift für Physik, 31*(1), 681–708.

Lacki, J., Ruegg, H., & Telegdi, V. L. (1999). The road to Stueckelberg's covariant perturbation theory as illustrated by successive treatments of compton scattering. *Studies in History and Philosophy of Science Part B: Studies in History and Philosophy of Modern Physics, 30*(4), 457–518.

Lamb, W. E., & Retherford, R. C. (1947). Fine structure of the hydrogen atom by microwave method. *Physical Review, 72,* 241–243.

Langer, R. M. (1929). Incoherent scattering. *Nature, 123*(3097), 345.

Oppenheimer, R. (1930). On the theory of electrons and protons. *Physical Review, 35*(5), 562–563.

Pawlak, A. (2004). Physik sakral. *Physik Journal, 3*(12), 36–37.

Peierls, R. (1934). The vacuum in dirac's theory of the positive electron. *Proceedings of the Royal Society of London A: Mathematical, Physical, and Engineering Sciences, 146*(857), 420–441.

Raman, C. V. (1928). A new radiation. *Indian Journal of Physics, 2,* 368–376.

Raman, C. V. (1929). Investigation of molecular structure by light scattering. *Transactions of the Faraday Society, 25,* 781–792.

Raman, C. V., & Krishnan, K. S. (1928). A new type of secondary radiation. *Nature, 121,* 501–502.

Rasetti, F. (1929). On the Raman effect in diatomic gases. *Proceedings of the National Academy of Science of the United States of America, 15*(3), 234–237.

Roqué, X. (1997). The manufacture of the positron. *Studies in History and Philosophy of Science Part B: Studies in History and Philosophy of Modern Physics, 28*(1), 73–129.

Rosenfeld, Leon. (1929). Über die longitudinalen Eigenlösungen der Heisenberg-Paulischen elektromagnetischen Gleichungen. *Zeitschrift für Physik, 58*(7), 540–555.

Rueger, A. (1992). Attitudes towards infinities: Responses to anomalies in quantum electrodynamics, 1927–1947. *Historical Studies in the Physical Sciences, 22*(2), 309–337.

Schweber, S.S. (1983). Some chapters for a history of quantum field theory: 1938–1952. In B. DeWitt & R. Stora (Eds.), *Les Houches, Session XL. Relativity, Groups, and Topology II*, (pp. 37–220). Amsterdam: North Holland Publishing.

Schweber, S. S. (1986). Feynman and the visualization of space-time processes. *Reviews of Modern Physics, 58*(2), 449–508.

Schweber, S. S. (1994). *QED and the men who made it: Dyson, Feynman, Schwinger, and Tomonaga*. Princeton, New Jersey: Princeton University Press.

Seife, C. (2000). *Zero. The biography of a dangerous idea*. New York: Penguin Books.

Singh, R. (2002). C. V. Raman and the discovery of the Raman effect. *Physics in Perspective, 4*, 399–420.

Singh, R., & Riess, Falk. (2001). The 1930 Nobel prize for physics: A close decision? *Notes and Records: The Royal Society Journal of the History of Science, 55*(2), 267–283.

Smekal, A. (1923). Zur Quantentheorie der Dispersion. *Die Naturwissenschaften, 11*(43), 873–875.

Tamm, I. (1930). Über die Wechselwirkung der freien Elektronen mit der Strahlung nach der Diracschen Theorie des Elektrons und der Quantenelektrodynamik. *Zeitschrift für Physik, 62*(7–8), 545–568.

von Meyenn, K., Hermann, A., & Weisskopf V.F. (Ed.). (1979). *Wolfgang Pauli. Scientific correspondence with Bohr, Einstein, Heisenberg a.o.. Volume I: 1919–1929*. New York, Heidelberg, and Berlin: Springer.

von Meyenn, K. (Ed.). (1985). *Wolfgang Pauli. Scientific correspondence with Bohr, Einstein, Heisenberg a.o.. Volume II: 1930–1939*. Berlin: Springer.

Waller, I. (1930). Die Streuung von Strahlung durch gebundene und freie Elektronen nach der Diracschen relativistischen Mechanik. *Zeitschrift für Physik, 61*(11–12), 837–851.

Weinberg, S. (1977). The search for unity: Notes for a history of quantum field theory. *Daedalus, 106*, 17–35.

Weisskopf, V. (1931). Zur Theorie der Resonanzfluoreszenz. *Annalen der Physik, 401*(1), 23–66.

Weisskopf, V. (1933). Die Streuung des Lichts an angeregten Atomen. *Zeitschrift für Physik, 85*(7–8), 451–481.

Weisskopf, V., & Wigner, Eugen. (1930). Berechnung der natürlichen Linienbreite auf Grund der Diracschen Lichttheorie. *Zeitschrift für Physik, 63*(1–2), 54–73.

Wentzel, G. (1933). Wellenmechanik der Stoß- und Strahlungsvorgänge. In H. Geiger & K. Scheel (Ed.), *Handbuch der Physik (2. Auflage). Band XIV, Erster Teil. Quantentheorie* (pp. 695–784). Berlin und Heidelberg: Springer.

Wentzel, G. (1938). Schwere Elektronen und Theorien der Kernvorgänge. *Die Naturwissenschaften, 26*(18), 273–279.

Wentzel, G. (1943). *Einführung in die Quantentheorie der Wellenfelder*. Wien: Deuticke.

Weyl, H. (1931). *Gruppentheorie und Quantenmechanik* (2nd ed.). Leipzig: Hirzel.

Wick, G. C. (1938). Range of nuclear forces in Yukawa's theory. *Nature, 142*(3605), 993–994.

Wood, R. W., & Diecke, G. H. (1930). The Raman effect in HCL gas. *Physical Review, 35*, 1355–1359.

Wüthrich, A. (2010). *The genesis of Feynman diagrams*. Dodrecht: Springer.

Chapter 16
Biography or Obituary? The Historiographical Value of the Death of the Ether

Jaume Navarro

On Obituaries and Biographies

In the preface to the second edition of his *A History of the Theories of Aether and Electricity*, issued in 1951, Edmund Whittaker explained why he decided to preserve the old title. First published in 1910, this book was intended as an encyclopedic account of the evolution of physics since the days of Descartes, although the main focus was optics, electricity and magnetism in the nineteenth century, and the argumentative backbone was the changing role of the ether(s) in this evolution. In words of an American reviewer of the first edition, the ether had, at the beginning of the twentieth century, achieved a "psychological triumph":

> difficulties which ... troubled everybody in the early days of the theory have by no means all been resolved; they have merely been ignored. The real triumph has not been physical but psychological; we no longer ask those awkward questions which are inimical to the theory. (Wilson 1913, p. 425)

It was only after the Great War that, according to Whittaker's 1951 preface, "the word 'aether' fell out of favour, and it became customary to refer to the interplanetary spaces as 'vacuous'". And yet, with the formulation of quantum electrodynamics

> the vacuum has come to be regarded as the seat of the 'zero-point' fluctuations of electric charge and current, and of a 'polarisation' corresponding to a dielectric constant different from unity. (Whittaker 1951, preface)

Whittaker's point was that "it seems absurd to retain the name 'vacuum' for an entity so rich in physical properties, and the historical word 'aether' may fitly be retained" (Whittaker 1951, preface).

J. Navarro (✉)
University of the Basque Country, Leioa, Spain
e-mail: jaume.navarro@ehu.eus

Ikerbasque, Bilbao, Spain

A few months later, Paul A.M. Dirac famously published a Letter to the Editor in *Nature* with the challenging title, "Is there an Aether?" in which he advocated for a revival of the elusive entity on similar grounds to Whittaker's. After 1905, it was "soon found that the existence of an aether could not be fitted in with relativity... [and] the aether was abandoned". But, according to Dirac, quantum mechanics made the re-establishment of *an* aether not only possible but also desirable. Possible, because "the velocity of the aether ... is subject to uncertainty relations", thus enabling the aether, at each point, to have a distribution of velocities, a wave function compatible with relativity. And desirable because the form of such wave function demands that an idealized perfect vacuum cannot exist: "we must make some profound alterations in our theoretical ideas of the vacuum. It is no longer a trivial state, but needs elaborate mathematics for its description" (Dirac 1951, p. 906). In other words, the vacuum turned out to be more complex than ideally thought of, and a vacuum with physical properties could very easily be named 'aether'.

Whittaker and Dirac's new aethers were indeed rather different objects, but both authors acknowledged that the ether was, by then and for most people, dead and buried. Their rhetoric was that of *retaining* or *reviving* a concept that had prematurely been abandoned: recent developments in physics, namely the complex properties of the so-called vacuum, demanded the resurrection of the name, albeit with properties different from those of the old Victorian object. Following the analogy of a "psychological triumph" used by the reviewer quoted above, Whittaker and Dirac possibly thought that the death of the ether had only been a hasty collective psychological delusion.

Dead or resurrected, the starting point of this paper is the assumption that by the early 1950s most physicists had declared the ether defunct. Only a generation earlier the panorama was different. In the recent collective work *Ether and Modernity* (Navarro 2018), a number of historians of science have described the status of the ether in the first third of the twentieth century, pointing at the fact that for many the ether was not only still alive but a sign of modernity. Contrary to the usual dichotomy between classical and modern physics and the portrayal of the ether as belonging only to the former, this volume argues that the ether worked as a kind of interstitial concept that bridged the classical and the modern, the cultural and the scientific, the material and the spiritual, the British and the Continental. Of course, belonging in the interstices of so many roles and audiences diminished the expected robustness of the ether qua epistemic object and eventually helped its general abandonment by the outbreak of World War II (Badino and Navarro 2018).

Scholars in what is normally known as historical epistemology often address the historicity of scientific objects in terms of their *biographies*. Far from being simply a catchy metaphor, the biographical approach addresses what Hans Jörg Rheinberger described as the 'robustness' of epistemic objects, the fact that they acquire a life and consistency of their own, be it in their experimental, logical or operational functions. In this sense, 'objects' are more than simply concepts in the mind of a person or a group of people. Following Hasok Chang, "it is important to recognize epistemic objects as objects; what we presume to be real functions as such, not only in our reasoning but also in our material practices" (Chang 2011, p. 414). From this point of view, the ether may be one epistemic object about which to write a biography: many people in the nineteenth and early-twentieth centuries *presumed* the ether to be real because it played a number of epistemic roles.

If deceased, however, a biography of the ether written after its death is likely to acquire the tone of an obituary. Indeed, biographical narratives of objects after their demise pose a challenge about their robustness. What might have been regarded as annoying inconsistencies yet to be resolved, may easily turn into absurd contradictions that explain the obit itself. This has been the usual strategy in the received histories of the ether. Lord Salisbury's famous statement in 1896 saying that the ether seemed only to be "the nominative case to the verb to undulate" (Salisbury 1894, p. 5), suddenly became part of a *common feeling*, and Michelson and Morley's interferometer null results were retrospectively given the status of *experimentum crucis*.

But while obituaries easily contain whiggish historical explanations to help the logical narrative, they may also play a role in reinforcing rather than challenging the robustness of the deceased object. The need to give a consistent account of what the object *was*, may easily turn into an exercise of demarcating its limits, highlighting its properties and unifying its history. That was the main criticism Simon Schaffer made of the classic 1981 volume, *Conceptions of Ether. Studies in the history of Ether theories, 1740–1900*, edited by Geoffrey Cantor and Jonathan Hodge (Cantor and Hodge 1981; Schaffer 1982). While acknowledging the diversity of ether *theories*, the editors ended up creating an object that might have not existed in a continuous way, except perhaps for the preservation of a name. As a matter of fact, the same happened with the title of Whittaker's work, namely *A History of the Theories of Aether and Electricity*. The emphasis seems to be placed on the changing dynamics of the *theories* explaining certain objects, not of the objects themselves. Moreover, in the 1951 edition, Whittaker traced that unity back to the ancient Greeks, as is commonly done.

Together with the ether, phlogiston is another well-known relinquished epistemic object. In Hasok Chang's reading, the abandonment of phlogiston created an explanatory void that only the electron filled a century later. But more importantly from an historiographical perspective, Chang argued that the contrast between a triumphant oxygen and a defeated phlogiston created two new objects significantly different from those of late eighteenth-century chemistry (Chang 2009). In contrast, Arabatzis' (2006) famous biography of the electron depicts an object with internal robustness or, as he calls it, recalcitrance, that would probably be explained otherwise if the term electron were to be dismissed. In other words, biographies of epistemic objects are, I would suggest, dependent on whether the term is still in use or not.

In this paper I thus intend to explore how the demise of the ether presents us with constraints and opportunities in the task of writing the biography of this particular object. I shall focus on three moments and will pay attention to very specific kinds of expositions. First, I shall take Maxwell's and Larmor's review articles in the Encyclopaedia Britannica as the guide to a time when an ether was supposed to exist, however contradictory or challenging this object was. From here, I shall move to the 1920s, a time in which, as noted above, the ether saw a peak of uses in physics and technology, as well as in culture at large. In retrospect, this could be seen as the swansong of an agonising object and a cause of its ultimate death, although contemporaneous accounts show a more complex array of reasons for such frantic presence. Finally, I pay attention to early accounts of the death of the ether as a door open to the way explanations of its disappearance created an object with specific properties that, though related, did not fully identify with previous descriptions.

Who Is the Ether … or Aether?

Where to start? *Who* are we talking about? Long, old-fashioned biographies tend to begin a few lines up the genealogical tree of the protagonist, and include the cultural, national and geographical landscape, as part of a causal account of the story to come. Obituaries are much simpler. Written soon after the obit, they are addressed to an audience for which the memory of the deceased person is still somehow alive. Recent episodes and major events become more relevant than origins and side-tracks. The ether that was pronounced dead throughout the first half of the twentieth century was not the old Greek's, nor Descartes' or Newton's. It was a nineteenth-century ether, an evolving object that belonged to a new physics that in the nineteenth century had reigned supreme among the sciences (Morus 2005). In David Wilson's words,

> throughout most of the century, optical evidence for its existence was more varied and more conclusive than ever before. Toward the end of the century, evidence pointed to its being a physical concept with extraordinary unifying qualities. (Wilson 1974, p. 220)

In his oft-quoted article for the ninth edition of the Encyclopaedia Britannica, written in 1878, James C. Maxwell made that break clear. "The only aether which has survived—he wrote—is that which was invented by Huygens to explain the propagation of light" (Maxwell 1878). The structure of the article would become canonical: it begins with the need to explain light in terms of waves, not particles; to dismantle any idea of light as a substance but to present it as a process. Only later does the aether appear:

> having so far determined the geometrical character of the process, we must now turn our attention to the medium in which it takes place. We may use the term aether to denote this medium, whatever it may be. (Maxwell 1878)

The article portrays a typical tension between the need to have a medium for the transmission of light and its conjectural character. What made Maxwell's aether more robust than Huygen's over a century earlier was, first, the triumph of the undulatory nature of light over Newton's corpuscular views and, more importantly, the link between light, electricity and magnetism postulated by Maxwell and which identified the 'electromagnetic medium' with the 'luminiferous ether'.

"What is the ultimate constitution of the aether? Is it molecular or continuous?" (Maxwell 1878). With this question, Maxwell introduced the last section of his article. If discrete, it could not be like a gas, where the then hypothetical atoms moved almost freely, but had to be very rigid; if continuous, it should allow for heterogeneous motions, like William Thomson's vortex theory of the atom seemed to imply (see below). In case the article might cast excessive doubts on the ether, Maxwell emphasised the need for it to exist:

> whatever difficulties we may have in forming a consistent idea of the constitution of the aether, there can be no doubt that the interplanetary and interstellar spaces are not empty, but are occupied by a material substance or body, which is certainly the largest, and probably the most uniform body of which we have any knowledge. (Maxwell 1878)

Two things are worth emphasising here, since they will come up again later in the argument. Maxwell finished his article in the same way he began it, namely, by appealing to the necessity of the ether. Not only because the wave theory of light necessarily called for the existence of the medium, but also due to an *horror vacui* in interstellar space.

Secondly, Maxwell talked about the ether as a "material substance" or a "body". As we shall later see, together with the inconsistencies in its structure and properties, the material ether presented another fundamental problem: that of its relationship with 'ordinary' matter. As I have elsewhere described, J.J. Thomson, Crookes and others studied electrical discharges in tubes because those were extreme situations in which the relationship between ether and matter might be unveiled. Eventually, this path led to the electron (or corpuscle) as the place where matter and electrification were most intimately united (Navarro 2012).

By 1911, Maxwell's tentative tone became more confident in the article written by Joseph Larmor, also for the Encyclopedia Britannica. For Larmor, the whole of "theoretical physics" consisted in "the science of the aether" which, in a way, had turned into the most fundamental substance, "a *plenum*, which places it in a class by itself; and we can thus recognize that it may behave very differently from matter". As is well known, Larmor's ether was so simple that the real problem appeared in the complexity of matter, not of the ether. Moreover, such complexity, namely "this doctrine of material atoms is an almost necessary corollary to the doctrine of a universal aether". Atoms became, in this picture, epiphenomena of the ether, manifestations of singularities, asymmetries, or vortices in the ether. An atom

> would be identical with the surrounding field of aethereal motion or strain that is inseparably associated with the nucleus. ... there are not two media to be considered, but one medium, homogeneous in essence and differentiated as regards its parts only by the presence of nuclei of intrinsic strain or motion. (Larmor 1911)

The dream of the unity of the physical world and the unicity of the ether was related to the conservation of energy, specifically, the transformation and quantitative equivalence between electric, magnetic, thermal and mechanical energies:

> This single principle of energy has transformed physical science by making possible the construction of a network of ramifying connexions between its various departments; it thus stimulates the belief that these constitute a single whole. (Larmor 1911)

The ether achieved further robustness by granting higher unification to physics, as the origin of matter and the seat for the transformations of energy. Rather than matter in movement, Larmor's physics was, as he well described, the science of ether.

As early as the 1850s, William Thomson, had used the expression 'universal plenum' to refer to the ether. The transformation of physics into the science of ether was fundamental to his dynamical project, the foundations of which were laid in his and Peter G. Tait's two-volume *Treatise on Natural Philosophy* (Thomson and Tait 1867). The intimate relationship between electricity and magnetism and the analogy of these phenomena with heat were the starting point to a grand project that also informed Maxwell's synthesis. It was also Thomson who developed and publicized the vortex theory of the atom, a "Victorian Theory of Everything", as Helge Kragh once put it (Kragh 2002). In a purely mathematical fashion Herman von Helmholtz had proved that vortex filaments in a perfect fluid would be absolutely stable and, thus, never destroyed. Thomson moved a step forward and turned Helmholtz's mathematical construct into a physical theory: assuming the perfect fluid was the ether, Thomson identified the stable vortex rings with the atoms. In this way, not only was he extending the possibilities of his dynamical program but also challenging atomism, or, as he put it, "the monstrous assumption of infinitely strong and infinitely rigid pieces of matter" (Thomson 1867, p. 94).

With the vortex theory one could advocate for atomistic physics without assuming that atoms were fundamental, metaphysical entities. In the same way heat had proved to be not a substance but the effect of matter in movement, atoms could be regarded as movements within the most fundamental of substances—the ether. Maxwell also helped to spread this theory since a fourth of his entry for Atom in the 9th edition of the Encyclopedia Britannica was devoted to it and with no small support. While the modern atomist

> may cut and carve his solid atoms in the hope of getting them to combine into worlds; the follower of Boscovich may imagine new laws of force to meet the requirements of each new phenomenon. (Maxwell 1875)

Over these two options, the vortex ring theory offered a simpler metaphysics, however mathematically complicated:

> [this] primitive fluid has no other properties than inertia, invariable density, and perfect mobility, and the method by which the motion of this fluid is to be traced is pure mathematical analysis. The difficulties of this method are enormous, but the glory of surmounting them would be unique. (Maxwell 1875)

That glory was never achieved. Maxwell soon retreated from any attempt to imagine the structure of the ether and opted instead for what he called a dynamical theory (meaning the use of Lagrangians and Hamiltonians without direct reference to a specific mechanical model). But the 1880s and 1890s saw a plethora of models for the ether, each applied to the resolution of particular problems: be it the "jelly" models of George Gabriel Stokes, the wheels and rubber bands of George FitzGerald or a few ether machines by Oliver Lodge. Even Lord Kelvin, who first rejected the modeling tradition of the ether for fear they might be misleadingly regarded as realistic, ended up constructing a number of models for pedagogical and heuristic reasons (Hunt 1991; Haley 2001). This point should be stressed: the ether qua epistemic object whose biography we are discussing was modelled in multiple ways, mostly heuristically, to explain optical, electric, magnetic, and even gravitational phenomena as well as an ultimate understanding of atoms and matter. The extent to which such ether models constituted competing and mutually contradictory expressions of the same object, different objects or heuristic manifestations of the ether qua research program is something that has no clear answer and for which the obituaries discussed in the last section might throw some light.

The ether was also modeled on the Continent. In M. Norton Wise's (1981) description, mid-century German physicists were reluctant to accept a purely mathematical notion of field, in reaction to the speculative excesses of *Naturphilosophie*. Wilhelm Weber epitomizes the shift from the French tradition of action at a distance to a more mathematical notion of field. As part of that process, Weber imagined an ether formed of neutral pairs of positive and negative electricity. After him, Bernhard Riemann moved to an explanation of gravitation and the transmission of light based on the inertia of a space-filling ether: resistance to change in volume would be the source of gravitation; to a change of shape, transmission of light. Finally, Herman von Helmholtz imagined the ether as an electrically and magnetically polarizable medium in which all forces acted successively in contiguous elements of such medium, also avoiding action at a distance. In any case, as Heaviside once put to Hertz, "My experience of so-called 'models' is that they are harder to understand than the equations of motion" (Heaviside to Hertz, August 1889, in Hunt 1991, p. 105).

Also in mid-century Germany, speculation on the ether was related to a broader interest on the relationship between physics, physiology and psychology. J. R. Mayer, for instance, distinguished between three substances, namely matter, force and *Geist*. The ether would be the seat of an all-pervading force (in opposition to action-at-a-distance forces) and also, by analogy, of mind, soul or *Geist*. Some such ideas were eventually appropriated and transformed by Energeticists and Vitalists, who thought less in terms of an ether and highlighted the centrality of energy or an *élan vital* as a substance (Wise 1981, pp. 270–276).

Finally, a consequence rather than a foundation of the ether was its role as the absolute reference framework. In the early 1880s, the need for global standards in several industries became a priority. Measurements linked to local conditions became useless for global projects in which precision was essential, like in the calibration of apparatus for the telegraph. In this context, the possibility of using the ether as the framework that would provide an absolute reference irrespective of the earth became prominent. As Richard Staley (2008) argued, this is the milieu of Michelson and Morley's experiments to measure the speed of the earth in the ether. Only in retrospect did this instrumental characteristic of the ether become central to its supposedly contradictory characteristics.

Splendour or Agony? The Early Decades of the Twentieth Century

In his 1928 book *The Ether and Growth*, the London-based American engineer and chairman of the European branch of Westinghouse Brake Company, John Wills Cloud, remarked that while some scientists repudiated the ether

> because it is intangible, and proof of its existence, satisfactory to Science, is therefore impossible... To the majority of Scientists, however, the existence and the function of such a medium have long seemed necessary... [Moreover] the modern development of Wireless communication has served greatly to strengthen this view, and the existence of an Ether medium which can be neither weighed nor measured is accepted by them through a sort of faith in the evidence of a thing not seen." (Cloud 1928, pp. 5–6)

The *Ether of Growth* is the speculative product of a mechanical and chemical engineer with spiritualist tendencies and, in it, the author developed a cosmic theory in which immaterial ethrons became the centres of growth of ponderable matter analogous to the growth of crystals. What interests me here is the idea that wireless technologies had come to the rescue of the ether in the face of attacks on its existence.

Indeed, as I have shown in another paper, at the time theoretical physics was abandoning the ether, the then emerging and increasingly popular wireless technologies came to the rescue of this elusive entity (Navarro 2016). Wireless technologies became crucial during the Great War, creating an army of wireless amateurs from the ranks of demobilized soldiers once the war ended. But it was in the early twenties, with the creation of the British Broadcasting Corporation in Britain, the BBC, that the public at large learnt about the ether for the first time.

Central to this strengthening of the ether was Sir Oliver Lodge, one of the major authoritative public figures of physics among electrical engineers and wireless amateurs. Indeed, in all his talks about the mechanisms of wireless, Lodge argued that the existence of wireless communications was the most material proof for the existence of the ether. In a lecture at the Wireless Society of London in June 1922, for instance, he used the opportunity to reassure "in the presence of those who are working with the ether and to whom we look for the discoveries of the future" that the ether did exist. "It is fashionable sometimes to disbelieve in the ether," he cautioned. Contrary to what "some of the newspapers" say, Lodge reminded that the only thing Einstein had done was to stop using the ether. But, he went on, "a thing does not cease to exist because you refrain from attending to it. Those who daily use the ether for signaling purposes can hardly disbelieve in its existence" (Lodge 1922, p. 408). This last sentence epitomises the core of the argument in favour of the ether in the context of wireless communications: the ether was obviously real since, so the argument went, without it wireless would not be even possible. Furthermore, when talks like this were broadcast live through the BBC, the reality of the ether turned psychologically unquestionable.

Lodge was also instrumental in keeping the ether alive through spiritualism. A convinced advocate of séances and mediums, Lodge struck a chord in 1916 with his best-selling *Raymond or Life and Death*, a book devoted to mourning the loss of his son in Gallipoli and the experiences of communicating with him after his death (Lodge 1916). Like Lodge, many people lost and would lose young loved ones in the war, and found solace, hope and comfort in the words of a scientist-mourning father giving ether-physics-based explanations of life hereafter. For large audiences already disappointed by traditional Christian churches, abandonment of the ether would deprive them from a last resource for hope in the aftermath of the war (Noakes 2018).

Contrary to the received view, the ether in the 1920s was not simply the stubborn remnant of an old concept resisting to accept its fate. In certain milieus, like the new practitioners and users of wireless, the ether was identified with modernity. So was among some modernist artists who, like the Italian futurist Umberto Boccioni, the French artist Marcel Duchamp, the abstract painter Wassily Kandinsky or the Hungarian dance artist Rudolf Laban, conceptualised their conceptions of space in terms of what they regarded as the modern theories of relativity and the ether (Henderson 1998, 2002, 2004 and 2014; Laemmli 2016). Indeed, while many would see the two (relativity and ether) as opposed, that was not the case in some cultural and artistic milieus of the early decades of the twentieth century where the lure of the images produced by X-rays, the modernity of wireless and the ether, the challenges to traditional notions of space by relativity and the existence of minute new particles like the electron were all part of the same modernist batch.

The formulation of General Relativity and the highly publicised eclipse observations of 1919 proving Einstein's predictions opened a new avenue for those wanting to preserve the ether. Einstein's (1920) famous lecture in Leiden, "Ether and the Theory of Relativity", was taken by some as the mending of the premature dropping of the ether fifteen years earlier (van Doengen 2012). Two things are worth stressing here. First, that the 1919 media campaign to create a "revolution in physics" opened up a space for relativity and the ether to be present in the public sphere (Sponsel 2002). The mystery and esoterism associated with the theory of relativity created a demand for public lectures, popular books and accessible explanations, and the ether, also mysterious and esoteric, became part of the argument, either by presenting it in contrast to and rejected by relativity or compatible with it.

Secondly, the 1920 lecture also gave authority to those wanting to preserve a new form of ether, even if distinct from the previous Victorian object. Acknowledging that "careful reflection teaches us, however, that the special theory of relativity does not compel us to deny ether", Einstein moved towards the interpretation of the space-time tensor as a new ether:

> According to this theory the metrical qualities of the continuum of space-time differ in the environment of different points of space-time, and are partly conditioned by the matter existing outside of the territory under consideration. ... But therewith the conception of the ether has again acquired an intelligible content although this content differs widely from that of the ether of the mechanical undulatory theory of light. The ether of the general theory of relativity is a medium which is itself devoid of all mechanical and kinematical qualities, but helps to determine mechanical (and electromagnetic) events. (Einstein 1920)

One of the most active popularisers of this new ether was Arthur Eddington. As early as 1918 Eddington stated that "The phenomena, electromagnetic as well as gravitational, will all be described by the $g_{\mu\nu}$, which represents the state of strain of this space-time. This space-time may be materialised as the aether, and the aether-theory does in fact attribute electromagnetic phenomena to strains in this supposed absolute medium" (Eddington 1918, pp. 79–80). What Einstein but mostly Eddington were doing here was to suggest a new transformation of the ether that, had it been successful, would now be part of the story line of this object. Certainly, the new ether had only a minor contact with a Victorian, material ether; but so did the new notions of matter, particle, force or action at a distance. Eddington was clear that not only the existence or non-existence of the ether was at stake, but rather a larger debate on realism in physics.

In 1920, Eddington wrote his first philosophical paper on "The Meaning of Matter and the Laws of Nature According to the Theory of Relativity", which was published in *Mind*. In it, he claimed that Einstein's equation did not relate matter to geometry but, actually, defined matter in terms of geometry: "matter does not cause an unevenness of the gravitational field; the unevenness of the field *is* matter" (Eddington 1920, p. 152). This subjective or idealist interpretation was made more explicit further on: "According to this view matter can scarcely be said to exist apart from mind. Matter is but one of a thousand relations between the constituents of the World", which begs the question as to "why one particular relation has a special value for the mind" (p. 153). Eddington faced this challenge with two arguments. First, that "matter" seemed to provide easier ways to relate to the only task of the senses, which was to perform measurements in space and time; and, second, that "matter" appeared as a permanent structure and, like with the conservation principles, "for some unknown reason the mind appears to have a predilection for living in a more or less permanent universe" (p. 154).

Although not explicit, the second argument could also be summoned in defending the ether, since it certainly gave more "permanence" than a non-ether, at least in the idealist way he understood it. As a matter of fact, preserving his ether and transforming the notion of matter were two sides of the same coin. The place where this is made most obvious was in his 1935 book *New Pathways of Science*. At a time when the ether had largely been abandoned, Eddington kept defending it in the following unequivocal terms: "As far as and beyond the remotest stars the world is filled with aether. It permeates the interstices of the atoms. Aether is everywhere" (Eddington 1935, p. 38). But his ether had clearly little to do with what he called the "materialist ether":

> How dense is the aether? Is it fluid like water or rigid like steel? How fast is our earth moving through it? Which way do the particles of aether oscillate when an electromagnetic wave travels across it? At one time these were regarded as among the most urgent questions in physics; but at the end of a century's struggle we have found no answer to any of them. We are, however, convinced that the unanswerableness of these questions is to be reckoned not as ignorance but as knowledge. What we have found out is that aether is not the sort of thing to which such questions should apply. Aether is not a kind of matter. Questions like these could be asked about matter but they could not be asked about time, for example; and we must reckon aether as one of the entities to which they are inappropriate (Eddington 1935, pp. 38–39).

Finally, while Eddington tried to preserve the ether as part of his convictions and defence of relativity, the ether also became the flagship of many anti-Einstein and/or anti-relativity representatives, the most outspoken of which were Philip Lenard in Germany and Arthur Miller in the US. While it would be anachronistic to dismiss these attempts at reclaiming the ether coming from the 'wrong side', this proliferation of ethers from contradictory stances may have easily been instrumental in its demise, which is the subject of the next section.

Early Obituaries

Who wrote the first obituary of the ether? The revival of positivism in the late nineteenth century pronounced many epistemic objects dead. With his theory of cognition based only on sense impressions or, as he would call them, "elements", Ernst Mach tried to purify physics from metaphysics, which he regarded as unnecessary and illegitimate constructions of the mind. Force, atoms, and even matter became subjective inventions that had to be abandoned. Yet not even Mach (1893) rejected the ether, "which scientifically is something more valuable than the doubtful notion of the absolute space" (in Kostro 2000, p. 22). More radical on this particular matter was Wilhelm Ostwald who thought the ether was an unnecessary hypothesis, especially when he embraced energeticism:

> after having recognised energy as a real substance—better, the only real substance of the so called external world—that we no longer have a need to search for a carrier of it... This allows us to recognise radiant energy as existing autonomously in space (Ostwald 1893, p. 1016).

One of the most influential optical physicists in Germany, Paul Drude kept a balance between a physical ether tradition and his positivistic environment. From his point of view,

> just as one can attribute to a specific medium, which fills space everywhere, the role of intermediary of the action of forces, one could do without it and attribute to space itself those physical characteristics which are now attributed to the ether. (Drude 1894, p. 9)

In what looks like a clear address to positivists, he would elsewhere defend that "the word 'ether' does not imply any new hypothesis, but is only the essence of space free of matter, which has certain physical properties" (Drude 1895, p. 9). This is an early example of a move towards changing the expression "ether of space", common mainly among Victorian physicists, for things like "physical space endowed with attributes".

If not the first, indeed the most famous obituary of the ether was written by Einstein in his 1905 founding paper on Special Relativity. In it, he claimed that

> the introduction of a 'luminiferous ether' will prove to be superfluous inasmuch as the view here to be developed will not require an 'absolutely stationary space' provided with special properties, nor assign a velocity-vector to a point of the empty space in which electromagnetic processes take place. (Einstein 1905, p. 38)

The ether that was here declared dead, or simply unnecessary, was a reduced version of the ethers we saw above, namely the notion of an absolute reference frame or stationary space. It was not until 1907, in his more pedagogical review article on relativity that Einstein named the Michelson and Morley results as the experimental trigger to the theory of relativity (Einstein 1907). This, as is well known, was a post hoc reconstruction and not a historically accurate account (Stachel 2002; Staley 2008; van Dongen 2009). Around 1909, with his further development of the theory of quanta of light, Einstein *killed* another ether: the one responsible for the transmission of light, or luminiferous ether, since light was now equated to particles and, like with Newton, in no need of a medium (Einstein 1909).

This obituary informed much of the pedagogy of the theory of relativity. Just to mention one example among many, in one of his popular accounts on modern physics, the physicist Edward Neville da Costa Andrade explained that the Michelson Morley experiment had killed the ether and paved the way towards the theory of relativity:

> No motion of the earth through the ether can, therefore, be detected, and this means that there cannot be an ether with the properties of any material body. It was to explain this extraordinary effect that Einstein originally devised the theory of relativity (Andrade 1930, p. 80).

As mentioned in above, Einstein resurrected the ether circa 1920 or, to be more accurate, imagined a new 'space-time ether'. However, by then, the association between relativity and the death of the ether had taken root. Moreover, the portrayal of a conflict between ether and relativity created a new public space for the former. As Milena Wazeck has described, some opponents to Einstein, to Relativity, or to both took the ether as the flagship for their attacks. Indeed, the international 'Academy of Nations' founded in 1921 to promote the restoration of 'true science' and the unification of knowledge against the specialisation and esoterism of relativity and modern sciences gave rise in Germany to the so-called 'German Society for Universal Ether Research and Comprehensible Physics' (Wazeck 2014, s. 4.3). Thus, an increasing acceptance of relativity involved a demise of the ether, not just because of Einstein's 1905 statement but also due to the embrace of such dichotomy by Einstein's opponents.

One of the most nuanced and influential popular obituaries of the ether came from James H. Jeans. One of the first converts to quantum physics in Britain, Jeans became a major populariser of modern physics highlighting the epistemological and ontological gap with a previous generation. In his 1929 *The Universe around us*, Jeans made it clear that

> It is now a full quarter of a century since physical science, largely under the leadership of Poincaré, left off trying to explain phenomena and resigned itself merely to describing them in the simplest way possible. (Jeans 1929, p. 329)

The example he chose to describe this shift towards what he called 'a simpler science' was the ether:

> the Victorian scientist thought it necessary to 'explain' light as a wave-motion in the mechanical ether which he was for ever trying to construct out of jellies and gyroscopes; the scientist

of to-day, fortunately for his sanity, has given up the attempt and is well satisfied if he can obtain a mathematical formula which will predict what light will do under specified conditions. (Jeans 1929, p. 329)

The indirect accusation of insanity and the link between mental health and simplicity cannot elude the reader. Jeans was advocating for a mathematical turn of physics, one in which "The formulae of modern science are judged mainly, if not entirely, by their capacity for describing the phenomena of nature with simplicity, accuracy, and completeness" (Jeans 1929, p. 329). It was with this approach in mind that Jeans pronounced the ether dead:

> the ether has dropped out of science, not because scientists as a whole have formed a reasoned judgement that no such thing exists, but because they find they can describe all the phenomena of nature quite perfectly without it. (Jeans 1929, p. 329)

Thus, the main question was not one of falsification of an idea but of abandonment as "not needed".

In a chapter on "Relativity and the Ether" in his 1930 *The Mysterious Universe*, Jeans made it clear that all ethers "are in all probability fictitious," meaning not that they do not exist at all, but that, if they do, "they exist in our minds" only (Jeans 1930, p. 79). This mental existence was the result of another conceptual distinction with no counterpart in modern physics, namely the distinction between space and time that Minkowski's metric had erased. This explained

> why the old luminiferous ether had inevitably to fade out of the picture—it claimed to fill 'all space', and so to divide up the continuum objectively into time and space... [With this] the laws of nature, not recognising such divisions as a possibility, cannot recognise the existence of the ether as a possibility. (Jeans 1930 p. 103)

Jeans also addressed and dismissed those who, like Eddington, wanted to preserve the word ether for the continuous space-time of relativity. From his point of view,

> as the hypothesis of relativity is the exact negation of the existence of the old ether, it is clear that any ether that relativity can allow to remain in being must be the exact opposite of the old ether...[and that is why] it seems a mistaken effort to call them by the same name. (Jeans 1930, p. 104)

With all this in mind, thus,

> it seems appropriate to discard the word 'ether' in favour of the term 'continuum', this meaning the four-dimensional 'space' we have already imagined, in which the three dimensions of ordinary space are supplemented by time acting as a fourth dimension. (Jeans 1930, pp. 109–110)

Ironically, Jeans' obituary gave a highly detailed description of the ether. He was addressing an audience that might have never heard of the ether and came to know about it through his extensive account: a whole chapter in a book on modern physics on something he regarded as non-existent looks a bit excessive. Moreover, this is a pattern we find in many popular books on physics in the 1920s and 1930s: negation of the ether came hand in hand with the creation of a space for it to be present, if only to be rejected.

More poignant than a dismissive obituary, however, is silence in the face of a loss. Explanations like Jeans' might induce some readers to hear about the ether for the first time and be convinced of its need rather than dissuaded. Another strategy was not to mention the ether at all. One example is particularly telling: John Ambrose Fleming, inventor of the thermionic valve and a key agent in the development of wireless technologies at the Marconi Company, gave a series of lectures at the Royal Institution in 1902. These were turned into a book that became very successful (re-issued a number of times in the following years), and which was often quoted in articles, lectures and books on wireless matters throughout the 1920s. In that book, the ether was a most real thing, since "there is abundant proof that it is not merely a convenient scientific fiction, but is as much an actuality as ordinary gross, tangible, and ponderable substances" (Fleming 1902, p. 192). The train of thought was the standard we saw in the previous section: light and radiation were waves in a medium, analogous to ripples in air and water.

In January 1922 Fleming gave the Christmas lectures at the Royal Institution for a second time and published them in extended form both in weekly instalments in the periodical *The Wireless World* and as a book with the title *Electrons, Electric Waves and Wireless Telephony* (Fleming 1923). Fleming gave a rather detailed account of the mathematics of waves, the constitution of matter, including the recent developments in atomic theory by Rutherford and Bohr, as well as the quantum theory of Planck, the basics of electricity, magnetism and hertzian waves, and then moved on to technical details about gramophones, telegraphy and wireless transmission. Surprisingly, in such a detailed book and contrary to what he had done in his Christmas lectures twenty years earlier, Fleming did not mention the ether, not even once, neither to defend nor to challenge its existence. Silence about the ether was the best way to pronounce it dead, without the need of a eulogy or an obituary.

Silence was also the tacit strategy that killed the ether in periodicals like *QST. An amateur radio magazine*, the American counterpart of *The Wireless World*. As we have seen in the previous section, British radio amateurs and engineers preserved the ether in many of their publications and popular accounts. The need for a medium to spread electromagnetic waves was the central argument. On the contrary, American accounts on wireless pivoted around rays, not waves, thus eluding the question of a medium for such waves. Similarly, the increasing acceptance of the quantum of light in the mid-1920s as a new corpuscular theory of light created a space for a robust abandonment of the ether as the medium for radiation. As the Liverpool physics professor James Rice very clearly stated in a popular book on relativity,

> nowadays we are so far from worrying about a material ether as a transmitter of a mechanical radiation that we actually regard radiation as an entity in itself, requiring no medium for its transmission, having an independent existence quite as real as that of an atom or electron. (Rice 1927, p. 48)

Perhaps equally piercing were the attempts to talk about an ether about which "its essential nature must for ever remain unknown, and, in fact, unknowable" (Corrigan 1928, p. 105). This was the stance of a few electrical engineers used to talking about the ether as a "scientific or philosophical necessity more than anything else" (p. 110).

The tone used by J.F. Corrigan, from the University of Manchester, testifies to the puzzling views about the ether that some wireless enthusiasts had towards the end of the 1920s:

> Einstein, it is said, with his now celebrated theory of relativity has stripped us of our conception of the ether. At least, that seems to be the popular opinion. But is such an opinion true? (Corrigan 1928)

As we have seen in the previous section, Lodge had addressed this audience with strong reassurance on the reality of the ether. Corrigan answers the question saying that

> It is more correct, perhaps, to say that Einstein has modified the ether into a four-dimensional 'continuum,' that is to say, a something which is ever-present, and which is continuous, permeating and transcending the material (Corrigan 1928, p. 104),

and about which nothing can be said and, therefore, it is better to ignore for any practical purpose. In a similar vein, William H. Bragg queried whether the question about the existence of the ether made sense at all:

> This is a question which we ask at once. And yet when we come to think what answer we shall give, we begin to doubt whether there is any real meaning to the question, whether in fact it is a proper question at all. We soon get into deep waters if we try to picture to ourselves what is meant by 'really existing'. Fortunately we need not try: and ought not to try (Bragg 1933, pp. 18–19)

and stick to empirical observations on light and their connections.

Another type of obituary for the ether appeared among Marxist ranks. In his book *Modern Science*, Hyman Levy described the old ether as a

> special fictitious substance… invented solely for the purpose of 'explaining' the special difficulties away… as a convenient dump into which all our difficulties could be flung or explained away by the simple expedient of saying 'The ether is like that'. (Levy 1939, p. 59)

The ether was simply a "verbal trick" because no experiment could be devised to "isolate" the ether. Thus, the ether was a "lying excuse"; it was "metaphysical" (Levy 1939, p. 59). But, if the ether did not exist, how could a Marxist explain action at a distance? Here, Levy used the interconversion of mass and energy in relativity to justify a notion of field that would enable mechanistic explanations without ad hoc entities like the ether:

> We have seen how mass is energy, and energy possesses mass form the relativistic standpoint, and so the transport of energy from A to B involves the passage of mass effects. Had we said that A affects B by bombarding it with the masses, the mechanistic habits of our minds would have not been violated. If we say it is energy that is being transported, our mechanistic habits again demand a picture of the material carrying energy. If we say that energy bears with it the imprint of mass, the mechanistic difficulty is resolved at the same time as the mind is released from the trammels of mechanistic thinking. (Levy 1939, pp. 592–593)

Indeed, not all Marxists tried to accommodate Einstein's theory within their ideological framework and many rejected what they regarded as an intrinsically idealist and bourgeois theory (Graham 1993). The thing to emphasise here is that, in contrast to the idealism of James Jeans, for instance, the ether could also be buried from a materialist conception. It is certainly ironic that while people like Lodge tried to preserve the ether as a way to fight against materialism, some Marxists saw the ether as a remnant of anti-materialism, while idealists Jeans and Eddington respectively rejected and rescued the ether from idealist perspectives.

One other way to kill the ether was by imagining an explosion in the number of ethers. Rather than falsification, rejection or simple uselessness, some obituaries explained the death of the ether as the result of having to imagine infinite ethers, one for each moving system. In their attempts to preserve an ether, some physicists had suggested the existence of local ethers as a way to explain the FitzGerald contraction and the null result of the Michelson Morley experiments. As philosopher and early defendant of relativity in Britain H. Wildon Carr put it, the principle of relativity rests on the absolute constancy of light propagation from any reference framework. In terms of ether, this would mean that

> instead of one absolute ether filling space, at rest and unalterable in relation to all systems of movement whatever, we must conceive the ether to be carried with and belong to every system of movement. (Carr 1913, p. 414)

This multiplication of ethers to infinity "is precisely the same thing as to suppose there is no ether" (Carr 1913, p. 414). Or, in words of physicist George P. Thomson in the face of suggestions that the de Broglie waves might be explained in terms of a sub-ether, with such inflation of ethers "space is becoming overcrowded" (Thomson 1930, p. 11).

Finally, not all obituaries of the ether relied on Relativity or the Michelson-Morley experiment. On occasion it was the discovery of the electron, as the first elementary particle, that had supposedly paved the way to the disappearance of the ether. In his amusing popular book *Within the Atom. A popular view of electrons and quanta*, John Mills, fellow of the American Physical Society, argued that

> with the discovery of the electron—the apparently indivisible particle of electricity—the ether rapidly lost its importance and finally with the work of Einstein it has ceased to be a necessary postulate in physical science. (Mills 1922, xii)

The reason behind this was that

> the vibrating systems of the electrons within an atom do not radiate energy continuously but emit it in definite quanta. According to the present accepted picture the electrons may vibrate in orbits without loss of energy to surrounding systems. (Mills 1922, p. 120)

To clarify things, the author explained the relationship between the quantum theory of the atom and the ether:

> This in itself is an argument against an all-embracing ethereal medium, for if it was capable of absorbing energy at all from a vibrating electron we should expect it to do so continuously. (Mills 1922, p. 120)

Incidentally, in this popular and, it should be said, rather humoristic book, the author warns against popularising the new physics through its history, since then "the terminology of the older physics of the ether is unavoidable" (Mills 1922, xii). Indeed, later on in the book, in a supposed dialogue between an electron, a proton, energy, the author, the general reader and the scientist, the latter two communicate through "ether waves", later referred to as "the same *hypothetical* medium" (p. 136).

Thus early accounts of the death of the ether used a diversity of explanatory strategies: from positivist and Marxist anti-metaphysics to instrumentalism and conventionalism; from a rejection of the old word ether to reference to modern relativistic space-time or dismissive silence about it; from an emphasis on the quantum aspects of radiation to mockery towards a potential inflation in the number of ethers; and, indeed, from experimental falsification to the claim of 'not needed' or the agnostic 'unknowable'. Interestingly, however, in most cases these obituaries worked as an instrument to create a space for the ether to be explained to the public, albeit in a way different to what Maxwell and Larmor had done in their articles for the Encyclopaedia Britannica. The strengths of the ether in those accounts, mainly being the necessary medium for the transmission of electromagnetic waves, became only secondary or even irrelevant to the identity of the deceased ethers. At the same time, a side property like the fact of it being a potentially absolute reference framework became centre stage. Obituaries were, thus, instrumental in changing the face of the ether.

Conclusion

In this paper I have tried to show how the first *obituaries* of the ether were instrumental in creating an object with specific and largely simplified properties related but different to nineteenth-century ethers. I suggest that writing the history of dead objects (or objects the author wants to be dead) does not seem to be epistemologically neutral but, on the contrary, it involves a reformulation of the object itself. That was indeed the case with the ether. Its most important property, being the medium for the transmission of electromagnetic waves, turned out to be irrelevant when explaining its demise. At the same time other properties of some previous ether(s) received full attention in order to prove its contradictions and justify its death.

To close these rather preliminary reflections on the historiographical role of obituaries, let me compare with two different objects: phlogiston and elementary particles. As the title of Chang's (2009) paper says, "we have never been whiggish about phlogiston". In his reconstruction of the so-called chemical revolution, the phlogiston and the oxygen used by late eighteenth century chemists were equally alien to the later notions of the same objects, only that the former was abandoned while the latter (name) remained. Similarly, the profound changes in the notion of elementary particles during the 1920s and 1930s created a totally new object, while the name remained (Navarro 2004). Indeed, history is largely contingent. I do not mean to argue that the ether should have remained; only to show that the process through which the ether disappeared included the creation of a non-existing object (the dead ether) significantly different from the exiting ether(s) of the nineteenth century.

Acknowledgements Ikerbasque Research Professor, University of the Basque Country. Part of this research was possible thanks to project HAR2015-67831-P MINECO/FEDER, EU of the Spanish Government as well as the Ikercambridge fellowship of the Basque Government.

References

Andrade, E.N.C. (1930). *The mechanism of nature. Being a simple approach to modern views on the structure of matter and radiation*. London: G. Bell & Sons.
Arabatzis, T. (2006). *Representing electrons. A biographical approach to theoretical entities*. Chicago: Chicago University Press.
Badino, M. & Navarro, J. (2018). *Introduction. ether—The multiple lives of a resilient concept* (pp. 1–13).
Bragg, W. H. (1933). *The universe of light*. London: G Bell & Sons.
Cantor, G.N. & Hodge, M.J.S. (1981). *Conceptions of ether. Studies in the history of ether theories, 1740–1900*. Cambridge: Cambridge University Press.
Carr, H. W. (1913). The principle of relativity and its importance for philosophy. *Proceedings of the Aristotelian Society, 14*, 407–424.
Chang, H. (2009). We have never been whiggish (about phlogiston). *Centaurus, 51*, 239–264.
Chang, H. (2011). The persistence of epsitemic objects through scientific change. *Erkenntnis, 75*, 413–429.
Cloud, J. (1928). *The ether and growth. A theoretical study*. London: Simpkin Marshall.
Corrigan, J.F. (1928). *Radio and relativity. How the Einstein theory affects our conception of the ether*. In Pitman's Radio Year Book, 100–4. London: Pitman & Sons.
Dirac, P. A. M. (1951). Is there an Æther? *Nature, 168*, 906–907.
Drude, P. (1894). *Physik des aethers auf elektromagnetischer Grundlage*. Stuttgart: Verlag von Ferdinand.
Drude, P. (1895). *Die Theorie in der Physik*. Leipzig: S. Hirzel.
Eddington, A. S. (1918). *Report on the relativity theory of gravitation*. London: Fleetway Press.
Eddington, A. S. (1920). The meaning of matter and the laws of nature according to the theory of relativity. *Mind, 29*, 145–158.
Eddington, A. S. (1935). *New pathways of science*. Cambridge: Cambirdge University Press.
Einstein, Albert. (1905). Zur Elektrodynamik bewegter Körper. *Annalen der Physik, 322*, 891–921.
Einstein, A. (1907). Über das Relativitätsprinzip und die aus demselben gezogenen Folgerungen. *Jahrbuch der Radioaktivität und Elektronik, 4*, 411–462.
Einstein, A. (1909). Entwicklung unserer Anschauungen über das Wesen und die Konstitution der Strahlung. *Physikalische Zeitschrift, 10*, 817–825.
Einstein, A. (1920). *Äther und Relativitätstheorie*. Berlin: Springer.
Fleming, J.A. (1902). *Waves and ripples in water, air, and aether*. Being a Course of Christmas Lectures delivered at the Royal institution of Great Britain. London: Society for promoting Christian knowledge.
Fleming, J. A. (1923). *Electrons, electric waves and wireless telephony*. London: The Wireless Press.
Graham, L. R. (1993). *Science in Russia and the Soviet Union*. Cambridge: Cambridge University Press.
Haley, C. (2001). *Envisioning the unseen universe: Models of the ether in the nineteenth century*. Unpublished Ph.D. dissertation, University of Cambridge.
Henderson, L. D. (1998). *Duchamp in context: Science and technology and related works*. Princeton: Princeton University Press.

Henderson, L. D. (2002). Vibratory modernism: Boccioni, Kupka, and the ether of Space. In L. D. Henderson & B. Clarke (Eds.), *From energy to information: Representation in science and technology, art, and literature* (pp. 126–149). Stanford: Stanford University Press.

Henderson, L. D. (2004). Editor's introduction: I. writing modern art and science—An overview; II. Cubism, futurism, and ether physics in the early twentieth century. *Science in Context, 17,* 423–466.

Henderson, L.D. (2014). Abstraction, the ether, and the fourth dimension: Kandinsky, Mondrian, and Malevich in context. In M. Ackermann & I. Malz (Ed.), *Kandinsky, Malewitsch, Mondrian: Der weisse Abgrund Unendlichkeit/The Infinite White Abyss* (pp. 37–55) (German); 233-44 (English). Düsseldorf: Kunstsammlung Nordrhein-Westfalen.

Hunt, B. J. (1991). *The Maxwellians*. Cornell: Cornell University Press.

Jeans, J. (1929). *The universe around us*. Cambridge: Cambridge University Press.

Jeans, J. (1930). *The mysterious universe*. Cambridge: Cambridge University Press.

Kostro, L. (2000). *Einstein and the ether*. Montreal: Apeiron.

Kragh, H. (2002). The vortex atom: a Victorian theory of everything. *Centaurus, 44,* 32–114.

Laemmli, W. (2016). *The choreography of everyday life: Rudolf Laban and the making of modern movement*. Unpublished Ph.D. dissertation, University of Pennsylvania.

Larmor, J. (1911). Aether. In *Encyclopædia Britannica ninth edition* (Vol. 11). https://en.wikisource.org/wiki/1911_Encyclopædia_Britannica/Aether.

Levy, H. (1939). *Modern science. A study of physical science of the world today*. London: Havish Hamilton.

Lodge, O. (1916). *Raymond or life and death*. London: Methuen & Co.

Lodge, O. (1922). Address to the wireless society of London. *The Wireless World, 13,* 407–415.

Mach, E. (1893). *The science of mechanics: A critical and historical exposition of its principles*. Chicago: Open Court.

Maxwell, J.C. (1875). *Atom. Encyclopædia Britannica ninth edition* (Vol. 3). https://en.wikisource.org/wiki/Encyclopædia_Britannica,_Ninth_Edition/Atom.

Maxwell, J.C. (1878). *Ether. Encyclopædia Britannica ninth edition* (Vol. 8). https://en.wikisource.org/wiki/Encyclopædia_Britannica,_Ninth_Edition/Ether.

Mills, J. (1922). *Within the atom. A popular view of electrons and quanta*. London: George Routledge and Sons.

Morus, I. (2005). *When physics became king*. Chicago: Chicago University Press.

Navarro, J. (2004). New entities, old paradigms: elementary particles in the 1930s. *Llull, 27,* 435–464.

Navarro, J. (2012). *A history of the electron*. In G.P. Thomson (Ed.), Cambridge: Cambridge University Press.

Navarro, J. (2016). Ether and wireless. An old medium into new media. *Historical Studies in the Natural Sciences, 46,* 460–489.

Navarro, J. (2018). *Ether and modernity. The recalcitrance of an epistemic object in the early twentieth century*. Oxford: Oxford University Press.

Noakes, R. (2018). Making space for the soul: Oliver lodge, Maxwellian psychics and the etherial body. *Navarro, 2018,* 88–106.

Ostwald, W. (1893). *Lehrbuch der allgemeinen Chemie* (2nd ed.). Leipzig: Engelmann.

Rice, J. (1927). *Relativity. An exposition without mathematics*. London: Ernest Benn Limited.

The Marquis of Salisbury. (1894). Presidential address. *British Association for the Advancement of Science Report, 64,* 3–15.

Schaffer, S. (1982). Conceptions of ether: Studies in the history of ether theories 1740–1900. *History of Science, 20,* 297–303.

Sponsel, A. (2002). Constructing a 'Revolution in Science': The campaign to promote a favourable reception for the 1919 solar eclipse experiments. *The British Journal for the History of Science, 35,* 439–467.

Stachel, J. (2002). *Einstein from 'B' to 'Z'*. Basel: Birkhäuser.

Staley, R. (2008). *Einstein's generation*. Chicago: Chicago University Press.

Thomson, G. P. (1930). *The wave mechanics of free electrons*. New York and London: McGraw-Hill Book Company.

Thomson, W. (1867). On vortex atoms. *Proceedings of the Royal Society of Edinburgh, 6,* 94–105.

Thomson, W., & Tait, Peter G. (1867). *Treatise on natural philosophy*. Oxford: Oxford University Press.

van Dongen, J. (2009). On the role of the Michelson-Morley experiment: Einstein in Chicago. *Archive for History of Exact Sciences, 63,* 655–663.

van Dongen, J. (2012). Mistaken identity and mirror images: Albert and Carl Einstein, Leiden and Berlin, relativity and revolution. *Physics in Perspective, 14,* 126–177.

Whittaker, E.T. (1951). *A history of the theories of aether and electricity* (vol. 1, 2nd ed). London: Nelson.

Wilson, E. B. (1913). Review of A history of the theories of aether and electricity. *Bulletin of the American Mathematical Society, 19,* 423–427.

Wilson, D. B. (1974). Aether studies. *History of Science, 12,* 220–227.

Wise, M. N. (1981). German concepts of force, energy and the electromagnetic ether: 1845–1880. *Cantor & Hodge, 1981,* 269–308.

Chapter 17
The Meaning, Nature, and Scope of Scientific (Auto)Biography

Thomas Söderqvist

> *The art of Biography is different from Geography.*
> *Geography is about maps, but Biography is about chaps.*
> (Bentley 1905)

Introduction

The theme of this volume is biography in the history of physics. In this chapter, I will go beyond the limitation to physics, however, and discuss aspects of the genre of biography and its relations to the history of science in general. My aims are, firstly, to remind historians of science that the genre of biography, including scientific biography, is about people, not institutions, concepts, or objects; and, secondly, to bring autobiography and memoir into the discussion.

I will begin with a discussion of the implications of taking the prefix bio- in the word 'biography' seriously. What is the subject matter of biographical studies, and what falls outside its denotation? More specifically, I will question whether the current extension of the use of the word 'biography' for historical studies of scientific institutions, theoretical entities, and material objects is sustainable. Can the use of phrases like 'biography of an institution', 'life of a concept', or 'biographies of objects' be justified? Why is the 'biography' metaphor so popular?

The main part of the chapter is motivated by the fact that autobiographies and memoirs (I use the two words synonymously throughout) are underestimated in the literature about scientific biography and history of science. For example, in two of the major collections of scholarly articles about scientific biography over the last decades (Shortland and Yeo 1996; Söderqvist 2007a) only two chapters out of 26 are devoted to autobiography (Outram 1996; Selya 2007). This neglect is to some extent

---

T. Söderqvist (✉)
University of Copenhagen, Copenhagen, Denmark
e-mail: ths@sund.ku.dk

understandable: self-centered accounts traditionally have had a bad reputation among historians of science for being subjective and self-congratulatory, and autobiography brings the old 'whiggish' approach (Jardine 2003) to the history of science into mind. But in the wider scholarly literature on life-writing, studies of biography and autobiography overlap; for example, one of the leading journals in the field is titled *a/b: Auto/Biography Studies* and most academic libraries similarly mix biographies, autobiographies and memoirs physically on the shelves and in the catalogues. In the main part of the chapter, I identify a number of existing and possible kinds of scientific auto/biographies and their relation to the history of science. I point out that writing scientific biography, autobiography and memoirs is not just an aid to history of science (an *ancilla historiae*), but has many other interesting aims as well, and suggest that an awareness of this variety of aims can qualify the discussion about auto/biography in the history of science, including the history of physics.

Auto/Biography Is About Individual Persons—not Institutions, Ideas, or Material Things

E. C. Bentley's famous clerihew in *Biography for Beginners* (Bentley 1905), quoted in the epigraph to this chapter, wraps up the definition of 'biography' succinctly: it's about chaps, not about maps, or anything else. While in Bentley's days, the word 'chap' referred to men only, a clerihew-poet of the twenty-first century would have to use a gender-neutral synonym that includes women (and other genders), for example, 'guys' or 'people'. The basic point of Bentley's whimsical verse is still valid, however: biographies are accounts of the lives of persons (in writing, pictures, speech, etc.). Similarly, an autobiography is the account of a person's life written by that very same person.

A person is an individual human being that possesses a number of defining features, such as cognitive abilities, self-consciousness, emotions, memory, morality, etc., and although the precise definition differs across ages and cultures, personhood is invariably attached to individual human beings (Carrithers et al. 1985). Institutions, ideas, material things, etc. are not individual human beings; in other words, university institutions are not persons, ideas are held by persons but are not persons, and things like cars do not have personalities (not even a driverless car). And—with the exception of some mammalian species, such as apes, dogs and perhaps dolphins—neither do animals seem to have personalities (Stamps and Groothuis 2010). As a consequence, institutions, ideas, material things, animal species, and so forth, cannot have their biographies written, unless the meaning of the prefix 'bio' is changed considerably.

Derived from the Greek noun $βίος$—usually translated as 'life' (German *Leben*, Latin *vita*)—it stands for a human mode of life or manner of living, for example in Homer, Aristophanes and Xenophon, or a person's lifetime, for example in Herodotus and Plato (Liddell and Scott 1897) in contrast to an animal life, or bare life ($ζωή$;

cf. the prefix zoo- in zoology). Plutarch even adopted βίος as a synonym for 'biography' in his comparisons between the lives of famous Greeks and Romans (Duff 1999). Traditionally and until recently, the use of the word 'biography' has therefore been restricted to accounts of the lives of individual human persons. In the last decades, however, there has been a growing trend to write about different kinds of non-human entities as if they had a life in the sense of βίος. Thus there are book length 'biographies' of cities, e.g., *Toronto: Biography of a City* (Levine 2015), of nations, e.g., *Australia: A Biography of a Nation* (Knightley 2000), of buildings, e.g., *Hearst Castle: The Biography of a Country House* (Kastner 2000), and of economically valuable animal species, e.g., *Cod: A Biography of the Fish That Changed the World* (Kurlansky 1999). The fact that most of such titles are trade books suggests that the use of the term 'biography' for non-human entities is primarily a marketing gambit—life-histories likely sell better than histories of entities—but it is also used increasingly in non-commercial scholarly publishing. A rapid survey of the literature through Google Scholar reveals the frequent use of phrases like "biography of a road", "biography of a blunder", "biography of an object", "biography of a thing", "biography of a concept", and so forth; the phrase "biography of an idea" alone results in around 1200 hits. Likewise some of the authors in this volume use the term 'biography' for historical accounts of institutions, scientific concepts, and technological objects.

The critical point I wish to make in this section of the chapter is that this proclivity to use the word 'biography' in historical analysis of entities that are neither individual persons nor express any of the features of personhood (consciousness, memory, morality, etc.) is at best the adoption of a superfluous metaphor and at worst a shoddy anthropomorphism.

Is It Meaningful to Speak About the Biography of an Institution?

For example, what does it mean that the history of a research institution, like the Brookhaven National Laboratory (Crease 1999), could be written as a 'biography'? As patterned and regulated collective outcomes of many interacting individuals, institutions are anchored in individual persons, but transcend these individuals by mediating their personal and intentional behavior. Each person can be described in biographical terms, but it is hard to see how the regulated interaction between aggregated individual life-courses can in any meaningful way be called a life-course at a higher organizational level, and accordingly, how an institution could have a 'biography'. The only defensible way to use the notion of 'biography' in histories of institutions without stretching the meaning of 'mode of life' (βίος) too far is to conceptualize institutions as collections of individual biographies, that is, writing the history of the institution as a collective biography (prosopography) (Pyenson 1977;

Werskey 1988; for a recent example of a prosopographical approach to the history of a scientific institution, see Svorenčík 2014).

In the sense of a collective biography, the term 'biography' can thus be defended for writing the history of scientific institutions.

Can Mental Constructs Be the Subject of Biographies?

It is more difficult, however, to see how the use of the term for historical studies of mental constructs, such as ideas, theories, concepts, memes, and so forth—for example the 'biography' of the mass–energy equivalence equation $E = mc^2$ (Bodanis 2000) or the 'biography' of the number zero (Seife 2000)—can be justified. Since these books were written by popular science writers for the general public one could argue that the word 'biography' in the title is just a marketing word, but scholarly authors, too, have employed it for historical accounts of mental constructs. Theodor Arabatzis' *Representing Electrons: A Biographical Approach to Theoretical Entities* is probably the best substantiated case in point. According to Arabatzis, theoretical concepts like the electron are "active participants" in science, they have "personalities" and "lives of their own", they are "born", have an "infancy", undergo "character formation", "gradually reach maturity", and eventually reach "death"—and can therefore "become the subject of biographies" (Arabatzis 2006, Chap. 2).

Surely, throughout human history, persons have entertained, disseminated and adopted ideas and memes, constructed, supported and criticized theories, and proposed, used and rejected concepts; the historical sub-disciplines of intellectual history, history of ideas and history of science are specialized in studying the institutionalized and intricate ways in which humans create, communicate and apply such mental constructs; writing biographies of the individuals involved in these collective mental processes is one of the many methods for this kind of studies. Yet mental constructs are not persons (or assemblages of persons) and do not have any of the properties of personhood; a concept does not literally have consciousness, memory or emotions, and thus does not have a life of its own. Arabatzis' and other historical studies of concepts and theoretical entities can not be called biographical studies in any meaningful way, unless the terms 'life' and 'life course' are defined so broadly that the denotation of 'biography' includes the description and analysis of the change of all kinds of mental constructs over time. But would it add anything to our cultural understanding to speak of 'a biography of Islam' (in contrast to a biography of Mohammed) or 'a biography of post-structuralism' (in contrast to a biography of Michel Foucault)?

Do Things Talk?

In my opinion, the most problematic use of the term 'biography' concerns the historical study of material objects. Drawing more or less explicitly on theoretical trends like actor network theory (Latour 2005), according to which not only humans but also non-humans and inanimate things (actants) have agency, and on works in anthropology that focus on objects themselves, their changing cultural careers and their lives as social markers rather than exclusively on their social functions and the networks surrounding them (Appadurai 1986), there has been an upsurge of attempts to write 'biographies of things'. Science writers and historians of science, technology, and medicine have contributed to this misuse of the notion of 'biography' into the non-human material world, as witnessed by book titles such as *The Microprocessor: A Biography* (Malone 1995), H_2O: *A Biography of Water* (Ball 1999), *Biography of a Germ* (Karlen 2000), *Asthma: The Biography* (Jackson 2009), and *The Emperor of All Maladies: A Biography of Cancer* (Mukherjee 2010). Even more philosophically trained historians of science have contributed to the meme of 'biography' of material objects; for example, Hans-Jörg Rheinberger has used the phrase "biography of things" for the historical analysis of material entities that embody concepts ('epistemic things') (Rheinberger 1997, p. 4), and Lorraine Daston has edited a whole anthology under the rubric of *Biographies of Scientific Objects* (Daston 2000).

With phrases such as 'evocative objects', 'things that talk', and so forth, some authors have even opened up for the implicit possibility of 'autobiographies of things'. In the Introduction to *Things That Talk* (Daston 2004), things do not just have a "life of their own", they also "talk to us". They are "eloquent" and "talkative":

> some things speak irresistibly, and not only by interpretation, projection, and puppetry. It is neither entirely arbitrary nor entirely entailed which objects will become eloquent when, and in what cause. The language of things derives from certain properties of the things themselves, which suit the cultural purposes for which they are enlisted. (Daston 2004, pp. 15, 24)

In the same vein, the organizers of an Austrian workshop in 2008 not only invited participants to bring objects to the meeting; they also arranged sessions where participants were encouraged to argue and discuss with the objects ("mit den Dinge zu argumentieren und diskutieren"), hoping that the objects, too, should have their say in the discussions ("die Dinge gleichsam selbst zu Wort kommen") (Wiener Arbeitsgespräche 2008). And when the German Society for Ethnography met in Berlin later the same year, the organizers not only wished to highlight things and their materiality but also gave things the status of agents and competent language users under the catch-phrase "Die Sprache der Dinge" (The language of things). What less clairvoyant scholars would have called inanimate things were, in the words of these ethnographers, "Handlungsträger und Akteure" (actors), "Vermittler und Übersetzer" (intermediary and translator) and "Produzenten von Bedeutungen, von sozialen Beziehungen und Praktiken, von Identitäten, Wertvorstellungen und Erinnerungen" (producers of meaning, of social relationships and practices, of identities, moral concepts, and memories) (Die Sprache der Dinge 2008). In other words, things were acknowledged to be speakers, actors, mediators, translators and producers of all possible social and cultural meanings. From there it is only a small step to argue that things can produce their own autobiographies and memoirs.

How shall we understand this viral meme that suggests that an object has a life of its own and can talk to us, maybe even tell us the story of its life? It seems unlikely that we are witnessing a collective expression of latter-day fetishism, a revival of the 'primitive' religious practice to attribute powers to inanimate objects, like stones or pieces of wood. Is the meme just bullshit (Frankfurt 2005), or a conceit, as Ludmilla Jordanova suggests in her devastatingly mocking review of *Things That Talk* when she lets her protagonist-thing bluntly end its soliloquy with the words "the idea that [things] talk, isn't that what's called a conceit?" (Jordanova 2006). A more generous interpretation is that it is 'just' a metaphor. Thing-theorists are usually aware of the metaphorical character of their vocabulary, as in the syllabus for a course on "thing theory" at Columbia University which claims that the new field of material culture studies "inverts the longstanding study of how people make things by asking also how things make people, how objects mediate social relationships—ultimately how inanimate objects can be *read as* having a form of subjectivity and agency of their own" (my emphasis) (Fowles 2008). This is a clear case of metaphorical understanding, namely, that intentional human beings read subjectivity, agency and language abilities into things, but that things themselves do not act. In the same way Arabatzis claims that his "biographical approach" is metaphorical only; the main historiographical advantage of this approach, he suggests, is that theoretical entities become explanatory resources:

> to explain the outcome of an episode in which a theoretical entity participated, one has to take into account the entity's contribution (both positive and negative) to the outcome of that episode. (Arabatzis 2006, p. 44)

The key word here is "participate", that is, concepts are seen as "active agents". Yet he does not want to attribute intentionality to concepts, or imply that they have "wishes or other anthropomorphic features"; he distances himself from Latour, "who obliterates completely the difference between human and nonhuman agents" (Arabatzis 2006, p. 46) and claims that he uses the term 'biography' in a metaphorical sense only: "my use of the biography metaphor aims at capturing the active nature of the representation of the electron." Daston, too, seems to agree, at least to begin with: things "do not literally whisper and shout"; but then again, even though she notes that those who are sceptical of talkative things will insist that all this talk is "at best metaphoric", she nevertheless seems to accept such sceptical doubts if only "for the sake of argument", before concluding that "there is still the puzzle of the stubborn persistence of the illusion [that things talk], *if illusion it be*" (Daston 2004, p. 12, my emphasis).

Why Is the 'Biography' Metaphor so Fashionable?

Why are parts of Academia currently obsessed with a vocabulary that suggests that objects are actors, have a life of their own, can think and talk, and can have biographies written of them, and maybe even write their own autobiographies? A possible answer (Söderqvist and Bencard 2010) is that the metaphorical phraseology that permeates the writing about 'biography of things' and 'things that talk' is a consequence of the persistence of the linguistic turn in the humanities. Terry Eagleton notes that the theoretical interest in the body during the 1980s and 1990s was a way of "having one's deconstructive cake and eating it too" (Eagleton 1998, p. 158); books on the history and culture of the body made the students wriggle under the emotional effects of reading about sex, death, torture and medicine, while at the same time explaining such effects away into the mists of language and cultural constructions; like Judith Butler, who addresses the biological materiality of the body and sex, only to translate it into a subset of problems about language and discourse (Butler 1993). The materiality of material bodies and things is both acknowledged and explained away. This linguistic turn continues unabated.

The current 'things that talk'- and 'biography-of-things'-vocabulary may thus be an expression of a wish to pay attention to the 'thingness' of things and yet keep one's language-centred approach to material culture intact. To allow things to become actors or actants with an uncanny ability to speak to us, can be seen as a license to maintain the set of scholarly tools and languages associated with the linguistic and cultural turns in the humanities, while still doing something apparently new. By suggesting that things have a life and can talk to us, scholars can maintain institutionally and traditionally enshrined ideas, while seemingly engaging with a new agenda. Rather than exploring the presence and effects of things *qua* things, things are turned into something which we, as academics trained in a discursive and cultural constructivist tradition, can relate to immediately. It is business as usual on a new subject matter, which still holds out the promise of being something different.

The Many Aims of Scientific Auto/Biography

Ever since Thomas Hankins' seminal article "In defence of biography" forty years ago, discussions about scientific biography have revolved around its usefulness for the writing of history of science. Hankins saw biography as a narrative about individual scientists that could shed light on the history at the macro-level: "We have, in the case of an individual, his scientific, philosophical, social and political ideas wrapped up in a single package" (Hankins 1979, p. 5). Since then scientific biography has become an increasingly acknowledged accepted subgenre of history of science. Several collected volumes (Shortland and Yeo 1996; Söderqvist 2007a) and special journal issues—for example on "Biography as cultural history of science" in the journal *Isis* in 2006—have been devoted to reflections about the genre. No serious historian of science today rejects the genre of biography out of hand.

Auto/Biography as an Ancilla Historiae

The acknowledgement of scientific biography is almost always confined, however, to it being a part of the historian's toolbox. To paraphrase Thomas Aquinas, who famously relegated philosophy to being an *ancilla theologiae* (a handmaid to theology; cf. van Nieuwenhove and Wawrykow 2005), scientific biography has acquired the identity of a handmaid of history of science—it is usually limited to being an *ancilla historiae* (Söderqvist 2007c, p. 255ff).

The lack of systematic reflections on scientific autobiographies and memoirs seems to suggest that self-life-writing has not been accepted by historians of science to the same degree as biography has. So far, no history of science journal has published a focus issue on autobiography, nor has the subgenre been the subject of a collected volume. One possible reason for this reluctance may be that autobiographies and memoirs are considered too subjective to count as serious historical research; this can, at least partly, explain the lack of attention, but does not justify the oblivious attitude to the subgenre among historians of science. After all, first-person accounts are a standard ingredient in mundane historical practices, and historians and biographers usually realize that bias and subjectivity is a matter of degree; few would claim that their texts are fully objective and free from ideological or other biases and interests. The alleged subjectivity of autobiographies and memoirs is thus just a matter of degree. Even though autobiographies and memoirs are often written from the standpoint of the author's interest to set the records straight and emphasize his/her importance, the historical factual matter is still, at least in principle, more or less verifiable. Both historians of science and scientific biographers rely more or less heavily on autobiographies and memoirs, or other pieces of self-writing, such as diaries, as source material, especially for events that have not generated other independent sources, thereby lending credibility to autobiographies and memoirs in the history of science.

Another argument in favour of paying more interest to autobiographies and memoirs in the history of science is that the voices of scientists, their first-person opinion about themselves and their colleagues, and the events they have experienced along their careers, are in themselves interesting aspects of the past. Scientific objects, theories, concepts and practices, social relations, institutions, and so forth are ordinary elements of the subject matter of history of science, but so are individual scientists and their personal opinions about themselves, their life trajectories and more or less idiosyncratic views of the world around them. Why should the views, opinions, self-understanding, and memories of individual scientists not be an integral part of the subject matter of history of science? Even if these views, opinions and memories can be unreliable sources for a more detached history of scientific institutions and practices etc., they are still part of the reality of the past. Thus scientific autobiographies and memoirs are part and parcel of the history of science.

But biography, autobiography and memoirs are more than an *ancillae historiae*. I think the distinction already made by Plutarch and other classical authors between βίος and ίστορία as two distinct ways of writing about the past (Momigliano 1971) is still valid (Söderqvist 2007b). History (ίστορία) originally meant 'an inquiry', but in the course of time such inquiries became restricted to historical studies of nations, classes, economic institutions, political movements, social interactions, cultural phenomena, etc., while βίος meant 'a life' in the sense of 'an individual life course' (cf. above). The classical distinction between βίος and ίστορία remains instructive for today's discussions about the uses of scientific biography. Even though most historians of science today think of scientific auto/biography as a handmaid of history, writings about the lives of scientists have other, and more independent, roles to play (Söderqvist 2006; Nye 2006). In the following, I extend my earlier typological analysis (Söderqvist 2011) of ideal-typical subgenres of scientific biography to include autobiographies and memoirs.

Auto/Biography as Case-Study of Scientific Work

Biography has been a preferred format for understanding the origin and construction of experimental findings, concepts, theories, and innovations. The idea is that scientific results should be understood, not primarily with reference to social, political or cultural circumstances, but with reference to individuals, their mental states and actions, such as motivations, ambitions, ideas, feelings, personality traits and personal experiences. One of the major motivations for writing about the life and work of individual scientists has actually been to understand science as a primarily individual achievement. This is not something particular to the historiography of science, but a methodology which historians of science share with literary historians, art historians, historians of music, and other historians of cultural artefacts. One of the most impressive examples is Frederick Holmes' fine-grained account in two volumes of how biochemist Hans Krebs came to the understanding of the citric acid cycle in the 1930s: relying on his subject's daily laboratory notebooks and many hours of interviews, Holmes follows the interaction between daily bench-work and biochemical ideas (Holmes 1991, 1993); this is 'science-in-the making' in painstaking detail.

Using life-writing to understand the development and psychological basis for creative work has its parallel in autobiography as well. Among contemporary writers, King's (2000) stands out as one of the best introspective observations of the creative process of a contemporary novelist. Most autobiographies of scientists contain elements of reflections on the creative process; a brilliant example is French molecular biologist François Jacob, who gives the reader a first-hand introspective insight into the thinking and passion behind his scientific work in *La statue intérieure*

(1987). The history of scientific work and creativity would benefit from more systematic introspective case-studies along these lines: but a book-length autobiographical counterpart to Holmes' detailed study of Krebs is still due.

Auto/Biography as Public Understanding of Science

Scientific biography is often used as a vehicle for popular science. One of the standard overviews of public understanding of science (Gregory and Miller 1998) covers books and magazines, mass media, museums, etc., but makes no reference to biography; likewise the *Routledge Handbook of Public Communication of Science and Technology* (Bucchi and Trench 2008) fails to include biography. These are amazing omissions given the fact that most scientific biographies have been written for a general public to create enthusiasm for science. British publishers like Longmans-Green, John Murray, and Macmillan poured out popular biographies about scientists around the turn of the last century, and some of the most impressive publications efforts were made in the German language area in the first half of the twentieth century with series such as "Grosse Männer" (Great Men) and "Große Naturforscher" (Great Scientists); likewise in the 1950s and 1960s the East German publisher Teubner issued hundreds of titles of popular biographies in the series "Biographien hervorragender Naturwissenschaftler, Techniker und Mediziner" (Biographies of Outstanding Scientists, Engineers and Physicians). Although few of them had scholarly ambitions, most were nevertheless based on earlier scholarly work. In fact, even scholarly scientific biographies have often taken the general educated audience into consideration. From the perspective of the authors and reviewers scientific biographies are seen as contributions to the history of science, but from the perspective of the publishers and readers they are also viewed as contributions to the public understanding of science; thus most scientific biographies occupy a broad middle ground between narrow scholarly history of science and popular understanding of science.

Autobiographies and memoirs, too, contribute to the public engagement with science and the history of science; in the same way as biographies make the history of science more appetizing to general readers by emphasizing the personal dimension of scientific practice, autobiographies and memoirs make history more approachable for the general reader. The first-person narrative voice is a traditional rhetorical device for creating emotional bonds between authors and readers, making the readers empathize with the lot of the author, and guiding them to see the world through the eyes of the author. Although it is difficult to quantify their impact on the public understanding of science, memoirs like Watson's (1968) and Feynman's (1985) became immediate bestsellers and have repeatedly been published in new editions and reprints. Similarly, the widespread positive reviews of Stephen Hawking's short autobiography *My Brief History* (2013) have undoubtedly contributed to the public interest in cosmology. Following the discovery of the structure of DNA through the eyes of Watson and the rise of quantum electrodynamics through the eyes of Feynman himself, or understanding the structure of black holes through the mind of

Hawking is a form of scientific *Bildung* (education), which can be compared to how medieval Christians understood God through the eyes of Saint Augustin of Hippo when reading *Confessions* (Augustin 2017).

Auto/Biography as Literature

A fourth subgenre of scientific auto/biography verges on literary biography. Although scientific biographies are probably rarely written primary for literary and aesthetic purposes, life-writing is nevertheless a genre in which literary features play a major role. In today's publishing world it is common knowledge that readers tend to choose biographies as substitutes for novels. Historians of science may be excused for mediocre writing skills if they dig up previously unknown archival material or construct new and interesting interpretations and explanations, but biographers of scientists can hardly get away with a lack of care for the literary qualities; it is difficult to imagine that a scientific biography that is a middling read becomes successful. Scientific biographies rarely match the highest literary standards of the biographical genre, but there are some good exceptions, for example, Janet Browne's two volumes on Darwin (Browne 1995, 2002), which received the History of Science Society's Pfizer Prize as well as two literary prizes: the National Book Critics Circle Award and the James Tait Black Award. Yet historians of science tend to underestimate such literary qualities as being just an extra bonus on top of the allegedly more important historical functions of the genre; accordingly the overlap between scientific biography and literature biography remains unacknowledged in the metabiographical literature. Maybe reviewers of scientific biographies are partly to blame for this ignorance of the literary aspects because they rarely mention the composition, style, or other aesthetic qualities of the book under review.

Autobiographies and memoirs are more frequently read and reviewed for their aesthetic qualities. Novelists have produced memoirs of high literary standards, such as Thoreau's (1854), Orwell's (1938), and Joan Didion's *The Year of Magical Thinking* (2005). Knausgård's *Min kamp* (*My Struggle*), published in six volumes 2009–2011, has set new standards for autobiographical novels. Yet there are only few examples of this kind of literary autobiography in the history of science. Franklin (1791) still stands out as one of the most well-written self-accounts of a scientist-engineer; Jacob's *La statue intérieure* gives not only a unique insight into the formation of a scientific mind, but is also a work of high literary quality. But Franklin's and Jacob's memoirs are rather exceptions than the rule; indeed the biography section in science libraries are filled with self-congratulatory and badly written autobiographies that often degenerate into mere listings of events and achievements. Readers of scientific memoirs are therefore looking forward to a Knausgård of scientific autobiography who will be able to win both a professional history of science award and a prestigious literary award.

Auto/Biography as (Self)Eulogy

To pay one's respect to a deceased person with 'good language' ($ε\dot{υ}λογ\acute{ι}α$) is the oldest use of biography and the function of the first *vitae* of natural philosophers in the seventeenth century (Söderqvist 2007c), and has remained a strong aspect of the genre of scientific biography. Most historians of science regard such explicit eulogistic aims as an embarrassing phenomenon of the past, which today are produced only at the margins of history of science by amateurs and scientists, who write about their heroes in scientific journals. But eulogistic commemoration is not at all absent from mainstream history of science and scientific biography; historians of science only need to look at their own practice of publishing praises of deceased famous members of their own profession to realize that the eulogistic tradition is strongly ingrained in the profession. Likewise the earlier tradition of writing eulogies for nationalistic purposes has given way to biographies written for gender or ethnic identity political reasons, for example, Linda Lear's hagiographical account of the famous biologist and conservationist Rachel Carson (Lear 1997) and Georgina Ferry's unashamedly eulogistic biography of biochemist Dorothy Hodgkin (Ferry 1998). Thus the eulogistic impulse as such has not disappeared from history of science and scientific biography, it has just changed focus: from 'dead white men' to women, ethnic minorities, and members of one's own profession.

The situation is quite different when it comes to autobiography and memoirs. Self-writing is still to a large extent characterized by eulogistic behavior (although they do not express 'good words' about another person, but about oneself, i.e., auto-eulogy). More often than not, scientist's autobiographies are self-congratulatory, smug and complacent textual selfies, which focus on the great achievements of its author, on accolades, prizes, important keynotes, prestigious grants and awards, highly cited publications in high-ranking journals, promotions to full professorships, election to academies—in other words narratives of professional success, in which failures and disappointments are passed over in silence, and spouses and children are mere decorations on the main theme.

The most common self-congratulatory autobiographical kind of text among scientists is the curriculum vitae (literally 'life's race'), a feature in the life of scientists, which so far has not been the subject of study from the side of historians or sociologists of science. As appendices to job applications and grant proposals and put on the web for the public gaze, the CV is continuously upgraded throughout a scientist's career. Scientists are thus well honed in writing in a complacent autobiographical mode throughout their whole career, and much autobiographical writing can thus be understood as a continuation and enlargement of the curriculum vitae. When retired scientists transmogrify into emeriti, they no longer have any need for updating their formal CV, but many of them still wish to look back on their careers in order to explain, display and legitimize their work and achievements. The scientific autobiography is the ultimate curriculum vitae.

Existential Auto/Biography

The ideal-typical subgenre of scientific biography in this exposé is that which Keynes' biographer Robert Skidelsky called "a new biographical territory, still largely unexplored": the story of "the life, rather than the deeds, the achievement" (Skidelsky 1988, p. 14), a form of life-writing that takes "us out of our old selves by the power of strangeness, to aid us in becoming new beings" (Skidelsky 1987, p. 1250). I call this type of biography 'edifying' and 'existential' (Söderqvist 1996, 2003a) with an eye to the use of biography that was founded by Plutarch in the *Parallel Lives* (Duff 1999). In the Plutarchian virtue-ethical tradition, biographies of scientists are written and read to explore the question: How to live a life in science in a good way? (Söderqvist 2001, 2003b). The subgenre also rests, implicitly, on the long philosophical tradition highlighted by the classical philologist Pierre Hadot, viz., the pronounced difference between philosophical practice as discourse on theories and conceptual systems, and philosophy as a mode of life based on the classical maxim $\gamma\nu\tilde{\omega}\theta\iota\ \sigma\varepsilon\alpha\upsilon\tau\acute{o}\nu$ (*nosce te ipsum*, know thyself) and Socrates' recommendation, in Plato's *Apology*, that the unexamined life is not worth living (Hadot 1981). Arguing that modern academic philosophy has largely gone astray in its attempt to objectify (externalize) its object of study, Hadot suggests that it should be more concerned about how its practice influences its practitioners. In Hadot's analysis, philosophy in the broad sense (that is, including the humanities and history) has always basically been a kind of intellectual self-therapy, a means for 'knowing oneself' or a care of self (*souci de soi*); a reading of the classical philosophers that had a seminal influence on the thinking of the late Michel Foucault and the third volume of his history of sexuality, subtitled *Le souci de soi* (Foucault 1984).

I think Hadot's argument for philosophy is applicable to scientific practice as well. One could say that it is a good and admirable thing to do science in order to understand the physical world, but another, and equally good and venerable thing, to be a scientist as a special mode of life. The same reasoning is also applicable to the history of science; it is a good thing to understand the history of, say, physics, but another, and equally good thing, to study the history of physics as a way of practicing *souci de soi*. Similarly, one could argue that it is a good thing to write about recent scientists in order to understand their work and their lives, but it is an equally good thing to write about them as a way of practicing the care of one's own scholarly self. Writing the history of science or $\beta\acute{\iota}o\iota$ of contemporary scientists are thus practices by which historians, biographers, and scientists can explore the perennial question of how to craft a worthwhile life-course out of talent and circumstances. Historians and biographers of science produce books, articles, lectures, etc., but from the point of view of the *souci de soi*-tradition, this is not the ultimate purpose of scholarship; according to Hadot, the basic aim of all humanistic writing is rather "to effect a modification and a transformation in the subjects who practice them" (Hadot 2002, p. 6).

The subgenre of existential and edifying biography described here has its counterpart in autobiographies and memoirs that aim to help their authors and readers to live better lives and prepare them for the inevitable death. This tradition for writing autobiography as an art of life (*Lebenskunst, ars vivendi*) and art of dying (*Kunst des Sterbens, ars moriendi*) can be traced back to classical antiquity too. In addition to the idea of 'know thyself' and 'care of self' mentioned above (Hadot 1981, 2002)—where the aim of autobiography and memoir writing is not to contribute to history, or understand the psychology of scientific creativity, or write well, or produce the final curriculum vitae and self-eulogy of one's life, but to undergo a personal transformation in the process of writing it—there is also a strand of existential autobiography which goes back to Augustine's *Confessions*, in which the church father portrays himself as a thief, a liar, and a lustful, adulterous sinner until his conversion to the Christian faith (Augustin 2017); as a guide to introspection for both religious and secular people, confessional autobiography has remained a paradigm for autobiographical writing for almost 1500 years, and is still reprinted and emulated, although today's confessional autobiographical writers are probably motivated more by a secular desire to shock their readers (Morrison 2015). A third strand of existential and edifying introspective autobiographical writing is the early fifteenth century *ars moriendi* (the art of dying) manuals which were written as instructions for one should deal with the last period before death; it was followed by a tradition of writing and reading death manuals throughout the following centuries, and has recently got the attention of scholars in the medical humanities (Leget 2007).

So far, none of these strands of existential autobiography has found its well-established practitioners among scientific memoirists. There are a few attempts: for example, *Surely You're Joking Mr Feynman!* (Feynman 1985) has some amusing passages with personal confessions, and the psychologist and notorious scientific fraudster Diederik Stapel does some apparently honest soul-searching in his attempt to atone for his massive fabrications of research data (Stapel 2012). But no truly confessional autobiography of an entire scientific career has yet been published. Similarly, to my best knowledge, no scientist in modern times has written an autobiography in the spirit of *souci de soi* or broadened the notion of *ars moriendi* to cover the whole scientific career. Thus, scientific autobiographers and memoirists still have some exciting and yet unexplored avenues to thread.

Conclusion

I have discussed two major aspects of the relation between the genre of biography and history of science (including history of physics). First, I analyzed what falls inside and outside of the genre; more specifically, whether the use of the word 'biography' for historical studies of scientific institutions, theoretical entities, and material objects is justified. My conclusion is that the notion of biography should be limited to accounts of the life courses of individual persons and avoided as an alternative term for histories of institutions, concepts, and objects. Then—after reminding the reader about the significance of autobiography and memoirs—I identified a number of kinds of scientific auto/biographies, thereby making the point that life-writing is not merely an aid to history of science (an *ancilla historiae*) but also has many other aims, and that an awareness of these can hopefully make future discussions about the relation between scientific auto/biography and the history of science more varied and interesting.

In other words, I believe that further discussions about scientific auto/biography and the history of science would benefit from a cognitive process of simultaneous restriction and expansion of the notion of biography. I suggest that the extension (denotation), i.e., the phenomena to which the notion can be applied, should be restricted to human life courses in order to avoid scholarly confusion. Vice versa, the restriction of the extension of the notion should go hand in hand with an expansion of its intension (connotation), i.e., its properties and qualities, in order to increase its conceptual richness. What is needed is a much sharper and simultaneously richer notion of what scientific auto/biography is and can do.

Acknowledgements I am grateful to Annelie Drakman (Uppsala University), Richard Staley (University of Cambridge), and Mark Walker (Union College) for comments on a draft manuscript, and to the participants in the workshop Biographies in the History of Physics: Actors, Institutions, and Objects at Physikzentrum, Bad Honnef, Germany, 22–25 May 2018, for lively discussions.

References

Appadurai, A. (Ed.). (1986). *The social life of things: commodities in cultural perspective*. Cambridge: Cambridge University Press.
Arabatzis, T. (2006). *Representing electrons: a biographical approach to theoretical entities*. Chicago: University of Chicago Press.
Augustin. 2017. *Confessions* (S. Ruden, Trans.). New York: Modern Library.
Ball, P. (1999). H_2O: *A biography of water*. London: Weidenfeld & Nicolson.
Bentley, Edmund Clarihew. (1905). *Biography for beginners*. London: T. Werner Laurie.
Bodanis, David. (2000). $E = mc^2$: *A biography of the world's most famous equation*. New York: Walker.
Browne, J. (1995). *Charles Darwin, vol. 1: Voyaging*. London: Jonathan Cape.
Browne, J. (2002). *Charles Darwin, vol 2: The power of place*. London: Jonathan Cape.
Bucchi, M., & Trench, B. (Eds.). (2008). *Handbook of public communication of science and technology*. London: Routledge.
Butler, J. (1993). *Bodies that matter: On the discursive limits of 'sex'*. New York: Routledge.
Carrithers, M., Collins, S., & Lukes, S. (Eds.). (1985). *The category of the person: Anthropology, philosophy, history*. Cambridge, UK: Cambridge University Press.
Crease, R. P. (1999). *Making physics: A biography of Brookhaven National Laboratory, 1946–1972* (p. 1999). London: University of London Press.
Daston, L. (Ed.). (2000). *Biographies of scientific objects*. Chicago: The University of Chicago Press.
Daston, L. (Ed.). (2004). *Things that talk: Object lessons from art and science*. New York: Zone Books.
Didion, J. (2005). *The year of magical thinking*. London: Fourth Estate.
Die Sprache der Dinge. (2010). kulturwissenschaftliche Perspektiven auf die materielle Kultur, Berlin, 21–22 November. In *H-Soz-Kult Newsletter*, 19 September 2008, www.hsozkult.de/event/id/termine-9883. Accessed 6 May 2010.

Duff, T. (1999). *Plutarch's lives: Exploring virtue and vice.* Oxford: Clarendon Press.
Eagleton, T. (1998). Body work. In S. Regan (Ed.), *The Eagleton reader* (pp. 157–162). Oxford: Blackwell.
Ferry, G. (1998). *Dorothy Hodgkin: A life.* London: Granta Books.
Feynman, R. (1985). *Surely you're joking, Mr. Feynman!: Adventures of a curious character.* In R. Leighton, & E. Hutchings (Eds.). New York: W.W. Norton.
Foucault, M. (1984). *Histoire de la sexualité, tome 3: le souci de soi.* Paris: Gallimard. English edition: Foucault, M. 1990. *The history of sexuality, volume 3: the care of the self* (R. Hurley, Trans.). London: Penguin Books.
Fowles, S. (2008). Thing theory course description, Columbia University. http://www.columbia.edu/~sf2220/TT2008/web-content/Pages/syllabus.html. Accessed 13 November 2018.
Frankfurt, H. G. (2005). *On bullshit.* Princeton, NJ: Princeton University Press.
Franklin, B. (1791). *Mémoires de la vie privée de Benjamin Franklin, écrits par lui-même, et adressées à son fils.* Paris (Many English editions.).
Gregory, J., & Miller, S. (1998). *Science in public: Communication, culture, and credibility.* London: Plenum Trade.
Hadot, P. (1981). Exercices spirituels et philosophie antique. Paris: Études augustiniennes. English edition: Hadot, P. 1995. *Philosophy as a way of life: spiritual exercises from Socrates to Foucault* (Ed: A. I. Davidsson, & M. Chase, Trans.). Oxford: Blackwell's.
Hadot, P. (2002). *What is ancient philosophy?.* Cambridge, MA: The Belknap Press.
Hankins, T. (1979). In defence of biography: The use of biography in the history of science. *History of Science, 17,* 1–16.
Hawking, S. (2013). *My brief history.* New York: Bantam Books.
Holmes, F. L. (1991). *Hans Krebs: The formation of a scientific life 1900–1933* (Vol. 1). New York: Oxford University Press.
Holmes, F. L. (1993). *Hans Krebs: architect of intermediary metabolism 1933-1937* (Vol. 2). New York: Oxford University Press.
Jackson, M. (2009). *Asthma: The biography.* Oxford: Oxford University Press.
Jacob, F. (1987). *La statue intérieure.* Paris: Odile Jacob.
Jardine, N. (2003). Whigs and stories: Herbert Butterfield and the historiography of science. *History of Science, 41,* 125–140.
Jordanova, L. (2006). Review of Lorraine Daston (ed.), Things that talk, 2004. *British Journal for the History of Science, 39,* 436–437.
Karlen, A. (2000). *Biography of a germ.* New York: Pantheon Books.
Kastner, V. (2000). *Hearst Castle: The biography of a country house.* New York, London: Harry N. Abrams.
King, S. (2000). *On writing: A memoir of the craft.* New York: Scribner.
Knausgård, K. O. (2009–2011). *Min kamp,* 6 vols. Oslo: Oktober. English edition: K. O. Knausgaard, 2014–2018. *My struggle* (D. Bartlett, Trans.). London: Penguin Random House.
Knightley, P. (2000). *Australia: A biography of a nation.* London: Jonathan Cape.
Kurlansky, M. (1999). *Cod: A biography of the fish that changed the world.* London: Vintage.
Latour, B. (2005). *Reassembling the social: An introduction to actor-network-theory.* Oxford: Oxford University Press.
Lear, L. (1997). *Rachel Carson: Witness for nature.* London: Allen Lane.
Leget, C. (2007). Retrieving the ars moriendi tradition. *Medicine, Health Care and Philosophy, 10,* 313–319.
Levine, A. G. (2015). *Toronto: Biography of a city.* Madeira Park: Douglas & McIntyre.
Liddell, H. G., & Scott, R. (1897). *A Greek-English lexicon.* Oxford: Clarendon Press.
Malone, M. S. (1995). *The microprocessor: A biography.* Santa Clara, CA: Telos.
Momigliano, A. (1971). *The development of Greek biography.* Harvard: Harvard University Press.
Morrison, B. (2015). The worst thing I ever did: The contemporary confessional memoir. In Zachary Leader (Ed.), *On life-writing.* Oxford: Oxford University Press.
Mukherjee, S. (2010). *The emperor of all maladies: A biography of cancer.* New York: Scribner.

Nye, M. J. (2006). Scientific biography: History of science by another means? *Isis, 97,* 322–329.
Orwell, G. (1938). *Homage to Catalonia.* London: Secker & Warburg.
Outram, D. (1996). Life-paths: Autobiography, science and the French Revolution. In M. Shortland & R. Yeo (Eds.), *Telling lives in science: Essays on scientific biography* (pp. 85–102). Cambridge: Cambridge University Press.
Pyenson, Lewis. (1977). 'Who the guys were': Prosopography in the history of science. *History of Science, 15,* 155–188.
Rheinberger, H. J. (1997). *Toward a history of epistemic things: Synthesizing proteins in the test tube.* Stanford, CA: Stanford University Press.
Seife, C. (2000). *Zero: The biography of a dangerous idea.* London: Souvenir.
Selya, R. (2007). Primary suspects: Reflections on autobiography and life stories in the history of molecular biology. In T. Söderqvist (Ed.), *History and poetics of scientific biography* (pp. 199–206). Aldershot: Ashgate.
Shortland, M., & Yeo, R. (Eds.). (1996). *Telling lives in science: Essays on scientific biography.* Cambridge: Cambridge University Press.
Skidelsky, R. (1987). Exemplary Lives. *Times Literary Supplement,* 13–19 November.
Skidelsky, R. (1988). Only connect: Biography and truth. In E. Homberger & J. Charmey (Eds.), *The troubled face of biography* (pp. 1–16). London: Macmillan.
Söderqvist, T. (1996). Existential projects and existential choice in science: science biography as an edifying genre. In R. Yeo & M. Shortland (Eds.), *Telling lives: Studies of scientific biography* (pp. 45–84). Cambridge: Cambridge University Press.
Söderqvist, T. (2001). Immunology à la Plutarch: biographies of immunologists as an ethical genre. In A. M. Moulin, & A. Cambrosio (Eds.), *Singular selves: Historical issues and contemporary debates in immunology,* pp. 287–301. Paris: Elsevier.
Söderqvist, T. (2003a). *Science as autobiography: The troubled life of Niels Jerne.* New Haven: Yale University Press.
Söderqvist, T. (2003b). Wissenschaftsgeschichte á la Plutarch: Biographie über Wissenschaftler als tugendethische Gattung. In H. E. Bödeker (Ed.), *Biographie schreiben* (pp. 287–325). Göttingen: Wallstein Verlag.
Söderqvist, T. (2006). What is the use of writing lives of recent scientists? In R. E. Doel & T. Söderqvist (Eds.), *The historiography of contemporary science, technology, and medicine: Writing recent science* (pp. 99–127). London: Routledge.
Söderqvist, T. (Ed.). (2007a). *The history and poetics of scientific biography.* Aldershot: Ashgate.
Söderqvist, T. (2007b). Plutarchian versus Socratic scientific biography. In J. Renn & K. Gavroglu (Eds.), *Positioning the history of science* (pp. 159–162). Berlin: Springer.
Söderqvist, T. (2007c). 'No genre of history fell under more odium than that of biography': The delicate relations between twentieth century scientific biography and historiography of science. In T. Söderqvist (Ed.), *The history and poetics of scientific biography* (pp. 242–262). Aldershot: Ashgate.
Söderqvist, T. (2011). The seven sisters: Subgenres of βίος of contemporary life scientists. *Journal of the History of Biology, 44,* 633–650.
Söderqvist, T., & Bencard, A. (2010). Do things talk? In H. Trischler, et al. (Eds.), *The exhibition as product and generator of knowledge* (pp. 92–102). Berlin: Max Planck Institute for History of Science.
Stamps, J., & Groothuis, T. G. (2010). The development of animal personality: Relevance, concepts and perspectives. *Biological Reviews, 85,* 301–325.
Stapel, D. (2012). *Ontsporing.* Amsterdam: Prometheus Books. English edition: D. Stapel, 2014. *Faking science: a true story of academic fraud* (trans: Brown, Nicholas). https://errorstatistics.files.wordpress.com/2014/12/fakingscience-20141214.pdf. Accessed 26 November 2018.
Svorenčík, A. (2014). MIT's rise to prominence: Outline of a collective biography. *History of Political Economy, 46*(suppl. 1), 109–133.
Thoreau, H. D. (1854). *Walden; or life in the woods.* Boston.

Van Nieuwenhove, R., & Wawrykow, J. (Eds.) (2005). *The theology of Thomas Aquinas*. Notre Dame: University of Notre Dame Press.
Watson, J. D. (1968). *Double helix: A personal account of the discovery of the structure of DNA*. New York: Atheneum.
Werskey, G. (1988). *The visible college: A collective biography of British scientists and socialists of the 1930s*. London: Free Association.
Wiener Arbeitsgespräche zur Kulturwissenschaft, Vienna 2008. http://www.trafik.or.at. Accessed May 6, 2010.

Index

A
Abbe, Ernst, 14
Adenauer, Konrad, 125, 126, 175
Airy, George, 222, 228–231
Andrade, Edward Neville da Costa, 292
Apel, Friedrich, 24, 28
Apel, Wilhelm, 25
Aquinas, Thomas, 308
Arabatzis, Theodore, 203, 262, 276, 283, 304, 306
Aristophanes, 302
Aristotle, 129
Arrhenius, Svante, 47
Asselmeyer, Fritz, 94

B
Bacher, Franz, 115
Becker, August, 25, 26
Benninghoff, Alfred, 101
Bentley, E. C., 302
Bergdolt, Ernst, 101
Bergmann, Peter, 188, 190, 194
Bernardi, Enrico, 252
Bessel, Friedrich Wilhelm, 21, 229, 230
Bethe, Hans Albrecht, 151, 152, 262, 270
Betz, Albert, 80, 82
Beurlen, Karl, 101
Beyler, Richard, 112
Bieberbach, Ludwig, 101
Blasius, Heinrich, 78
Blum, Alexander, 180
Boccioni, Umberto, 289
Boepple, Ernst, 90
Bohr, Margarethe, 176
Bohr, Niels, 69, 118, 121, 122, 164–166, 174, 176
Boltzman, Stephan, 148
Bondi, Hermann, 185, 186, 190–194
Bonhoeffer, Dietrich, 132
Born, Max, 67, 112, 115, 123, 125, 134, 266
Bosch, Carl, 163
Bothe, Walther, 170
Boyle, Robert, 20
Bradley, James, 229, 230, 232
Bragg, William H., 295
Bragg, William Lawrence, 81
Braginsky, Vladimir, 195
Brewster, David, 16
Brooks, Harriet, 48
Browne, Janet, 311
Browning, Robert, 37
Bühl, Alfons, 97
Butler, Judith, 307

C
Cabrera, Blas, 61–65, 69, 70
Cabrera, Nicolás, 70
Cantor, Geoffrey, 283
Carl, Philipp, 15
Carmeli, Moshe, 190
Carnap, Rudolf, 118
Carpenter, James, 226
Carr, Frank, 227
Carr, Wildon H., 296
Carson, Cathryn, 270
Carson, Rachel, 312
Cassidy, David, 213, 215
Catalán, Miguel, 63, 69
Chang, Hasok, 282, 283, 297

Chisholm, Grace, 43
Clay, Jacob, 125
Clegg, Samuel, 237, 249
Cloud, John Wills, 288
Compton, Arthur Holly, 175
Corrigan, J.F., 295
Crookes, William, 285
Cross, Charles, 41
Curie, Marie, 66

D

Dahn, Ryan, 112
Darwin, Charles, 311
Daston, Lorraine, 223, 305, 306
Davis, Natalie Z., 16
Davy, Sir Humphrey, 242
Debye, Peter, 271
Delambre, Jean Baptiste Joseph, 230
Descartes, René, 59, 284
De Unamuno, Miguel, 59
DeWitt, Bryce, 192
Dickens, Charles, 222
Dicke, Robert, 186
Didion, Joan, 311
Diederichs, Carl, 28
Dieke, Gerhard, 268
Dingler, Hugo, 98, 101, 102, 105, 116–119
Dirac, Paul A.M., 69, 149, 150, 262–272, 276, 282
Domeier, Ernst, 112
Drude, Paul, 291
Dubček, Alexander, 191
Duc de Chaulnes, 18
Duchamp, Marcel, 289
Dukas, Helen, 125
Dunkin, Edwin, 226
Dunnington, Waldo G., 21
Dyson, Freeman, 145, 153, 275

E

Eagleton, Terry, 307
Eddington, Arthur, 66, 289–291, 296
Einstein, Albert, 62, 64–67, 71, 99, 122, 124, 125, 129, 130, 135–137, 139, 153, 154, 166, 179, 180, 182, 188, 289–292, 295
Eisenstaedt, Jean, 179
Engels, Friedrich, 137, 138, 141
Euler, Hans, 262, 269, 272, 276

F

Faraday, Michael, 14
Fermi, Enrico, 167, 170, 175, 262, 263, 270
Ferry, Georgina, 312
Feynman, Richard, 145–147, 149–157, 261, 262, 273–276, 310
Finkelnburg, Wolfgang, 97
Finlay-Freundlich, Erwin, 139
FitzGerald, George, 286, 296
Fleck, Ludwik, 147
Fleming, John Ambrose, 294
Flexner, Abraham, 64
Flügge, Siegfried, 164, 165
Fock, Vladimir, 184, 190, 192, 195
Föppl, August, 76
Forbes, George, 226
Foucault, Michel, 313
Fowler, Alfred, 65
Fox, Tobias, 261
Franco, Francisco, 64
Franklin, Benjamin, 237, 238, 240–242, 244, 250, 252, 254, 311
Frank, Nathaniel H., 149
Frank, Philipp, 118
Frank, Walther, 98
Frayn, Michael, 176
Frenkel, Yakov Illich, 265, 266
Freundlich, Erwin, 67
Frick, Wilhelm, 163, 164
Frisch, Otto, 164, 166
Fry, Roger, 38
Fuhrmann, Georg, 79

G

Gaßmann, Heinrich, 30
Galileo, Galilei, 21
Galison, Peter, 206
Gauß, Carl Friedrich, 14, 19–22, 28, 29
Gerlach, Walther, 95, 121, 171
Gill, Jane, 53
Gill, Laura, 48
Gill, Raymond, 50, 53
Gil Santiago, Eduardo, 69, 70
Glaser, Ludwig, 102
Gleick, James, 154
Goebbels, Josef, 171
Goeppert-Mayer, Maria, 266, 276
Goethe, Johann Wolfgang, 59
Göring, Hermann, 77, 82, 84, 85
Goudsmit, Samuel, 172, 173
Grimsehl, Ernst, 98, 102
Grunder, Hermann, 204–206, 208, 210, 212, 214, 215

Index

H
Hadot, Pierre, 313
Hager, Kurt, 134
Hahn, Hans Peter, 277
Hahn, Otto, 121, 163, 164, 168, 174
Halley, Edmond, 232
Halpern, Otto, 271
Hankins, Thomas, 307
Harich, Wolfgang, 134
Harig, Gerhard, 134
Harteck, Paul, 168, 171
Hasse, Helmut, 120, 121
Havemann, Robert, 134
Hawking, Stephen, 310, 311
Heaviside, Oliver, 287
Heidegger, Martin, 65
Heilbron, John, 204
Heilmann, Peter, 134
Heisenberg, Werner, 80, 96, 112, 118, 121, 123, 125, 136, 165, 166, 168–176, 213, 265–267, 269–274
Heitler, Walter, 263, 273
Hemings, Sally, 55
Henle, Jakob F. G., 15
Herodotus, 302
Hertz, Gustav, 140
Hertz, Heinrich, 287
Hildebrandt, Kurt, 101, 115
Hipparchus, 228
Hitler, Adolf, 82, 83, 111, 113, 115, 121, 123, 125, 166, 173–176
Hodge, Jonathan, 283
Hodgkin, Dorothy, 312
Hollitscher, Walter, 134, 138, 139, 141
Holman, Silas, 41
Holmes, Frederick, 309, 310
Homer, 302
Honig, Georg, 25
Hugenberg, Alfred, 113
Hulme, H.R., 268
Hume, David, 135
Husserl, Edmund, 65
Huygen, Christiaan, 284

I
Infeld, Leopold, 140, 188
Ingold, Tim, 223
Iske Brothers, 252
Ivanenko, Dmitri, 195, 196

J
Jacob, François, 309, 311

Jaeger, Werner, 65
Janés, Clara, 60
Jauch, Josef M., 275
Jeans, James H., 292–294, 296
Jefferson, Thomas, 55
Jehle, Herbert, 150
Johnson, Benjamin, 153
Jordanova, Ludmilla, 306
Jordan, Pascual, 112–126, 136
Jungk, Robert, 166, 174, 176
Jungnickel, Christa, 21

K
Kaiser, David, 270
Kandinsky, Wassily, 289
Kapferer, Norbert, 134
Kaufhold, Karl-Heinrich, 24
Kekulé, Friedrich A., 150, 153
King Alfonso XIII, 62
King Ernst August, 20
King, Stephen, 309
Klaus, Georg, 134
Klein, Felix, 43, 76, 77, 79
Klemperer, Viktor, 116
Knausgård, Karl Ove, 311
Kockel, Bernhard, 262, 269, 272, 273
Kohlrausch, Friedrich, 38
König Ludwig II, 29
Kopfermann, Hans, 267
Kragh, Helge, 130, 286
Kramers, Hendrik Anthony, 265, 267, 269
Krebs, Hans, 309, 310
Kubach, Fritz, 101, 103, 104, 117, 118
Kuhn, Thomas, 156

L
Laban, Rudolf, 289
Ladenburg, Rudolf, 267
Lamb, Willis, 273, 274
Landis, Israel, 252, 254
Larmor, Joseph, 283, 285, 297
Latour, Bruno, 223
Lavoisier, Antoine, 242
Lawrence, E. O., 175, 204, 208, 211
Lear, Linda, 312
Lederman, Leon, 210, 215
Lenard, Philipp, 89, 90, 94, 96, 98–100, 102, 104–106, 111, 291
Lenin, Vladimir, 132, 137, 141
Lenz, Wilhelm, 124
Leroi-Gourhan, André, 223

Leslie, Sir John, 237–239, 242, 244–247, 249
Levy, Hyman, 295
Ley, Hermann, 134
Lichnerowicz, André, 185, 187
Lindemann, Frederick, 66, 67
Listing, Benedikt, 30
Litten, Freddy, 101, 106
Lodge, Sir Oliver, 286, 288, 295, 296
Loewenherz, Leopold, 21
Lord Kelvin, 286
Lord Salisbury, 283
Lorentz, Hendrik A., 148, 272
Luxemburg, Rosa, 132

M
Mach, Ernst, 135, 291
Maltby, Margaret, 37–55
March, Arthur, 66
March, Hilde, 66, 67
Martius, Ursula, 123
Marx, Karl, 137
Maskelyne, Nevil, 230
Matthiessen, August, 249
Maunder, Walter, 226
Maxwell, James Clerk, 69, 272, 283–286, 297
May, Charles, 222
May, Eduard, 102
Mayer, J. R., 287
McCormmach, Russell, 21
McFadden Orr, William, 80
Meißner, Walther, 93, 95
Meitner, Lise, 125, 163, 164
Melloni, Macedonio, 247
Mendelsohn, Nathan, 18
Mercier, André, 182, 183, 185–191, 194, 196–198
Merz, Georg, 16
Merz, Sigmund, 16
Meyer, Philip Randolph, 47–51, 53–55
Meyerstein, Moritz, 13–26, 28–30
Michelson, Albert A., 148, 283, 287, 292, 296
Michling, Horst, 21
Milch, Eberhard, 169
Miller, Arthur, 291
Millikan, Robert A., 133
Mills, John, 296
Minkowski, Hermann, 293
Möglich, Friedrich, 140
Moles, Enrique, 62, 63

Møller, Christian, 187, 192, 196, 197
Moore, Walter John, 60
Morley, Edward W., 45, 148, 283, 287, 292, 296
Müller, Wilhelm, 97, 101, 105

N
Nairne, Edward, 240
Nernst, Walther, 38, 43, 53–55
Neurath, Otto, 118
Newcomb, Simon, 228, 230
Newton, Isaac, 16, 130, 157, 228, 255, 284, 292
Niels, Bohr, 294
Nietzsche, Friedrich, 132
Nobili, Leopoldo, 247
Novikov, Igor D., 195
Noyes, Arthur, 43

O
Of Hippo, Saint Augustin, 311, 314
Oppenheimer, J. Robert, 146, 175
Ørsted, Hans Christian, 19
Ortega y Gasset, José, 59, 60
Orwell, George, 311
Ostwald, Wilhelm, 43, 291

P
Palacios, Julio, 63, 69
Papapetrou, Achilles, 139, 140, 142
Pauling, Linus, 149
Pauli, Wolfgang, 112, 125, 182, 183, 269–271
Pearson, Karl, 232
Peierls, Rudolf, 166, 269
Peña Serrano, Fernando, 69–71
Penborn, Charlotte, 49
Peng, Huanwu, 273
Peres, Asher, 190, 191
Petrov, Alexei, 140
Pistor, Carl Philipp Heinrich, 16, 18
Pistor, Lutz, 95
Planck, Max, 118, 120, 148, 294
Plato, 302, 313
Plutarch, 303, 313
Poggendorff, Johann Christian, 15
Popper, Karl, 136
Prandtl, Ludwig, 75–85

Q
Queen Victoria, 132

Index

R
Rabi, Isidor I., 146
Raman, C.V., 262, 266–269
Ramón y Cajal, Santiago, 63
Ramsden, Jesse, 18
Ransome, Robert, 222
Rathkamp, Wilhelm, 26
Reiche, Fritz, 69
Reichenbach, Georg Friedrich, 18
Reichenbach, Hans, 118
Renn, Jürgen, 180
Repsold, Johann Georg, 16, 18
Retherford, Robert, 273
Rheinberger, Hans-Jörg, 282, 305
Rice, James, 294
Richards, Ellen Swallow, 41, 42, 46, 53
Richter, Liselotte, 134
Rieck, Alex, 136
Rieke, Eduard, 44
Riemann, Bernhard, 287
Ritter, Johannes, 19
Robinson, Ivor, 194
Robison, John, 241
Rohrlich, Fritz, 275
Rompe, Robert, 123, 124, 132, 139–141
Roosevelt, Franklin D., 166
Rosbaud, Paul, 120
Rosenberg, Alfred, 122
Rosenfeld, Léon, 187, 270
Rosen, Nathan, 190, 192, 194
Rotblatt, Joseph, 175
Rumford, Benjamin Thomson, 237, 242–245
Rumpf, Johann Philipp, 24, 28
Runge, Carl, 76
Rutherford, Ernest, 48, 294

S
Sachse, Christian, 135
Salinas, Pedro, 64
Salleron, Jules, 238
Sánchez Ron, José Manuel, 60, 70
Sartorius, Florenz, 24
Satterthwaite, Gilbert, 229
Sauter, Fritz, 91, 93, 95–97, 105
Schaefer, Clemens, 21
Schaffer, Simon, 231, 283
Scheel, Gustav Adolf, 101, 103
Scherrer, Paul, 66
Scherzer, Otto, 97
Schlichting, Hermann, 81
Schlick, Moritz, 118, 134

Schrödinger, Annie, 59, 62, 66, 67
Schrödinger, Erwin, 59–62, 64–72, 119, 125
Schumacher, Heinrich Christian, 15
Schwinger, Julian, 145, 152–154, 156, 274, 275, 277
Seife, Charles, 262
Shakespeare, William, 153
Shapin, Stephen, 20
Simms, William, 222
Simon, Leslie E., 84
Singh, Sohan, 94
Slater, John C., 148, 149
Smekal, Adolf, 267
Smyth, Henry, 167
Sommerfeld, Arnold, 65, 70, 80, 81, 91, 105, 125, 148
Speer, Albert, 169, 171
Spinoza, Baruch, 59
Sponer, Hertha, 67
Staley, Richard, 287
Stalin, Joseph, 137, 138, 141, 184
Stapel, Diederik, 314
Stark, Johannes, 89, 90, 111, 120
Steck, Max, 102
Stokes, George Gabriel, 286
Strassmann, Fritz, 163, 164
Strauß, Franz Josef, 125, 175
Stueckelberg, Ernst, 277
Sturmius, John Christopher, 242
Sullivan, Miles, 252
Szilard, Leo, 166, 167, 175
Szöllösi-Janze, Margit, 130, 141

T
Tait, Peter G., 286
Tamm, Igor, 271
Taylor, Geoffrey Ingram, 81–83
Teller, Edward, 166
Terradas, Esteve, 69
Thomson, George P., 296
Thomson, J.J., 48, 285
Thomson, William, 285, 286
Thoreau, Henry David, 311
Thorne, Kip, 196
Thüring, Bruno, 97, 98, 101, 102
Timoféef-Ressovsky, Nikolai, 121
Tollmien, Walter, 81
Tomaschek, Rudolf, 89–91, 93–106
Tomonaga, Sin-Itiro, 145, 153, 154, 156, 274, 275
Tonnelat, Marie-Antoinette, 140, 185
Traynor, Elizabeth, 50, 53

Traynor, Mary, 53
Treder, Hans-Jürgen, 129, 130, 132–142
Treder, Max, 132
Troughton, Edward, 222

V
Valente, Bacelar, 261
Van der Waerden, Bartel L., 174
Van Helmont, Jan Baptist, 242
Van Vleck, John H., 64
Vögler, Albert, 169
Von Bobers, Baron, 25
Von Helmholtz, Herman, 286, 287
Von Kármán, Theodore, 77, 81
Von Laue, Max, 67, 68, 71, 120, 122
Von Neumann, John, 146, 154
Von Weizsäcker, Carl Friedrich, 97, 137, 165, 166, 168, 169, 173, 174, 176

W
Watson, James D., 310
Wazeck, Milena, 292
Weber, Wilhelm Eduard, 14, 19–21, 28, 29, 287
Webster, Arthur, 38, 50
Wegner, Peter, 120
Weisskopf, Victor, 267
Weiss, Pierre, 64, 65

Wentzel, Gregor, 265, 270, 276
Weyl, Hermann, 64, 65, 67, 71, 140
Wheeler, John Archibald, 146, 150, 154, 164, 187, 188, 192, 273
Whittaker, Edmund, 281–283
Wigner, Eugene, 166, 167, 267
Wildangel, Ernst, 132
Wilson, David, 284
Wilson, Bright E., 149
Wilson, Robert, 204–208, 210–215
Winston, Mary, 43
Winthrop, John, 238
Wise, Norton M., 112, 287
Wolf, Karl Lothar, 101–103
Wollaston, William Hyde, 237, 241
Woodhull, Mariane, 49
Wood, Robert, 268
Woolf, Virginia, 38
Woolley, Sir Richard, 228

X
Xenophon, 302

Z
Zeiss, Carl, 14
Zenneck, Jonathan, 90, 93–95
Zubiri, Xavier, 60–62, 65, 66, 70, 71
Zweiling, Klaus, 134–138, 141

CPSIA information can be obtained
at www.ICGtesting.com
Printed in the USA
LVHW011200260720
661559LV00002B/157